CHEWING TOBACCO TIN TAGS 1870 - 1930

Louis Storino

77 Lower Valley Road, Atglen, PA 19310

Published by Schiffer Publishing Ltd.
77 Lower Valley Road
Atglen, PA 19310
Please write for a free catalog.
This book may be purchased from the publisher.
Please include $2.95 for shipping.
Try your bookstore first.

We are interested in hearing from authors
with book ideas on related subjects.

Library of Congress Catalog Card Number: 95-74743

Printed in Hong Kong

ISBN: 0-88740-857-5

Contents

Preface 4

Acknowledgments 4

Introduction 5

Collecting Tin Tags 13

Tobacco Advertising 18

Chewing Tobacco Manufacturers 21

TinTags Described 28

Tin Tags Illustrated 96

Price Guide 200

Preface

In collecting and researching chewing tobacco tin tags, I have found that they are one of the least known collectibles, with little written about them. Here we have a collectible that is truly part of our American history and culture. Most of these beautiful little pieces of art are over 100 years old and come in various sizes and shapes, many very colorful.

In the late 1870s and the early 1900s the collecting of chewing tobacco tin tags was an active, vital hobby, enjoyed by the young and old. It enjoyed a popularity possibly equaling the popularity baseball cards enjoy to day, yet very little information on or relating to the chewing tobacco tin tag is available to the collector or dealer.

This reference will be welcomed by collectors, old and new, by dealers, and by others who enjoy America's colorful past. With a listing of over 6000 tin tags described and priced, 2000 illustrated tags, plus the many other illustrated and related features, this new work will fill the void and bring hours of pleasure to tobacco tag fanciers.

Correspondence concerning unlisted tags or additional information is welcomed. Please write the author at Tags, P.O. Box 189, Los Altos, CA 94022

Acknowledgments

This book represents the accumulated knowledge of many people, exchanging information and discoveries with me. Without their help this book would not have been possible. The author wishes to acknowledge the kind assistance of the following people:

Sal and Barbara Falcone, Alice Evans, George Hilligoss, Clement Zambon, Barbara Edmonson, John Stech, Wayne Wormsley, Pete French, (Duke) R.W. Nolte, Randy McDonald, (Tagger) Lee Jacobs, Tony Hyman, Peggy Dillard, and a special thanks to Chris Cooper.

Thanks also to Ron Willis and everyone at Willis Photo Lab for their help and suggestions and dedicated print work on many of the photographs in this book. Audrey Whiteside at Schiffer Publishing photographed the tags. We also wish to acknowledge the use of the historical sources found at the following: the Toledo Public Library, Toledo, Ohio; the New York Public Library, New York City, N.Y.; the Mountain View Public Library, Mountain View, California; and The Bancroft Library of the University of California at Berkeley, Berkeley, California.

INTRODUCTION

THE ORIGIN OF CHEWING TOBACCO TIN TAGS

In the early 1870s, the country was overstocked with cheap grades of manufactured chewing tobacco. Manufacturers of some of the better grades or brands felt the necessity of adopting some means of marking or branding each plug of their goods. Up to this time the branding of their name on the box had been all the identification that was needed. But in the environment of surplus, it was no protection against unscrupulous retail dealers. They would buy one box of the good grade tobacco and after it was sold they would slip a box of the cheap goods into the empty box and represent it to be the higher grade. The consumer and the manufacturer of the good grade were cheated at the same time.

THE WOODEN CHEWING TOBACCO TAG

In the late 1870s, at least two plug tobacco manufacturers marketed wood-tagged brands of plug chewing tobacco. One was P. Lorillard & Co. of New York, and the other was the Pioneer Tobacco Co. Manufacturers, with a factory in Brooklyn, New York.

The Pioneer Tobacco Co. Factory.

Pioneer made plug tobacco exclusively. They described their wood-tag brands as "MATCHLESS finest in the world", "FRUIT CAKE a delightful chew", and "PIONEER a splendid chew".

In a circular mailed to their customers in 1877, Pioneer emphasized the advantages of their 16 ounce Pound Plug. It noted, "to help enable the dealer to secure every possible advantage, we have placed the Wooden Tags or Trademark at intervals throughout the entire length of the plug, which permits the retailer to cut the lump into small pieces to suit his customer, each piece holding its identity as though a perfect plug in itself."

The circular also stated that "retailers would be rewarded with a FREE Gift of a Champion Improved Tobacco Knife, with the purchase of 60 pounds or more of their Matchless or Fruit Cake Brands of wood-tagged plugs of chewing tobacco."

Research failed to reveal how long the wooden tobacco tags were in use. Disappearing with the advent of the tobacco tin tag, it was most likely only a two to three year period, which would account for their scarcity.

Above: Wooden tags for the three brands: Matchless, Fruit Cake, and Pioneer. Below: The illustration from the circular shows how they are situated on the plug ready for the merchandiser to cut them for his customers.

THE ADVENT OF TIN TAGS

The first company to use tin tags was P. Lorillard & Co. of New York. At its Jersey City, New Jersey plant it applied a tin tag with the word "Lorillard" to the inside of the outer leaf wrapper of the plug. It was less than a successful experiment. The tag was hidden from sight and the consumer, being unaware of its existence, he was almost sure to bite on it, injuring his teeth and gums. It is said the Lorillards were responsible for a great share of the swearing done by tobacco chewers about that time, as well as the loss of a great many teeth. As this made their tobaccos unpopular, the use of the inside tag was discontinued.

At about the same time, the Ben Finzer Tobacco Company of Louisville, Kentucky had the distinction of being the first to introduce the outside tin tag. The company attempted to patent the devise, but the courts ruled that tin tags were not patentable, soon after tin tags were in use by most plug makers.

The brand names used for plug chewing tobacco ranged from the whimsical to the unusual, but in the 1880s & 1890s there was no great deterrent to imitation or out right theft of a name, the average plug maker offered anywhere from 40 to 120 different brands, there were at least six manufacturers with a Legal Tender plug, three with Honey Suckle, six with Strawberry, four with Rosebud, and five with Pine Apple, to name a few.

Plug chewing tobacco was sold everywhere, and because tin tag labels were very inexpensive custom brands could be ordered. As a result, brands were created by manufacturers, wholesalers, and by many retailers. Factories often

Flyer for Horse Shoe Plug claims "over one hundred twenty-five million ten cent pieces sold annually." The slogan says it all: "Good all through...Even the tags are valuable."

made two or three types of plug tobacco and sold them under hundreds of different brand names. Because of this, there are more then 12,000 different tin tags.

Chewing tobacco tin tags pronged into a plug served two purposes. First, a manufacturer's name, origin or trademark identification gave some assurance to the buyer that the same quality would be present at all times. Second, as competition among plug makers grew in the early 1900s, tags became premium tokens redeemable for prizes and cash.

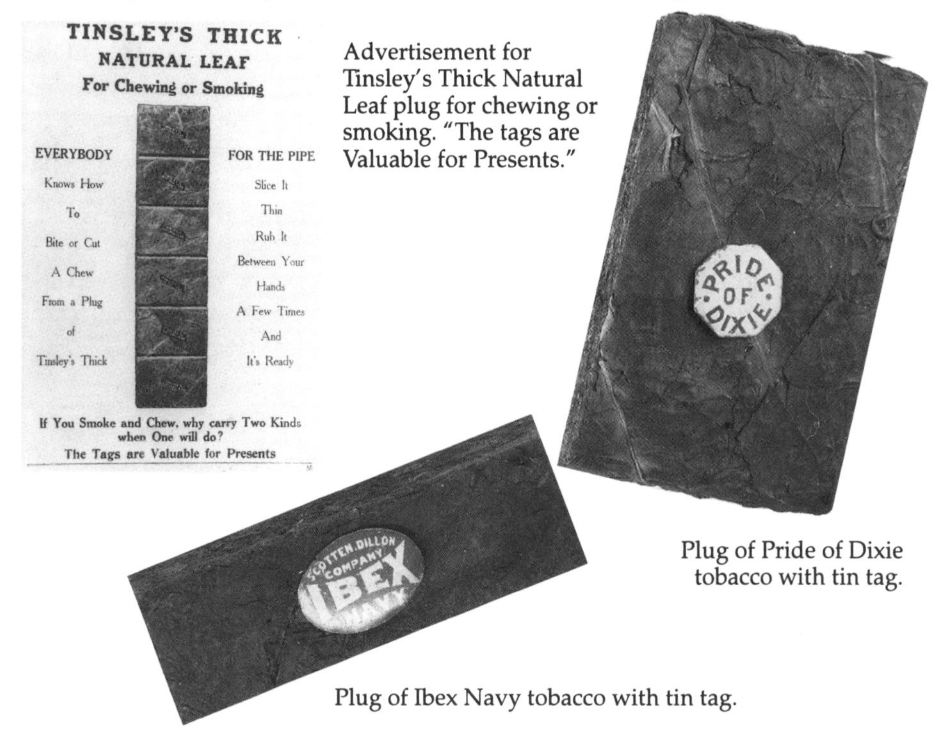

Advertisement for Tinsley's Thick Natural Leaf plug for chewing or smoking. "The tags are Valuable for Presents."

Plug of Pride of Dixie tobacco with tin tag.

Plug of Ibex Navy tobacco with tin tag.

TYPES OF CHEWING TOBACCO - SALES AND VOLUME

Four major types of chewing tobacco were manufactured from the 1870s through the early 1900s. They were:

"Flat plug," a compressed rectangular plug of bright tobacco, sometimes lightly sweetened

"Navy," also a compressed flat rectangular plug made of highly flavored burley leaf

"Twist," a tobacco rope made of burley leaf braided by hand then compressed. It was also known as "pigtail"

"Fine-cut," a chewing tobacco (not a plug and not compressed) made of shredded stripped leaf, comprised mainly of burley leaf with the addition of a little dark Green River leaf. This was the lowest cost and least popular for chewing, although among those who smoked a pipe or rolled their own cigarettes it was very popular, because of its low cost.

Plug chewing tobacco should not be confused with cut plug, which is a pipe or cigarette smoking tobacco. Frontier America chewed tobacco and most of America was frontier until after 1870. A man could not smoke when he was plowing his field, driving a wagon, or riding horseback but he could chew. Because of its low price, plug tobacco was popular well into the 1900s.

Chewing tobacco sales and plug volume reached their peak from 1897 through 1917, averaging 200,000,000 pounds per year. At that time ten leading manufacturers accounted for 60% of the Nation's plug tobacco production. Liggett & Myers and National, were the two largest, producing 26,000,000 pounds annually. Lorillard, Reynolds, Hanes, Scotten, Sorg, Finzer, Butler, and Drummond had average annual productions of from 8,000,000 to 10,000,000 pounds.

This Lorillard flyer claims a production of 33,148,937 lbs of tin tag plug during the years 1877-78-79.

A chewing tobacco plug factory, where workers are making up one pound plugs of tobacco, measuring 3" x 16". After they were marked into seven to twelve "cuts" of about 1 1/4" to 2 1/2", the tin tags would be pressed into each marked piece by young boys. The retailer in turn could then cut and sell the plugs for 5 to 15 cents each depending on the size of the cuts.

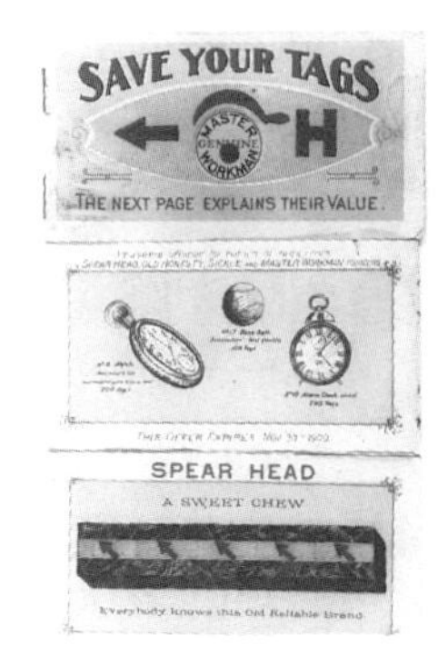

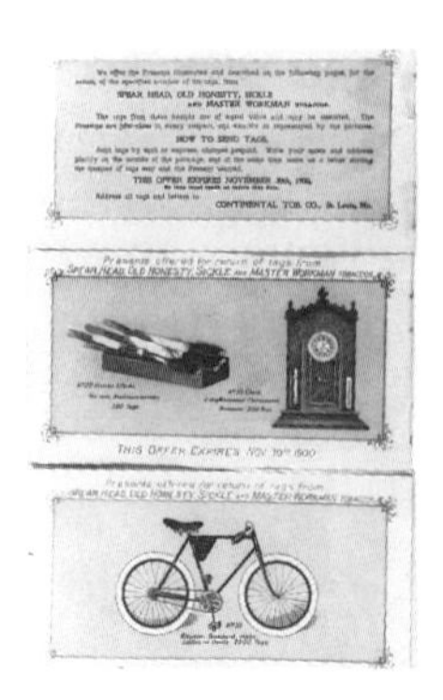

Tags were redeemable for cash and prizes. As illustrated here in Continental Tobacco Co.'s November, 1900, premium book, "Save Your Tags,"some tags were worth as much as one cent a piece in trade-in-value. The average tag was worth closer to one-eight of a cent. In 1902 alone, one manufacture spent well over $1,500,000 redeeming chewing

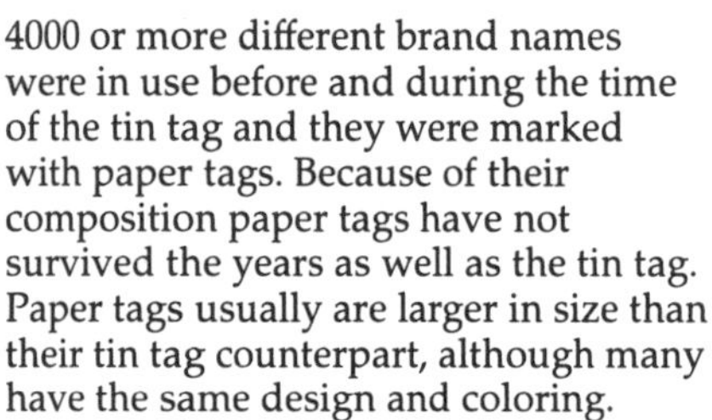

4000 or more different brand names were in use before and during the time of the tin tag and they were marked with paper tags. Because of their composition paper tags have not survived the years as well as the tin tag. Paper tags usually are larger in size than their tin tag counterpart, although many have the same design and coloring.

THE FLAVORING OF PLUG CHEWING TOBACCO

As competition among chewing tobacco manufacturers grew, plug makers became more elaborate, adding some type of flavoring to the cured tobacco leaf, including sugar, rum, honey, licorice, or other sweetener.

Harry Weissinger's 1885 advertisement describes "Prune Nugget" plug. "Prune Nugget is made from the highest grade of White Burley leaf and Fruit. By a process known only to us the fruit becomes a subsistent part and delicate flavor to the tobacco, and yet does not destroy the identity of the tobacco itself, giving it a delicately delicious flavor never heretofore obtained ".

TIN TAG MANUFACTURERS

The following is a partial list of tin tag manufacturers of the 1880s.

Hamilton, Lilley & Co., tin, Brooklyn, New York
Hasker & Marcuse Mfg. Co., tin & paper, Richmond, Virginia
Lanier Printing Co., paper, Winston, North Carolina.
G. Moser & Sons, tin, Cincinnati, Ohio
Raible, Smith & Co., tin, Louisville, Kentucky

Illustrated is a page entitled "SPECIAL BRANDS" taken from the sales catalog dated 1892 of The Wertheimer Company, Tobacco Manufacturers of San Francisco, California. It offers the opportunity for the purchaser of large amounts of chewing tobacco to have his own specially designed tin tag. This helps explain why there are so many different chewing tobacco tin tags.

45

The Wertheimer Company, - - - San Francisco, Cal.

OUR recognized standing with the various extensive factories whose goods we control for this Coast, and our EXCLUSIVE AGENCIES in connection therewith, permit us to extend to our trade an advantage that no other house can offer—that of furnishing shipments direct from factory under private brands which become the sole property of the customer without fear of the article being sold by any other firm or individual, all rights being guaranteed and protected by ourselves and the factories.

For **Plug Smoking** *and* **Chewing Tobaccos** *special tin tags can be made up, either a design of the party placing the order, or selected from many original ones of our own not hitherto made use of. All such special factory orders, of course, will have to be of sufficient a quantity to warrant their exclusive use, and to cover the attendant expense of special manufacture, tags, etc., etc.*

WE SOLICIT CORRESPONDENCE

upon this subject, and can promise best of care to all such orders, as this feature has been a distinct one of our business for years.

When writing to us, state the grade of tobacco you desire a special brand of your own for, selected from any of the styles, on pages 41 to 54 inclusive, with the quantity you wish manufactured as an initial order, and we will promptly respond, with all other particulars necessary to an intelligent understanding of the matter.

An advertisement from The Wertheimer Company of San Francisco, for Special Brands. Customers could design their own tin tag or select one from a group the manufacturer has produced. They may have had it put on the grade of tobacco they chose.

THE LAST FAMILY RUN INDEPENDENT

Taylor Brothers Tobacco Company was founded in Winston, North Carolina in 1883, by Bill Taylor, aged 32, and his brother, Jack. At this time Winston was home to more then thirty tobacco companies. Despite this Bill Taylor prospered and grew, acquiring brand names like Ripe Peaches, Red Coon, Foot Prints, Stars and Bars, Rose Bud, and Bull of The Woods. These, added to his already successful brands, kept his tobacco company competitive.

Over the years Bill turned down many offers to merge with larger firms, as was the custom with many small plug firms. Bill's stubborn streak lent strength to his little company. Taylor Brothers Tobacco Co. remained an independent plug firm and Bill remained his own boss throughout his life. In 1952, nineteen years after Bill's death Taylor Brothers Tobacco Co., the last family run independent plug firm, was sold to a large corporation.

COLLECTING TIN TAGS

TIN TAG COLLECTING IN THE 1880s

In researching material for this book I had the good fortune of acquiring the chewing tobacco collection and correspondence of Mr. Arthur Bohanna of Wellington, Kansas, for the years 1884 to 1888. At that time he was advertising for the buying and trading of tin tags with other collectors. The following are excerpts from some of the more then 200 cards and letters in this collection, illustrating the popularity of tin tag collecting in the 1880s.

Samuel Mac, card printer
Philadelphia, Pennsylvania, Aug. 22, 1887
"I have 1245 rare all different tin tags to exchange"

Edward T. Stoeber, wine-grower
Columbia, South Carolina, Sept. 25, 1888
"With this I mail you 200 different samples of tin tobacco tags, please send me an equal number in exchange"

Smith P. Galt, attorney at law
St. Louis, Missouri, July 20, 1887
"I have a collection of 145 all different tin tags in good order. I will trade them to you for your tags if suitable to me"

George Wolfenden
Detroit, Michigan, Sept. 18, 1886
"I saw in "Harper's Young People" that you would exchange tin tags. I have a collection of 469 tags which I will exchange for others."

R.J. Sawyer
Bangor, Missouri, Nov. 3, 1888
Enclosed 173 tags to exchange.

Arthur Paul
Alton, Illinois, Oct. 30, 1884
"I received your letter with the tags and was glad to hear from you. In looking over the tags I found 9 tags which I had in my collection and I return them with this letter, the other 62 I will keep and send the same number of different ones for them, please find enclosed the above mentioned. If you have any others to exchange please send list."

T.J. Krom
Sayre, Pennsylvania, July 12, 1885
"I have a few tin tags that I would like to exchange. Big H, Little H, Spear Head, Hot Shot, Boot, Climax Grade, Violet, Elephant, K of L Plug, Rooter, Rebecca, Chocolate Cream, Hub red, Hub blue, Hub brown, Green Turtle, J.T., Crusher, Magpie, Rum, Happy Nig, Silver Coin, Key Note, D.S. st Co., Panel Tag. Please send me list of your tags. From, Charlie Krom."

Mr. Arthur Bohanna's tin tag correspondence and a small sample of his collection.

OLD HONESTY TAG COMPANY

The popularity of tin tag collecting in the late 1880s, led some companies to capitalize on the hobby. Not all tin tags came from plugs of chewing tobacco, as tin tag manufacturers began to sell them in sets for collectors.

A letter dated July 27, 1887, from F.W. Finzer & Co. Louisville, Kentucky, proprietors of the Old Honesty Tag Company and publishers of "The Tag and Stamp Herald", offers collectors the opportunity to purchase tags, both tin and paper. "Our prices of fine unused Tin & Paper tags are as listed below. 50 varieties new tin tags 15¢, 300 tin tags (60 varieties evenly mixed) 55¢, 100 all different paper tags 15¢, 200 all different paper tags 25¢." The letter further notes that "all of our tags are very fine, the tin are plain & pictured, embossed & Japanned, about equally mixed, the paper tags very colorful...We believe you could double your money by ordering 1000 of our tin or paper tags. Each purchaser of 1000 either tin or paper gets a set of 4 fine picture cards or photographs of 4 famous beauties."

Sets included subjects like American Generals, Flags of Nations, Presidents, and Cartoons made to satisfy the demand from the collector, and were made in addition to brand name tin tags made for the plug tobacco manufacturers.

The Four Famous Beauties picture cards offered by W. Duke & Sons as premiums for Honest Long Cut Chewing and Smoking Tobacco tags. They are Alba, Jennie Ricci, Lillian Russell, and Miss Marion Hood.

This trade card from Climax Plug, shows the various picture card premiums that were available for tin tag redemption.

COLLECTING TODAY

What does the beginning collector look for? Collecting tin tags is often likened to collecting pin-back buttons. The tin tag should be in good to very fine condition. The embossed tin type should have little or no rust and the painted or lithographed type should not be scratched or chipped. While condition certainly affects value, condition matters less if the tin tag is very rare. Tin tags from the early 1880s are now well over 100 years old. Because of their rarity, the attractiveness of an older tag must be weighed against the chances of ever finding another example.

The most convenient way to collect and store tin tags is to flatten the prongs and put them in 2 x 2 clear mylar coin slips, insert them into 20 pocket vinyl coin pages, and keep them in three-ring binders.

Tin tags can also be mounted in collector's cotton backed display frames, which I find most suitable. The larger frames can hold 100 or more tin tags all in view at one time. Tin tags also look attractive mounted in a frame to hang on a wall. Whether in one frame hung on a wall or in a large collection, the tin tag is a reminder of a colorful era in the chewing tobacco industry which is now long past.

A group of tags in a collector's cotton backed display frame.

TOBACCO ADVERTISING

On the following pages are photos of just a few of the many signs, trade cards, advertisements & premiums, offered by chewing tobacco manufacturers to the merchant and tobacconist to help sell their brand of chewing tobacco.

These colorful old signs and premiums tend to enhance one's collection when displayed with it. Many are very rare and highly prized by collectors. All are very collectible pieces of Americana.

Battle Ax Plug Tobacco. Lithographed cardboard hanging sign with image of Uncle Sam as a peacemaker. 12" x 12" framed.

Old Honesty Plug Tobacco. Lithographed cardboard hanging sign. 12" x 12" framed.

Sweet Tip Top Tobacco. Lithographed cardboard sign with striking image of horse drawn fire engine moving at full speed. 23" x 36".

Hold Fast Cut Plug. Paper sign picturing black children sledding in tobacco boxes, 23" x 12".

Happy Thought Plug Tobacco. A very large paper sign 39" x 28", featuring Wave Line Plug Tobacco The Sailors Choice, the Wilson & McCallay Tobacco Co.

Newsboy Plug Tobacco. Lithographed paper sign, 31" x 23", on cardboard, featuring a group of news boys wrestling for a plug of tobacco while the tobacconist and his wooden Indian watch with smiling faces. "Scrapping Fer De Plug" is the caption of this sign, made for the National Tobacco Works, Louisville, Kentucky.

Lucky Strike Tobacco Clock. 24" tall school clock , made for the R.A. Patterson Tobacco Company.

Bull Dog Cut Plug. A colorful match holder resembling the Bull Dog tobacco tin tag, die-cut, 6.5" x 3.25". Very rare.

CHEWING TOBACCO MANUFACTURERS

The following list consists of tobacco manufacturers who made or sold chewing tobacco in the late 1890s, alphabetically listed by state and city.

ALABAMA

Huntsville
D.T. Harrison

ARIZONA

Tuscon
Victor Griffith
Tin Wo Chan
Albert Steinfeld
Phoenix
Sam'l. N. Seip

ARKANSAS

Bentonville
Geo. M. Hancock
Berryville
Berryville Tobacco Co.
Decatur
Decatur Tobacco Co.
Evening Shade
Edwin F. Smith
Jonesboro
H. Kortenber
Little Rock
J.W.G. Wierman
Osage Mills
Jas. N. Whitesides

CALIFORNIA

Los Angeles
J.H. Borum
F. Wm. Flemming
San Francisco
H. Bohls & Co.
A. Newman
R. Penny
A. Podesta
The John Bollman Co.
The Wertheimer Co.
J. Wierner

COLORADO

Denver
Colorado Tob. Works
Jacob Kerkis
C.J. McLaughlin
Nagely Tobacco Co.

DELAWARE

Yorklyn
W.E. Garrett & Son
Wilmington
Juda Barber

GEORGIA

Atlanta
Augustus De Voe
C.E. Lloyd
N.S. Lloyd
Ball Ground
Jewell & Hendricks
Fairmount
John M. Patton
Jos. A. Riddle
Greely
Jas. P. Mahan
Salaca
Geo. W. Jefferson
Walesca
Geo. M. Harmon

ILLINOIS

Bloomington
E.C. Richardson
Breese
Geo. Schwarz
Carmi
W.S. Rice & Co.
Centralia
Morey & Eis
Chicago
American Tob. Co.
August Beck & Co.
H. Bolander
F. Bujack
Thos. Bujack
Victor Cohn
Cuban Tob. Fact'y
Aug. Delke
Finkel & Silpske
Gradle & Strotz
Herman Gray
H.O. Hansen & Co.
Albert Hert
H.M. Hess
P.O. Johnson
F. Jungkard
Albert Klemendt
Morris Krost
Jos. Kuszewski
D.J. Landfield
John P. Larsen
Northwestern Tob. Fact'y
D.J. O'Connell
C.W. Peterson
M. Pritikin
Reimbold & Co.
J. Saslawaky
John Sokup
Spaulding & Merrick
E.J. Struska
C. Van Lenners
Decatur
E.C. Webster
Flora
C.O. Stanford
Freeport
John E. Tuckett & Sons
Galesburg
Nels. Peterson
Girard
Frederick W. Rothgeber
Peoria
Sander Bros
Quincy
Eagle Tobacco Co.
Wm. Heilhake
J.B. Kohl
Wm. Richmiller
Wellman & Dwire Tobacco Co.
River View
Desplains Tobacco Factory
Rockford
Rockford Tobacco Co.
A. Vernier
Springfield
Leonard Denkel
Thomsonville
John McFarland

INDIANA

Boonville
Flihrer Tobacco Co.Inc.
Crawfordsville
Schweitzer & Son
Dale
John Anderson
Samuel F. Johnson
Dresden
J.M. Records
Evansville
John Beatty
R.W. Harper
Geo. Huck
Jacob Kissel
M.W. McCoy
W.L. Tucker
Jeffersonville
C.O. Davenport
Lebanon
John H. Hoy
New Albany
Feuger Bros
Progress Tobacco Works
Newburgh
J.H. Foster
Hampton Norwood & Co.
Tell City
Columbia Tobacco Co.
Terre Haute
Fred J. Biel
C.H. Biel
Vincennes
H. Stahlschmidt
Wadesville
Albert Endicott

IOWA

Albia
Benj. Strasburger
Burlington
A.S. Cook
Gabriel & Beck
Council Bluffs
F. Luchow
Dubuque
Cox Myers & Co.
Tice Myers & Co.
Davenport
Great Western Tob. Co.
Keokuk
A.J. Siebert
Longgrove
C.F. Jacobson
Ottumwa
Emil Fecht

KANSAS

Abilene
Kathe Lenze

Atchison
Julius Seitz
Beman
John E. Sample
Ellinwood
John Hofman
Fort Scott
W.W. Dillard
Holton
Haist & Ellis
Humboldt
A.A. Miller
Lawrence
Geo. F. Leward
Leavenworth
C.A. Gercken
Okete
G.B. Watson
Topeka
Curry Tobacco Co.
L.H. Holtwick
Wichita
Fern. N. Hussey
Loveland & Cook

KENTUCKY

Bethany
J.W. Blythe
J.W. Cassity
Bowling Green
R.T. Garvin
Ben S. Perkins
H. B.Scott Tob .Co.
Starn & Lasley
Brodhead
Martin & Perkins
Calhoun
Johnson & Landrum
Campbellsville
G.W. Gowdy
Carrollton
White Burley Tobacco Co.
Covington
Mary Arnold
The Capitol Tobacco Co.
E.O. Eshelby Tob. Co.
John Flavin
J.S. Hudson
Theo. Kenneweg
The Kentucky Rail Road Tobacco Co.
Lovell & Buffington Tobacco Co.
Noonan-Dorset Tob. Co.
Perkins & Ernst
Senour & Gedge
Ellis Spilman & Co.
John A. Thompson & Son
I.L. Walker Tob. Co.
The McNamara Whiteman Tobacco Co.
Dixon
Dorris & Rice
Samuel D. Trice
East Bernstadt
J. McNeill & E. S.
Elizabethtown
Samuel L. Morris & Co.
Flemingsburg
The Million-Kendall Co.
Fordsville
Shapero Bros.
Frankfort
Capital Tobacco Manufacturing Co.
Abe Gum
Franklin
Sloss-Groves Tob. Co.
Georgetown
The Kentucky Tob. Co.
Kinzea Stone
Glasgow
T.J. Samson
Gradyville
John E. Moore
Greensburg
Wilson Tobacco Co.
Greenville
Buran Martin & Co.
C.E. Martin & Co.
C.Y. Martin & Co.
H.N. Martin & Co.
E. Rice
Hanley
Henry Nichols
Hanson
Weir Bros
Henderson
Henderson Tob. Ext. Wks.
Hodge Tobacco Co.
Rohards Tobacco Co.
George Knake
Hopkinsville
Greenville Tob. Mfg. Co.
Hopkinsville Tobacco Mfg. Co.
L.D. Reese
Junction City
R.S. Martin
Kirksey
Samuel Cain
Leitchfield
Martin & Co.
Lexington
Blue Grass Tobacco Co.
Livermore
Peay & Atherton
Louisville
American Tobacco Co.
Axton-Fisher Co.
J.D. Baldridge & Co.
Belle of Meade Tob. Co.
Jno. Brumback
J.M. Buckner Jr. Co.
L. M. Burford
Jno. Burrell
Geo. Dannenhold
Robt. Daugherty
J.K. Deictrich
John Finzer & Bros
Jas. Frohle
Greenville Tob.Works
Hail & Cotton
The Harry Weissinger Tobacco Co.
W.P. Helt
W.J. Jenkins
Kentucky Tob. Products
Leora Landrm
Louisville Spirit Cured Tobacco Co.
Manufacturers Tob. Co.
Thomas A. Mann
H.N. Martin & Co.
Samuel B. McGill
Monarch Tobacco Co.
Nall & Williams Tob. Co.
National Tob. Works, Branch of the American Tobacco Co.
John Noonan
E.J. O'Brien
Ohio River Tob. Co.
Owen Tobacco Co.
Wm. E. Peck
Peerless Tobacco Co.
Rothert & Co.
S. Sensbach
John T. Skeldon
Stanley & Otto
Jos. Stienberg
Strater Bros.
D. K. Tate
Taylor Bros
Wm. J. Trent
H. G. Trompeter
A. J. Turpin & Co.
Ludlow
Hardy Bros
Madisonville
S. H. Hollornan & Son
Mayfield
Pigram Tobacco Co.
Maysville
Green River Tob. Co.
J.H. Rains & Son
Monterey
Jones & Smithers
Mote
William Gerrett
Mount Sterling
Kentucky Leaf Tob. Co.
Murray
B. F. Clayton
Gilbert Son & Co.
J.D. Rowlett
L.C. Stubblefield
Valentine Bros & Wilson
Wilson & Son
New Concord
P.M. Rowlett & Son.
Newport
Chas. H. Lansdale
Owensboro
American Tob. Co.
O'Flynn & Scott
Paducah
Smith & Scott Tob. Co.
Pembroke
W. M. Bronough
Rockfield
Ben. S. Perkins
Taylorsville
Charles P. Polk & Bros.
Trammel
J.A. Dowell
Waynesburg
Gooch McHenry & Bro
Winchester
W.T. Baldwin
Wingo
West Kentucky Wheel and Alliance Tob. Mfg.

LOUISIANA

Convent
G.H. Dugas
A. Lartigues
Florian Poche
Hester
F. Guisalph
Paul Louque
F. Michel
Claude Petit
Ozemee Poche
Augustave Roussel
Norbert Roussel
Lutcher
Elisee Jaubert
Ed Malancon
F. Reynaud
Ernest Roussel
Victor Roussel
New Orleans
American Tobacco Co.
R. Beauvais
D.R. Carroll
S. Heinsheim Bros & Co.
W.R. Irly Tobacco Co.
Obershmidt & Schuman
C.E. Sarrazin & Bro
Joseph Wackerbarth
Paulina
H.O. Bourgeois
J.W. Bourgeois
Prudent Bourgeois
Arthur Brignac
F.M. Brignac
Lezimme Deslattes
Optimer Frederick
F. Guglielmo
J.B. Jaubert
John Killburn
L.G. Laiche
A. LeBlanc
Clement LaBlanc
Adrien Louque
Ernest Louque
Jos. Louque
Louis Louque
Ulysse Louque
C. Madere
Eug. Matherne
Oct. Matherne
John Martin
Leopold Martin
Maurice Martin
Van Buren Martin
Evariste Millette
Agricole Poche
Geo. Poche
J.W. Pugh
W. John Rome
Armand Roussel
Art. Roussel
Augustin Roussel
Charles Roussel
Numa Roussel
Octave Roussel
Valery Roussel
Arthur St. Pierre
Floriestain St. Pierre
Theodule St. Pierre
Moise Schexnaidre
Welcome
Norbert Louque

MARYLAND

Baltimore
Jon. B. Adt
H. Arnd & Co.
Barton-McEvoy & Well
Paul Braym
C.M. Benninghaus
Chas. Becker
Francis Dreves
F.W. Felgner & Son
Gail & Ax, Branch American Tobacco Co.
Raphael Goodman
Solomon Kann
Marburg Bros, Branch American Tobacco Co.
Jos. Nadisch
Frederick Schultze

R. Starr & Co.
Cambridge
John Foble

MASSACHUSETTS

Adams
Theo. Koehler
Athol
F.J. House

Boston
Sol Berlowitz
Samuel Brickel
Harris Clarfine
Estabrook & Eaton
Fillppo Goduti
Theodore Koehler
Leavitt & Pierce
N.E. Tob. Mfg. Co.
National Tobacco Co.
Oriental Tobacco Co.
Jos. Tenenblott
Brockton
Peter Hohner
Byfield
Byfield Tob. Co.
Lawrence
Israel Davis
Saugus
Larken & Merrill
Benj. Pearson Jr.

MICHIGAN

Adrian
Michigan Tob. Works
Detroit
American Eagle Tob.Co.
John J. Bagley & Co.
Banner Tobacco Co.
Hans Danielson
Globe Tob Co.
Michigan Tob. Co.
Daniel Scotten & Co.
Scotten-Dillon Co.
Scotten Tob Co.
Ignatz Wolff
Grosse Pointe
Henry Ellspass
Ishpenning
Ishpenning Tobacco Co.
Ithaca
Ithaca Tobacco Co.
Ludington
A.W. Hogland & Co.
Petoskey
Boyington & Corbett Co.

MINNESOTA

Forest Lake
Robt. Bronberger
Minneapolis
Scandia Tob Co.
Swedish Crown Tob Co.
St. Paul
The Ware Tob. Works

MISSOURI

Brookfield
M.P. Kinsey
Boonville
John N. Gott
Brunswick
Brunswick Tob. Co.
A.G. Kennedy
Cape Girardeau
M. Routh & Co.
Clarksville
W.P. Boone & Son
Major-Mackey Tob. Co.
Collins
P.S. Arnold
Edina
Henry Koh
Greenfield
W.G. Porter
Hannibal
Hannibal Tobacco Co.
W.H. Robison
Kansas City
Thos. J. Casey
Geo. M. Foley
Midland Tob. Works
A.D. Weills Jr.
Licking
W.S. Nichol
Louisiana
Tinsley Addison Tob.Co.
John W. Glen
Western Tob. Co.
Montgomery City
The Brown Tob co.
Nevada
Nevada Tob. Co.
Ozark
Jas. F. Graber
Pine
M.S. Dale & Sons
Poplar Bluff
Chas. F. Duke Jr.
St.Charles
Wright Bros. Tobacco Co.
St. Genevieve
Wm. F. Cox
St. Joseph
Anton Kloss
St. Louis
American Tob. Co.
Catlin Tobacco Co.
Central Tobacco Works
Chas. W. Dieterichs
Drummond Tobacco Co.
Joseph Fleig
Wm. G. Hills
E.D. Holthaus
James G. Butler Tob. Co.
Christian Laupp
Liggett & Myer's Tob. Co.
Henry McCabe
Miller & Worley
Phil. Muehlhaeusler
Fred. Mueller
Christian Peper Tob. Co.
Rassfeld & Albers
Michael Rautz
Weisert Bros. Tob. Co.
John Weisert Tob. Co.
H.D. Westerheide
Salisbury
Best Hit Tobacco Co.
Sedalia
Herman Guerrant Tob. Co.
Springfield
G.W. Anthony
Geo.H. McCann & Co.
Robt. L. Pate
W.G. Porter
Webb City
Southwest Tob. Co.
Wentzville
Hunter & Carr
J.A. Thompson
Wellsville
The Wellsville Tob. Co.
Zebra
W.J. Payne

NEW JERSEY

Brookford
James M. Parsons
Caldwell
Lane & Lockwood Co.
Changewater
Bowers Tobacco Co.
Helmetta
The George W. Helme Co.
Jersey City
John Botthof
F.A. Goestxe & Bro. Co.
Gottlieb B. Herbst
P. Lorillard & Co.
Peter Spieles
Newark
Campbell Tobacco Co.
North Brunswick
James M. Parsons
Patterson
Allen & Dunning Co.
Spotswood
Augusts De Voe Co.
Skinner & Co.
Westfield
Maurice Weingarten

NEW YORK

Albany
Constantinople Tob. Co.
Dongan Tob. Mfg. Co.
Payn's B. Sons Tob. Co.
Francis Shields
Auburn
Anton Bros.
Binghamton
Lyman Clock & Son Co.
Brooklyn
Herman Abrahams
Jacob Bahr
John Bassler
Ernest Beringer
John Bramm
Conrad Breininger
F.J. Britt
Buchanan & Lyall
Sandel Conen
E.E. Falke
Edward Frankel
Garcia & Son
Eugene Gerard
Gross & Lampe
Fred. Heeg
Frank Hummel
Adolph Jaeger
Philip Jung
Theo. Kanschsa
J.C. Knaup
G.L. Krapf
Lenz & Gottlieb
Franz Marggrof
Pincus Marks
Geo Mayer
Jesus Mendez
Henry Merdes
H.W. Meyer
L. Miller & Sons
W.N. Moench
Patrick Murray
C.J. Nielson
Max Peyser
H.H. Ruhl
John Sehey
Samuel Wise
Buffalo
John Baum & Co.
Fred. Bresling
Andrew Ignatz
Jos. Jansowski
Julius Swartz
Castleton Corners
Catherine Mayer
John Nicks Tob. Co.
Elmira
John Nicks Tob. Co.
Glenham
Samuel Marsh
F. & M. Herbs
Hudson
F. & M. Herbs
Kingston
Egbert Mullen
Newburgh
John R. McCullough
New York City
J. Abramowitz
Alexander Bros
R. Alexander
Allen Tobacco Co.
American Tob. Co.
Anargyros & Co.
Lasar Aretjksy
P. Ascher
Baker & DuBois
M. Barranco & Co.
Michael Berger
D. Buchner
Butler & Butler
Esther Cohen
Wm. J. Connelly
J.B. Day & Co.
A. Diamond
E.J. Donigan & Co.
Chas. Dress
W. Duke Sons & Co.
C. Erdt
The Ferdinand Hirsch Co.
Fernandez & Saxby
E Fleischauer
P. & J. Frank
E. Freiman
F. Frieman
A. Gershel
B. Goldsmith
Sol. Gutterman
Jacob Heimberger
G. Hillen
Cras. Hlavac
Kelly R. Horace
Horwitz & Goodman
Chas. Jacobs
P. Kassel
Kinney Bros., Branch of American Tob. Co.
J. Krinsky
Kugelmann & Co.
M. Langenzen
J. Levin
Sarah Levy
M. Mahler
S.C. Mariun
G. Mayer
D.H. McAlpin & Co.
A.J. Mellor
L. Miller & Sons

Mrs. G.B. Miller & Co.
J.S. Molins
Monopol Tob. Works
A. Nestler
Onarga Sigaret Co.
Oriental Tob. Co.
C.S. Philips
A. Pinner
E. Pons & Co.
M. Reichert
Jos. Resnick
M.Ritter
E.G. Rohrberg
J. Rose
Chas. Rosen
R. Rosenthal
M.J. Sanchez
Sanchez & Haya
S. Scharlin
Schenker & Co.
F.A. Schleiff
J.C. Schramm
G.A. Schnitzler
Chas. Sillenberg
Sklamberg & Abramson
Sloebeder & Wa-towitch
J.E. Stake
C. Stenberg
J. Stein & Co.
J.W. Sunbreg Co.
L. Sylvester & Son
Turco-American Tob. Co.
V. Vallauri
A. Wadro
A. Wasserman
A. Weingarden
S. Weingarden
I. Weissenhorn
R. Xiques
V.M. Ybor
F. Zaloon & Sons

Red Hook
Hoffman & Co.

Riverhead
Elijah Griswold
Newins & Son

Rochester
British Amer. Tob. Co.
G.F. Hess & Co.
Wm. S. Kimball & Co., Branch of the American Tobacco Co.
Jacob Weissager
R. Whalen & Co.

Tompkinsville
Theo. Scholle

Utica
Warwick & Brown Co.

Winfield
Sim. Scharlin

NORTH CAROLINA

Abshers
Joshua Spicer

Advance
H.T. Smithdeal
C.D. Ward

Arnold
R. Everhard & Co.

Asheville
Asheville Tobacco Works
E.J. Holmes & Co.

Ayersville
D.S.R. Martin

Belew Creek Mills
B.S. Brown

Bethania
O.J. Lehman

Blackwells
King Bros.

Boonville
J.M. Speer

Cana
E. Frost

Catawba
Catawba Tobacco Co.

Clingman
J.G. Segravis

Clyde
L.V. Rogers

Cobbs
J.R. Martin

Copeland
W.R. Doss Tob. Co.

Crossroads Church
D.I. Reavis

Cullers
Mack D. Boyd
Culler & Schaub
C.J. Watson & Co.

Dabney
Vance County Farmers Alliance Tob. Co.

Dalton
D.N. Dalton

Dobson
Samuel Falkner
Samuel Falkner & Co.

Dulin
Thos. F. Atkinson

Durham
Blackwell's Durham Tobacco Co.
Durham Farmers Alliance Tobacco Co.
W. Duke Sons & Co., Branch American Tob.Co
Faucett-Durham Tob. Co.
Z.I. Lyon & Co.
R.F. Morris & Sons Mfg.
Swift & Brown
The Whitted Tob. Co.

East Bend
John A. Martin
R.G. Patterson
Morse & Wade

Efland
S.T. Forrest

Elbaville
J. G. Peebles

Elkin
R.G. Franklin

Farmingtion
J.R. Cornelison
W.F. James
S.A. Jarvis & Co.
Jas. M. Perry
O.L. Williams

Fayettevile
J.L. Allen
<st3a>
Forbush
Vestal & Wooten

Fort Church
A.M. Conatza

Francisco
D.W. Dodd
Milton Smith

Franklinton
R. R. Homes

Fulton
L.A. Peebles & Bros.

Gamewell
L.H. Tuttle

Greensboro
J.L. King & Co.
S.J. Pegram & Co.
R.H. Stanley
E.J. & A.G. Stafford

Grogansville
W.P. Grogan

Henderson
Burgwyn Bros. Tob. Co.
J.H. Prichard & Co.
Carolina Tobacco Co.

Hickory
A.W. Marshall & Co.
N. Martin

High Point
J.H. Jenkins & Co.
W.P. Pickett & Co.
E.D. Steele

Hillsboro
R.C. Hill
H.P. Jones & Co.

Huntsville
L.G. Hunt

Kernersville
Beard & Roberts
B.A. Brown
John M. Greenfield
Lowery, Son & Co.
Adkins Shore & Co.

Kinston
Hamlin & Abbott

Leaksville
Dillard & Moin
Dyer B.H. & G.D.
Guerrant T.D.F. & Co.
Benj. F. Ivie
D.F. King
Martin Tobacco Co.
J.B. Taylor Tobacco Co.

Lexington
A.A. Springs & Co.

Madison
Fleming Goolsby

Marion
Crawford & Morgan
J.L. Morgan

Mars Hill
Wm. O'Connor

Mevane
J.A. Tate

Milton
C.J. Allen Jr.
W.B. Lewis & Co.
Nettie E. Oliver
E.D. Winstead

Mocksville
Kelly & Johnston
Jas. L. Sheak

Mooresboro
Jos. T. Bland

Mooresville
J.G. Benson
W.L. Cadwell & Co.
J.C. Frost
A.D. Plyle

Moravian Falls
Rufus A. Slainhour

Mount Airy
L.W. Ashby & Sons.
Geo. M. Booker
Bower & Co.
Olive Forkner & Co.
Forkner & Fulton
Fulton & Brother
Saml. W. Gentry
R.L. Gwynn & Bro.
W. McKinney & Bro.
W.E. Patterson & Co.
Prather & Whitlock
Rucker-Witten Tob. Co.
Jno. D. Satterfeld & Co.
Sparger Bros.

Mount Tirzah
J.I. Cothran

Newton
Marcus M. Rowe

Oxford
Cooper Tobacco Co.
Granville County Farmers Alliance Tob. Co.
Hicks Tobacco Co.
Oxford Tobacco Co.

Panther Creek
T.W. Poindexter

Pilot Mountain
Dodson Bros. & Stockton
Daniel Marion
Redmen Bros.

Poindexter
G.L. Mathews
J.C. Nicholson

Price
R.P. Price

Raleigh
Harvey & Taylor
J.E. Pogue

Redland
H.W. Dulin
Reaves, Poplin & Co

Reidsville
J.W. Dameron
Robt. Harris & Bro.
Johnston Bros.
Wm. Lindsey & Co.
J.H. Lyle
A.H. Motley Co.
F.R. Penn& Co.
R.P. Richardson Jr.

Republic
W.E. Bovender

Ridgeway
J.D. Scott

Roaring River
W.H. Reves

Rockford
H. Holyfield
Thos. B. Holyfield

Rocky Mount
Emelius Stone

Salem
Elbert Payne & Co.

Salisbury
Beall & Co.
D.L. Gaskill
Homes & Miller
Robertson & Peebles
Thompson Tobacco Works

Sandy Ridge
R.A. Deshaso
John R. Martin

Siloam
A.H. Marion

Statesville
Chas. Adams & Co.
Benjamin Ash
L. Ash
H. Clark & Sons
L. Harrill
Irvin & Poston
J.H. McElwee
Miller & Brown
Powell Turner & Co.
Ramsey & Maxwell

Stoneville
R.T. Stone
Swancreek
W.M. Sparks
Tally Ho
D.W. Wheeler
Taylorsville
Taylorsville Tobacco Mfg. Company
Smith & Beckham
Tilden
J.W. Warren
Tootville
T.A. Steelman
Stem
D.C. Farrabow
Summerfield
Ogburn & Co.
Geo. W. Parrish
Walkertown
Thos. A. Crews
N.D. Sullivan
Walnut Cove
G.H. Crews
Fair & Waddell
J.G. Fulton
Walnutgrove
White & Fowler
Wilkesboro
Hall & Davidson
J.V. Hall
Wilson
Kentucky Tob. Co.
S.B. Rierson
Wells-Whitehead Tob. Co.
Jos. J. Wilson
Wilton
E.L. Harris
Alonzo Mitchell
R.E. Strother
Winston
Bailey Bros.
Bitting & Hay
Blackburn Dalton & Co.
Brown & Bro.
Brown & Williamson
S. Byerly & Son.
Bynum Cotten & Co.
R.L. Candler & co.
W.B. Ellis & Co.
Griffith & Bohanmon
Hamlin Lipfert & Co.
B.F. Hanes
P.H. Hanes & Co.
Hodgkin Bros. & Lunn
Johnson & Humphrey
Leak, Beall & De Vane
Leak Bros. & Hasten
T.F. Leak Tobacco Co.
Lipfert-Scales Co.
Vaughn Lockett & Co.
M.L. Ogburn
S.A. Ogburn
Hill Ogburn & Co.
H.H. Reynolds
Reynolds Bros.
R.J. Reynolds Tobacco Co.
O.J. Shepperd & Co.
Taylor Bros. Co.
T.L. Vaughn & Co.
Walker Bros.
W.A. Whitaker
Whittaker-Harvey Co.
T.F. Williamson & Co.
Williamson Tobacco Co.
N.S. Wilson & T. J. W. Wood & Co.
Yadkin College
T.S. Dale & Co.
Rea Green & Co.
J.A. Hartley & Co.
Ed. L. Owen
Yadkinville
C.F. Dunnagan
J.D. Hamlin
W.L. Kelly
J.E. Zachary
York Institute
L.M. Davis

OHIO

Cincinnati
Otto Autenreith
H. Bade & Son
August Beckmeyer
J. Berger & Co.
Meyer Besuner
Fenner Cottrell & Co.
Day & Night Tob. Co.
B. Durrell & Bros.
Chas. E. Halley
Henry Moorman
Queen City Tobacco Co.
Parmelia C. Robbins
L. Rothert
Casper Ruthmeyer
Schroer Tobacco Co.
The Spence Bros. Co.
Jos. Steidel
L.B. Tenent Jr. & Co.
M.V.B. Weighell
Cleveland
Block Bros.
Julius Borsh
F. M. Federman
S. Greenspon & Co.
Aug. H. Kokanner
Benj. Klein
Chas. A. Kohl
Louis Lewis
Mason Tobacco Co.
Max Meretzky
Ohio Flyer Tobacco Co.
B. Rohrheimer & Sons.
E. Rosenfeld
Chas Sauer
Adolph Seidman
Henry H. Serrer
The Standard Tobacco Co.
Columbus
Columbus Tobacco Co.
Solomon Loeb
Dayton
Gem City Tobacco Co.
The Terry & Porterfield Tobacco Company
Findlay
I.E. Hapner
Middletown
Luhrman & Wibern Tobacco Co.
The P.J. Sorg Company
The Wilson & McCallay Tobacco Company
Norwalk
F.B. Case
The Lake Erie Tob. Co.
Sandusky
F.A. Riedy
Toledo
Edward Frieder
Isherwood Chase & Co.
Toledo Tobacco Works Company
J. F. Zahm Tobacco Co.
Urbana
J.B. Hitt & Co.
Charles Shaul
Zanesville
The Pinkerton Tobacco Company

OKLAHOMA TERRITORY

Hennessey
W.J. Schiffer
Kingfisher
J. A. Eggleston
Oklahoma City
H. L. Haines

OREGON

Portland
Gerson & Hart
E. Schiller
Sichel Co.

PENNSYLVANIA

Beartown
Henry B. Hoffman
Boyerstown
R.R. Engle
T.D. Ochsenford
Fleetwood
O.A. Hoch
Geryville
Charles Zipf
Gilbertsville
J.W. Custer
Hanover
J.G. Fisher & Co.
Jacob H. Hostetter
Harrisburg
Smith & Keffer
Hazleton
John Schwartz's & Sons
Union Tobacco Co.
Huntingdon
Robt. K. Miller
Lancaster
Fred. J. Bradel
Henry C Demuth
Geo. Fisher
Chas. C. Herr
Iona Tobacco Co.
J.L. Matzger Tobacco Co.
Isidor H. Neuman
Morris Rosenstein
Geo. J. Sheld
Rudolph Shultz
Milon B. Weidler
Philip Zimmerman
Lebanon
B.P. Biecker
Lewiston
Geo. G. Frysinger
Lititz
E.K. Bear
J.R. Bricker & Co.
Ambrose E. Furlow
A.H. Hershey
C.J. Keller & Co.
Marietta
John W. Smith
Mechanicsburg
I.F. Barkley
Mohn's Store
Jno. A. Bohler
Neffs
Neffsville Tobacco Co.
Norristown
S.T. Babham & Bro.
Jno. K. Thomas
Philadelphia
B. Appelbaum
J.G.S. Beck & Co.
L.M. Blitzstein
H. Bolevsky
Crabtree Tob. Co.
Jacob Flet
Frishmuth Bros & Co.
W.E. Garrett & Sons.
Geo. Geisler
L. Halpern
Hipple Bros.
Marks Kaban
Jacob Kerkis
Bertha Millen
Wm. M. Orr
J.W. Palmer
Parham & Duff
Pioneer Tobacco Factory
Lena Rappaport
M. Rappaport
Stephano Bros.
W.H. Sterner
Ralph Stewart & Co.
Theodore Teuscher
Geo. W. Weaver
Wetmore Mfg. Co.
N. Yuleman
F. Zimmerman
Pittsburgh
R. & W. Jenkinson Co.
A. Kirkpatrick
F. Manns
Geo. Steurnagle
Weyman & Bro.
Pottstown
J. W. Evans
Jno. A. N. Schuyler
Reading
Atlas Tobacco Co.
C. Breneiser & Son.
Cork Tobacco Works
Joel F. Darrah
Jno. W. Fehr
Morris Goldman
P.H. Hautsch
Hautsch & Rhein
Keystone Tobacco Co.
Isaac C. Koch
E.M. Luden
Sam'l. Milmore
Reading Tobacco Mfg.Co.
J.M. Roland
L.R. Romig
Triple Name Tobacco Co.
Benj. Weiss
Wentzel & Maurer
H.L. Winters
Scranton
Isaac Allabach
J.D. Clark
Clark & Scott
The Clark & Snover Co.
Adam Starkman
Spring City
Jacob Dorean
Springmount
Samuel S. Wolford
Terre Hill
Isaac M. Richmond
Wilkes-Barre
Penn. Tobacco Co.

Windsor
Hiram F. Martin
York
J.G. Bergdoll
David Forry
B.F. Gable & Co.
Weaver Tobacco Co.

SOUTH CAROLINA

Cherokee
J.R. Dover
Fairbanks
Dutch Fork Tob. Mfg. Co.
Florence
Florence Tobacco Co.
Thos. E. Gregg
Springfield
J. W. Jumper
Varnville
J. W. Colson
Walhalla
Norton & Ashworth

TENNESSEE

Alexandria
Eaton Bros.
Allisona
Andrews & Ladd
U.J. Owen
Tatum & McAllister
Arno
W.R. Miller
Belleville
W.H. Ladd Jr.
Bloomingdale
J.F. Howard
Bristol
A.D. Reynolds
Buchanan
Coleman Winchester
Burt
W.D. Smith
Camden
R.W. Ayers
Ayres & Morris
A.H. Gibson & Co.
I.W. Gibson
Carthage
Carthage Tob. Works
Ford & Flippen
Christiana
W.T. Smotherman & Co.
Chattanooga
Chattanooga Tob. Works
Phillips Co.
Clarksville
American Tob. Co. Merriwether Br.
Clarksville Tob. Mfg. Co.
College Grove
J.S. Ogilvie
Columbia
C.F. Brittain Tob. Works
Conyersville
J.M. Hooper
Cookeville
Ford & Elston
Crossland
Henry Co.
Crossland Mfg. Co.
Dixon Springs
Dixon Springs Tob.Works
Eagleville
P.H. Elmore
R.C. Owen
W.J. Owen
Elmwood
W.B. Ford & Co.
Franklin
J.W. Corlett
Gainsboro
A.C. Washburn
Greenville
Campbell Tob. Co.
East Tennessee Tob. Mfg. Company
Greenville Tobacco Mfg. Company
Merchants Tob. Mfg. Co.
J.R. Noell Tob. Co.
Unaka Tob. Works
Holt's Corner
T.C. Brittain
Huntington
Huntington Tob. Works
Inda
R.Y. Glover
Johnson City
Crandall, Harris Tobacco Works
Knoxville
W.R. Johnsons
Knoxville Tob. Co.
H. Levy & Bro.
Lick Creek
W.H. Dean
Martin
West Tennessee Tobacco Works
McIllwain
Cain Bros.
McIllwain & Cain
Morristown
Morristown Tobacco Mfg. Company
W.T. Rippetoe
Nashville
Belle Meade Tob Works
Bruton & Condon
Crescent Tob. Co.
Cumberland Tob. Works
W.W. Ford Tob.Works
R.H. Lee & Co.
Wm. Morrow
Nashville Tob. Works
Jesse J. Seawell
E.T. Smotherman & Co.
Newbern
N. Porter
H.C. Porter
Paris
Empire Tobacco Works
W.H. Hudson & Son
J.S. Mossman
Williams & Hudson
Pekin
D.H. Thompson
J.B. Edwards
Portland
William Buntin
Pulaski
J. Bugg Tob. Co.
Reed's Store
W.C. Mosley
Rockbridge
Rockbridge Tob. Works
Rossview
E.B. Ross
Saint Bethlehem
E. B. Ross & Co.
Sharon
Thomas & Brother
Smithville
Eaton & Foster
Springfield
Elliott & Watson
Strawberry Plains
J. W. Trent & Son
Sugar Tree
Wesson Bros.
Sweet Water
Thomas Forkner
Trentville
J. W. Trent & Sons
Wayland Springs
Ayers & Pierce
Willard
Willard Tob. Works
TEXAS

Alice
Geo. Hobbs & Sons
Bonham
Martin Bros.Tob. Co.
Dallas
P.P. Martinez
Denison
Geo. Hog
El Paso
Kohlberg Bros.
Laredo
Villegas Mercantile Co.
San Diego
L. Levy
Shiner
Louis Ehlers
Whitesboro
Choice & Frazell

VIRGINIA

Abingdon
W. Ingraham & Co.

Alexandria
Frank E. Corbett
Ararat
Scales Bros.
Bedford
Berry-Suhling Tob. Co.
Bolling, Wright & Co.
Clark & Co.
Gish & Smith
Jud, Hurt & Co.
W.P. Hurt Tob. Co.
Bristol
Critz & Reynolds
Brooklyn
B. Barksdale & Co.
Brook Neal
Elder & Wingfield
Brosville
Berry Bros.
J.W. Burton & Bro.
Centralia
Brown-Holland Tob. Co.
Chase City
Chase City Mfg. Co.
Chatham Tobacco Co.
J.H. Hargrave & Co.
T.E. Roberts Tob. Co.
Christiansburg
Wake Kasey & Co.
Clarksville
Flanks & Griffin
Danville
Harvey Bendall & Co.
J.H. Cosby & Bro.
A.G. Fuller & Co.
Gravely & Miller
P.B. Gravely & Co.
R.H. Herndon & Co.
W.C. Hurt Tob. Co.
Lyon & Goodson
Charles Norborn
Talbot Pace & Co.
Penn, Green
Penn & Rison
J.E. Perkinson & Co.
C.A. Raine & Co.
Schoolfield & Watson
Snelling & Sparrow
L.P. Stovall & Co.
Stultz, Lisberger & Co.
Thomas & McAdams
Traylor & Spencer
Geo. E. Tuckitt & Co.
Ellerson Wemple & Co.
J.N. Wyllie & Co.
Drake's Branch
Payne & Jackson
Dyer Store
G.W. Lester & Son
Edgwood
F.R. Brown
Fork Union
D.R. Norvell
Glasgow
Rockbridge Tob. Co.
Graham
Union Tob. co.
Graystone
R.C. Payne & Co.
Hopper
D.F. Haislip
Leatherwood
B.F. Gravely & Sons
Lynchburg
E.A. Allen & Bro.
L.L. Armistead
C.G. Bruning Tob.
J.W. Carroll
W.S. Carroll
J.B. Evans & Bro.
J.H. Flood
W.A. Ford
F.C. Ford & Co.
Hancock Bros. & Co.
Highlander Tob. Co.
Owen Irey & Co.
William King Jr. & Co.
Lewis-Johns Mfg. Co.
S.T. Labby
H.A. Robinson
G.W. Smith & Co.
Timberlake, Woodson & Snead
Virginia Tob. Co.
West-Winfree Tob. Co.
Winfree & Loyd
Manchester
Manchester Tob. Co.
Marrowbone
W.T. Deshazo & Sons
Martinsville
Bevan & Co.
J.R.& F. R. Brown
W.A. Brown Tob. Co.
Brown Bros. & Dudley
Belcher English & Co.
B.F. Gravely & Co. Limited
Henry Tobacco Co.
H.C. Lester
Penn, Watson & Co.
Rucker & Witten Tob. Co.

Sparrow & Gravey Tob. Company, Inc.
J.H. Spencer
D.H. Spencer & Son, Inc.
Stanley & Finley
Stultz, Sparrow & Co.

May's Forge
Thos. H. Penn

Mendota
Dorton & Benham

Nottaway Court House
E.H. Witmer

Olympia
D.T. Davis
W.B. Spratt

Petersburg
Bland Bros. & Wright
Wm. Cameron & Bro.
The Cameron Tob. Co.
Campbell & Co.
David Dunlop
J.H. Maclin & Son
S.W. Venable Tob. Co.
Watson & McGill
Williamson & Routh
Zimmer & Co.

Portsmouth
C. Oeins

Richmond
Allen & Ginter, Branch American Tob. Co.
Seddon Boykin & Co.
W.A. Bragg
Butler & Bosher
Cameron & Cameron
Alex Cameron
Carrington & Co.
W.H. Connell
E.T. Crimp
J.N. Cullingworth
James G. Dill
The Durham Company
Chas. E. Ellison
J.H. Goddin
Gould Tob. Co.
W.T. Hancock
Hardgrove & Co.
J.J. Hickok & Co.
Kentucky Tob.Prod.Co.
Laidlaw, Mackill & Co.
James Leigh Jones
Larus & Bro.
L.H. Lightfoot
Lottier Tob. Co.
A. Maupin & Co.
T.T. Mayo
P.H. Mayo & Bro.
Myers Bros. & Co.
J.B. Pace Tob. Co.
R.A. Patterson Tob. Co.
E.T. Pilkinton Co.
Rucker & Witten Tob. Co.
C.F. Russell & Co.
W.W. Russell
Spicer Son & Co.
D.O. Sullivan
H.T. Thomas
Thornton & Co.
United States Tob. Co.
L.B. Vaughan
T.C. Williams & Co.
J. Wright & Co.
W.J. Yarbrough & Sons

Ridgway
H.A. Deshazo & Co.
George O. Jones

Roanoke
Fishburne Bros.

Rocky Mount
Perry Harrison
Tazewell Helms
Cannaday Noel & Co.

Scottsburg
Hudson & Mosely

Shady Grove
M.F. Parcell & Co.

South Boston
Edmondson Tob. Works
Virginia Boss Tob. Co.

Spencer
Job Deshazo
J.W. King
D.H. Spencer & Son

Taylorsville
W.H. Rangeley & Co.

Union Hall
Parker Bros. & Co.

Whitmel
F.A. Swanson

Wilmington
Belle & Perkins
Dunkum Bros.

WASHINGTON

Seattle
John Lindblom

Spokane
Fred Siegenthaler

WEST VIRGINIA

Charlestown
D.S. Hughes

Clarksburg
J.H. Fye Co.

Fairmount
Antonio Furans

Fetterman
R.N. McClonkey

Forest Hill
Mann, Woodson & Co.
Roberts Middleton & Co.

Wheeling
Beer Tobacco Co.
The Bloch Bros. Tob Co.
Sam'l S. Bloch
C.F. Miller Jr.
J.F. Miller
Augustus Pollack & Co.
John Schneider & Co.
The West Virginia Tob. Co.

WISCONSIN

Dodgeville
Jno. A. Hahn

Edgerton
The Edgerton Tob. Mfg. Company

Green Bay
Felix Freix

La Crosse
Pamperin & Wigenhorn Co.

Milwaukee
The Adams Tob. Co.
J.G. Flint Jr.
Gottleib Graber
Hanson & Schmidt Co.
Aug. C. Kurz
Edwd. R. Lange
B. Leidersdorf & Co.
Ernst Ludorf Jr.
Mathew Pelk
F. Reuter & Co.
Sternemann Bros. & Haydon
Chas. Spangenberg
J. Wicke & Co.

Oshkosh
Wm. Wichman

Platteville
M.S. Sickles & Co.

Racine
Jensine Petersen

Richland Centre
Chas. H. Hyatt

Sheboygan
Herman Schuelke

Watertown
Chas. H. Miller

Wausau
Theodore Buller

Wequiock
H. C. Moeller

TIN TAGS DESCRIBED

NUMBERING SYSTEM AND DESCRIPTION

(1) Each tin tag is assigned an individual number for means of identification, and to help communication between collectors or dealers.
(2) Name of brand.
(3) Wording in [ ] is information on or related to the tin tag.

Example:

B486 BILLY [round black on red, picture a goat]

B486 is the tags number, **BILLY** is the brand name, **round** is the tags shape, **black on red** is the color of the tags lettering on its back ground, **picture a goat** is other descriptive feature on tag.

A

A101	**A**	**[die-cut]**
A102	A	[round, embossed A]
A103	A.	[round, black on yellow, J.W. Trent]
A104	A	[round, embossed, D.F. Kings]
A105	**AA**	**[die-cut]**
A106	AAA	[rectangular, embossed]
A107	A.A.A.A.	[round, gold on red, Finks]
A108	AAAA	[triangular, Henry Co.]
A109	**AAAA**	**[die-cut, red, green, yellow, orange]**
A110	AAAA	[rectangular, black on yellow, Extra Fine Powder]
A111	AAAA	[die-cut multicolored, Mfd. by A.D. Reynolds, Bristol, Tenn.]
A112	AAAA	(a) [rectangular, white on blue, The F.R. Penn Tob Co.] (b) [red on yellow]
A113	A.A. FIG	[round, black on yellow, picture of a fig, H.C.L.]
A114	**ABC**	**[die-cut block, red, green, yellow]**
A115	A.B.C.	[round black on red, Crews]
A116	A.B.C.	[round, embossed]
A117	**A.B.C. 5 cent**	**[round, embossed]**
A118	A C-Co.	[odd shape, gold on red]
A119	A.C. FULLER & CO.	[rectangular, odd shape, embossed]
A120	A.C.H.	[rectangular, embossed]
A121	**A D**	**(a) [rectangular, embossed, red]** (b) [gold] (c) [silver]
A123	A D A	[rectangular, embossed]
A124	A D A	[oval, yellow on black]
A125	**A 1**	**[round, embossed]**
A126	A S W	[rectangular, black on blue]
A127	**AT**	**[diamond, cut-out]**
A129	AT & Co.	[round, embossed]
A130	A.W.C.	[round, embossed]
A131	**A.Y.P.**	**[rectangular, embossed]**
A133	A BIT	[banner embossed]
A134	A GENTLEMAN'S CHEW	[rectangular, yellow on black, Spaulding & Merrick]
A135	**A. No. 1**	**[round, Trade Mark, orange on white]**
A136	ABOVE ALL	[rectangular, black on red]
A137	ABOVE ALL	[rectangular, embossed]
A138	ABOVE ALL	[rectangular, black on red, Berry Bros]
A139	**A CHEWER**	**[round, black on pink, picture of a man chewing]**
A142	**ACME**	**[die-cut shield, Sullivan &**

		Early]
A143	**ACME**	**[rectangular, embossed, C.H. & S.]**
A144	ACME	[diamond, black on red]
A145	ACME	[round, multicolored picture of a tomato, Liipfert, Scales & Co.]
A146	**ACORN**	**(a) [small die-cut shape, red]**
		(b) [silver]
		(c) [gold]
A147	**ACORN**	**(a) [large die-cut shape, red]**
		(b) [silver]
		(c) [gold]
A149	**ACORN**	**[round, embossed, Dick Middleton & Co.]**
A151	**ACTOR**	**[rectangular, white on red]**
A153	ACTUAL V CENT PIECES	[round, embossed, S.F. Hiss & Co.]
A155	ADA BRYAN	[rectangular, picture of a lady's face]
A157	ADAMS APPLE	[round, red on yellow, picture of an apple]
A158	ADAMS TOBACCO	[round, embossed]
A159	ADAMS TOBACCO COMPANY	(a) [round, red on yellow]
		(b) [round, black on red]
A161	ADMIRAL DEWEY	[octagonal black on yellow, picture of a man]
A162	ADMIRAL TOGO	[oval, picture of a man]
A163	**ADVANCE**	**[oval, embossed]**
A165	AFTER DARK	[rectangular, black on red]
A167	**AFTER TEA DESSERT**	**[round, embossed]**
A169	AGATE	[oval, embossed]
A170	AGGIE	[oval, embossed]
A171	A GRADE	[round, yellow on light brown]
A174	AH THERE	[oval, black on yellow, picture of a boy]
A175	A.H. MOTLEY & CO.	[round, black & red on yellow]
A176	AIR CURED	[round, multicolored, an eagle on quarter moon]
A177	AIR CURED	[round, brown on green & yellow]
A178	AIR MAIL TWIST	[die-cut large T, black on red]
A179	**AJAX**	**[round, black on red]**
A180	**ALABAMA**	**[diamond, white on black]**
A181	**THE ALABAMA**	**[rectangular, black on orange]**
A182	ALABAMA BELLE	[oval, embossed]
A183	**ALABAMA COON**	**[round, black man, G. Penn Sons Tob. Co.]**
A184	ALABAMA COON	[round, black on yellow & red, black man & top hat]
A185	ALAMO	[octagonal, black on green]
A186	**ALAMO**	**[hexagon, black on red]**
A187	ALEX CAMERON	[rainbow shape, black on red]
A189	ALEX CAMERON	[heart shape, black on red]
A191	ALEX CAMERON	[round, shape, black on red]
A193	ALEX H. STEPHENS	[oval, black on yellow]
A199	**ALL ROUND**	**[small round scalloped, black on orange, picture of a man]**
A200	ALL THE GO	[oval, black on red, picture of men]
A201	ALL TWIST	[rectangular, black on yellow]
A203	ALL UNIONS	[round, red & blue on white]
A205	ALL OK	[oval, black on red]
A206	ALL WOOL	[round, white on red]
A207	C.W. ALLEN	[round, black on red]
A208	**ALLEN COUNTY**	**[round, black on red, picture of a twist of tobacco]**
A209	**ALLEN'S**	**[round, brown on yellow, chewing smoking]**
A210	ALLEN'S	[oval, black on red & yellow, Suncured Tobacco]
A211	ALLEN'S JEWEL 5¢ PLUG	[round, black, blue & red on yellow]
A213	ALLEN'S TWIST	[round, picture, chewing-smoking]
A215	ALLEN'S C.W. FIRST GRADE	[round, black & red on yellow]
A217	**ALLIANCE GIRL**	**[round, embossed, by Taylor]**
A219	**ALLIGATOR**	**[die-cut embossed, K & R]**
A221	ALLIGATOR	[die-cut embossed]
A223	ALPINE	[round, green & blue on white, picture of mountains]
A225	ALTO	[die-cut shape, cow]
A227	ALTO	[die-cut shape, camel]
A229	ALTO	[die-cut shape, horse]
A231	ALTO	[die-cut shape, elephant]
A232	**ALTON TOBACCO**	**[round, embossed, Drummond Randle Tobacco Co.]**
A233	AMERICAN BEAUTY	[round, black on white, picture of red flower, S.B. Brown Co.]
A234	AMERICAN BOY	[rectangular, black on red & blue, picture of soldier & flag]
A235	**AMERICAN EAGLE**	**[round, red, white & blue, A.T. Co]**
A236	AMERICAN EAGLE	[round, red, white & blue, C.T. Co]
A237	AMERICAN EAGLE TOBACCO	[rectangular, black on red, British-Australian Tob. Co.]
A238	**AMERICAN FLAG**	**[round, red white & blue on white]**
A239	AMERICAN FLYER	(a) [die-cut large T, black on red]
		(b) [black on yellow]
A242	**AMERICAN NAVY**	**(a) [small rectangular, red, white & blue, picture of ship]**
		(b) [large]
		(c) [on back of tag: redeemable]
		(d) [on back of tag: not redeemable]
A244	AMERICAN NAVY	[round, white on blue]
A247	AMERICAN PLUG	[round, black on red]
A249	AMERICAN TWIST	[round, black on yellow]
A251	A.M. LYON & Co.	[round, embossed]
A252	ANCESTOR	[rectangular, black on yellow]
A253	**ANCHOR & CHAIN**	**[die-cut, large]**
A255	**ANCHOR**	**[round, embossed]**
A256	ANCHOR	[round, black on yellow, picture of anchor, tag made in U.S.A.]
A257	**ANCHOR**	**[die-cut, O.H.]**
A259	**ANCHOR**	**[die-cut, with embossed rope]**
A261	**ANCHOR**	**[die-cut, with embossed chain]**
A263	ANCHOR	[die-cut, black on red]
A264	**ANCHOR**	**(a) [round, small, black on yellow, picture of a anchor]**
		(b) [black on white]
		(c) [red on yellow]
A265	ANCHOR	[die-cut small, black on red, The P.J. Sorg Co.]
A266	ANCHOR	(a) [die-cut large, black on red, The P.J. Sorg Co.]
		(b) [black on yellow]

A267	**ANCHOR PLUG**	**[die-cut anchor, black on red]**
A268	ANCHOR MEDIUM	[die-cut anchor, red on yellow]
A269	ANDERSON RED PLUG TWIST	[round, black & yellow on red]
A271	ANHEUSER BUSCH	[rectangular, odd shape, red on yellow]
A272	ANNA CAMP	[round, black on red, picture of a girl, West Winfree Tob. Co.]
A273	ANNA CARTER	[round, black on yellow, picture of girl]
A275	**ANNIE JONES No. 1**	**[oval, black on red & yellow, Ridgeway, Va.]**
A276	ANNIE LEE	[round, embossed]
A277	ANNIE LISLE	[rectangular, embossed]
A279	ANNIE McLEOD	[rectangular, black on yellow, Henry Co.]
A280	ANNIE PEARLE	[octagonal black on yellow, The F.R. Penn Tob. Co.]
A281	ANNIE RANGELEY	[rectangular, black on red]
A282	ANNIEKE'S	[round, black on red]
A283	A No. 1, MAYO'S TOB.	[rectangular, black on yellow, Trademark Reg'd. 1878]
A285	**ANVIL**	**[rectangular, black on red, picture of anvil, O. O. T. W.]**
A287	**ANVIL**	**[die-cut embossed, large]**
A289	**ANVIL**	**[die-cut embossed, small]**
A291	ANVIL	[die-cut embossed, Trade Mark]
A293	AOUW	[round, black on red]
A294	APEX	[oval, embossed]
A295	**APEX**	**[rectangular, black on red]**
A296	APEX	[round, white on blue]
A297	APPLE	[die-cut, black on red]
A298	**APPLE JACK**	**[gold & black]**
A299	**•APPLE JACK**	**[hexagon, black & red on yellow, picture of apple]**
A301	**APPLE JACK**	**[hexagon black & red on yellow, picture of apple, P.H. Hanes & Co.]**
A302	APPLE JACK	[rectangular, gold on green]
A303	**APPLE SUN CURED**	**(a) [die-cut small, black on red]**
		(b) [die-cut larger, black on red]
A307	APPLE TRADEMARK	[round, black & red on yellow]
A308	APOLLO	[round, black on yellow, Cameron]
A311	APRIL FOOL	[octagonal red on yellow]
A313	**APRIL KICK THIS**	**[hexagon black on pink, picture of man with sign on back end]**
A315	APRIL PLUG	[round, embossed]
A317	**ARAB**	**[octagonal, red on gold]**
A319	ARABIC	[round, gold on black]
A321	**ARCH**	**[die-cut shape]**
A323	ARCHER	[rectangular, black on yellow, picture of man with bow & arrow]
A325	ARCADIA	[oval, embossed]
A327	ARK	[round, black on red, picture of boat]
A328	ARM CHAIR	[rectangular, black on yellow, picture of chair]
A329	**ARM & HAMMER**	**[die-cut small, embossed Trade Mark]**
A331	**ARM & HAMMER**	**[die-cut large, red]**
A333	ARMSTRONG'S TOBACCO	[round, red on yellow]
A335	ARMSTRONG'S TWIST	[round, black & red on yellow]
A336	ARMY	[oval, black on yellow]
A337	**ARMY**	**[rectangular, die-cut crossguns, embossed]**
A338	ARMY PLUG	[round, black on red, white & blue]
A339	**ARMY & NAVY**	**(a) [round, blue on white, picture of two sailors, P. Lorillard Company]**
		(b) [black on orange]
A340	**ARMY & NAVY**	**[round, small black on white, P. Lorillard & Co.]**
A341	ARMY & NAVY	[round, small, black on blue, P. Lorillard]
A342	**ARND'S BEST**	**[die-cut flag, embossed]**
A343	ARNETT'S NATURAL LEAF TOBACCO	[die-cut shield, white on red & blue]
A345	**ARROW**	**[die-cut, E. Leidy]**
A347	ARROW	[round, white on red, picture of arrow]
A349	ARROW	[rectangular, embossed, picture of arrow]
A351	**ARROW HEAD**	**(a) [white on blue, die-cut, N. & W.T. Co.]**
		(b) [white on red]
		(c) [white on yellow]
		(d) [white on dark blue]
A352	**AUGUSTA EVANS**	**[round, embossed]**
A354	AUGUSTA EVANS	[round, black on yellow, picture of girl, West, Winfree Tob. Co.]
A357	**AUTO CHEW & SMOKE**	**[round, black on yellow & orange]**
A359	**AUTUMN**	**(a) [round, small blue]**
		(b) [green]
		(c) [yellow]
A361	A WHOLE TOWNSHIP	[round, multicolored, picture of town]
A363	AXE	[die-cut shape, black on red]
A365	AXE HEAD	[die-cut axe head, red on yellow, Dick Middleton & Co.]
A368	AXLE TWIST	[oval, black on yellow, picture of an axle, Lovell & Buffington Tob. Co.]
A369	AXTELL	[round, red on black]
A371	**AXTON'S NATURAL LEAF**	**[diamond, black on red]**
A373	AXTON'S TWIST	[rectangular, black on yellow]
A375	**AYRES**	**[die-cut shield, embossed, Geo. C & D]**

B

B101	**B**	**[round, small embossed]**
B102	B	[round, black on yellow, smoke, chew, Eastern Tob. Co.]
B103	**B**	**[round, large embossed]**
B104	B	[rectangular, octagon, black on red, Rowlett's Twist]
B105	**B**	**[die-cut buffalo embossed]**
B106	B	[rectangular, black on red, J.W. Trent, Strawberry Plains, Tenn.]
B107	B	[rectangular, thin black on red, J.W. Trent]
B108	**B**	**(a)[die-cut B embossed]**
		(b)[smaller]
B109	B-B	[rectangular, black on yellow, J.W. Trent, Strawberry Plains, Tenn.]
B110	**BB**	**[round, black on white]**
B111	**BB**	**[rectangular, embossed]**

B112	**BB**	**[round, Smoking, black on red]**
B113	**BB**	**[round, embossed]**
B114	**B and B**	**[Hazel Kirke]**
B115	B B B B	[round, white on red & white, Trade Mark]
B116	B B B B	[round, black on yellow]
B117	B.C. HOME SPUN	[round, black on red]
B118	B.C.R.R.	[rectangular, embossed]
B119	**BD**	**[cut-out oval, orange & yellow]**
B121	B.F. GRAVELY & SONS	[round, embossed]
B123	B.G.	[round, red on black]
B125	**B.G. TWIST**	**[round, red on yellow]**
B126	B J	[rectangular, cut-out]
B127	**B.L.**	**[round, embossed, Chic, K.Y.]**
B129	B-L	(a) [round, white on red, chromo B & L]
		(b) [white on green]
		(c) [white on blue]
B131	**B.G./N.G.**	**[square, embossed]**
B132	**B M**	**[cut-out embossed]**
B133	B-M	[round, red on yellow, light, pressed]
B134	B on N	[rectangular, embossed]
B135	**B & O**	**[diamond, embossed]**
B136	B. & R.R.	[rectangular, black on yellow]
B137	**B S**	**[round, embossed]**
B138	B.S. & CO.	[round, scalloped, gold on black]
B139	**B & S TRADE MARK**	**[die-cut crescent embossed]**
B140	B.T.	[rectangular, white on red]
B141	B.T.	[oval, embossed]
B142	**B V**	**[rectangular, embossed]**
B143	B & W	[two rounds rectangular, white on blue, Sweet Chew, trade mark, Reg. Dec 30, 1884]
B145	**B L**	**(a) [large blue round cut-out, Buchanan & Lyall]**
		(b) [small blue]
		(c) [large brown]
		(d) [small brown]
B151	BABY BOY	[round, black on red]
B153	BABY BRAND	[round, picture of a baby, Trade Mark]
B155	BABY	[round, black on yellow, picture of baby]
B157	**BABY HEAD**	**(a) [die-cut embossed red]**
		(b) [gold]
		(c) [silver]
B158	**BABY RUTH**	**[round, black on yellow, picture of little girl]**
B159	BABY'S BOTTOM	[round, black on yellow, picture of baby's back side]
B161	BACHELOR'S COMFORT	[rectangular, black on red, picture of man sitting, Iredell Tob. Co.]
B163	BACHMANN'S DELIGHT	[die-cut embossed, heart shape]
B164	BACHMANN'S FIDDLE	[die-cut fiddle, black on red]
B165	**BACK TO DIXIE**	**[round, embossed]**
B166	BAD BOY	[round, picture of boy's head]
B167	BAGDAD	[round, red on yellow]
B168	BADGER FIGHT	[die-cut embossed badger]
B169	BAGLEY	[die-cut drum, blue on gold, Union Maid]
B170	**BAGLEY'S NAVY**	**[oval, black on white]**
B171	BAILEY BROS.	[rectangular, embossed]
B173	BAILEY BROS, BIG 20	[rectangular, odd shape, yellow on red]
B174	BAILEY BROS. #5	[round, red on black]
B175	BAILEY'S OX	[die-cut & picture of ox, black on red]
B176	BALD EAGLE	[round, gold on red, E.A. Saunders & Sons]
B177	BALD EAGLE TOBACCO	[round, black & red on yellow, picture of eagle, E. Pluribus Unum]
B179	BALL & BAT	[round, black & yellow on red, picture of ball & bat]
B180	BALL PIN HAMMER	[die-cut hammer embossed]
B181	**BALLOON**	**[die-cut shape, black on red]**
B183	BALLOON	[round, embossed, picture of balloon]
B185	BALTO	[round, embossed]
B187	BANG UP	[round, red on black]
B188	BANGLE	[die-cut scroll, black on red, J.B. Pace Tob. Co.]
B189	**BANGLE**	**[banner embossed]**
B190	BANGOR	[oval, black on yellow, red spot]
B191	BANKER	[rectangular, white on red, tags worth one cent cash]
B192	BANKER	[rectangular, white in red, The Clean Tobacco]
B193	**BANKO**	**[rectangular, black on yellow, Crescent Tob. Co.]**
B194	**BANNER**	**(a) [die-cut shield, embossed blue]**
		(b) [silver]
		(c) [gold]
B195	**BANNER TWIST**	**[rectangular, octagon, black on red]**
B196	BANNO	(a) [large round white on red, Hampton Tob. Co.]
		(b) [small round]
B197	BANTON'S NO.1 NAVY	[rectangular, black & blue on white]
B199	BARBED WIRE PLUG	[rectangular, black on yellow, picture]
B200	BARE CITY	[round, black on red, P. & W.]
B201	BARNEY'S B&O EXPRESS	[round, black on red]
B202	BARLOW TWIST	[round, black on yellow, picture of pocket knife]
B203	BARROW'S CHOICE	[round, black on yellow]
B204	**BARS**	**[die-cut prison bars, black on red]**
B207	BASEBALL	[rectangular, embossed]
B208	BASEBALL CLUB BOSTON	[rectangular, picture of baseball player & bat]
B209	**BASKET**	**[round, black on red]**
B210	**BAT**	**[round, embossed]**
B211	BAT FIRED	[round, black on red, picture of bat]
B212	BAT NATURAL TWIST	[round, black & brown on yellow, picture of bat]
B213	**BATTLE AX**	**[die-cut ax, black on red]**
B214	**BATTLE AX**	**[round, white on red, A.T. & Co.]**
B215	BATTLE AX	[round, embossed, A.T. & Co.]
B216	BATTLESHIP	[round, black on yellow, picture of ship]
B217	**BATTLESHIP**	**[round, white on black & yellow, picture of ship]**
B218	**BAXTER'S**	**[rectangular, embossed]**
B219	BAY LINE	[round, black on green, picture of ship]
B221	BAY STEAMER	(a) [large round, black on yellow, picture of ship]
		(b) [small round]
B223	B-B TWIST	[octagonal, black on red]
B224	BEAM WEIGHT	[round, black on orange, J.B. Taylor Tob. Co.]
B225	**BEAR**	**[die-cut shape, small embossed]**
B226	**BEAR**	**[die-cut shape, large]**

B227	BEAR	[round, embossed bear head]
B229	BEAR	[die-cut embossed, S. Bear & Sons]
B230	BEAR & EYE	[round, black on yellow, picture of dressed bear & eye, Allen Bros. Tob. Co.]
B231	BEAUREGARD	[round, black on yellow, picture of man, J.N. Wyllie & Co.]
B232	BEAUREGARD	[octagonal black on yellow, picture of man, J.N. Wyllie & Co.]
B233	BEAUTY	[round, black on yellow, picture of girl, P. Lorillard & Co.]
B235	**BEAUTY**	**[round, embossed, A. Ethridge & Co.]**
B237	**BEAUTY**	**[rectangular, embossed]**
B238	BEAUTY	[diamond, embossed]
B241	**BEAVER**	**[die-cut shape, embossed]**
B242	BEAVER	[rectangular, shield, brown on yellow, picture of beaver]
B243	BEAVER	[round, embossed]
B244	BEAVER	[rectangular, black on yellow, picture of beaver]
B245	RED ROCK	[die-cut banner embossed]
B247	**BEE**	**[round, embossed**
B249	BEE	[round, yellow on red, picture of bee]
B251	BEE	[die-cut beehive shape, embossed]
B253	**BEECH-NUT PLUG**	**[oval, red, white & blue]**
B255	BEEHIVE	[round, black on yellow & tan, picture of beehive & bees]
B256	BEE-WAX	[round, black on yellow, picture of beehive]
B257	BEFORE THE JUDGES	[rectangular, black on red, picture of three heads]
B258	BELFAST	[round, embossed]
B259	BELGIUM FLAG	[oval, blue, yellow & red on white]
B260	BELL	[die-cut shape, large, black on gold]
B261	BELL	[die-cut embossed, Macon]
B262	**BELL**	**[die-cut small embossed]**
B263	BELL	[rectangular, picture of bell, black on yellow, Statesville]
B266	BELL'S OF GREENVILLE	[die-cut embossed]
B267	BELLE	[round, black on red, picture of girl's head]
B268	BELLE	[hexagon black on yellow, picture of girl's head]
B269	BELLE BRANDON	[rectangular, white on black]
B270	BELLE OF ALABAMA	[oval, embossed]
B272	BELLE OF ALABAMA	[oval, embossed]
B273	BELLE OF HENRY	[rectangular, black on gold, L. B. & T]
B274	BELLE OF PINNACLE	[rectangular, black on green, picture of girl]
B275	BELLE OF PINNACLE	[octagonal, Culler, N.C.]
B276	BELL OF PINNACLE	[octagonal embossed, Culler, N.C.]
B277	BELLE OF N.C.	[round, black on yellow, picture of girl]
B278	BELLE OF NEW ENGLAND	[round, gold on green]
B279	BELLE OF THE SOUTH	[round, black on yellow]
B281	BELLE OF WASHINGTON	[rectangular, black on light blue]
B283	**BELLE OF WEST VA.**	**[rectangular, embossed]**
B285	**BELLE OF WINSTON**	**[rectangular, odd shape, embossed, C.H. & S. Co.]**
B286	BELLE OF WINSTON	[rectangular, odd shape, embossed, H L & Co.]
B287	**BELLMAN'S OWN**	**[rectangular, red on yellow]**
B282	BELWOOD	[rectangular, red on yellow]
B288	BEN BOW	[round, embossed]
B289	BEN FRANKLIN	[oval, green on white, picture of man]
B290	BENGAL	[round, black on yellow & brown, picture of tiger head]
B291	BENGAL 43	[round, red on yellow]
B292	BEN HILL	[rectangular, yellow on black, Dalton & Ellington]
B293	**BEN HUR**	**[round, black on yellow & red]**
B294	**BEN HUR**	**[round, embossed]**
B295	BENJAMIN HARRISON	[round, center photo, red, white & blue]
B297	BERKSHIRE	[rectangular, black on red, picture of a sow]
B298	BERTA GRAVELY	[oval, black on yellow, picture of a girl]
B299	BESSIE BIRCH	[round, black on yellow, picture of a girl]
B300	BEST	[round, black on yellow, Z. Gravely, Henry Co. Va.]
B301	**BEST**	**(a) [round, gold on green, Taylor Bros.]**
		(b) [gold on black]
		(c) [gold on red]
B302	**BEST**	**[die-cut shield, Hancock, black on red]**
B303	**BEST**	**[diamond, R.A.D., white on red]**
B304	BEST	[oval, red on yellow]
B305	**BEST**	**[round, black on orange, B. Schuster & Co.]**
B306	BEST	[rectangular, scalloped, picture of plug, Gravely, Henry Co.]
B307	**BEST OF ALL**	**[round, embossed, H. Callender & Co.]**
B308	BEST OF BEST	[rectangular, black on red, Anderson]
B309	BEST CHEW	[rectangular, black on yellow, Lester, Barrow & Towne]
B310	**BEST DOG**	**[rectangular, embossed]**
B311	**BEST GRADE**	**[round, Lennert, black on red]**
B312	BEST NAVY	[rectangular, banner embossed, Butter & Bosher]
B313	**BEST OF ALL**	**[round, embossed, H. Callender & Co.]**
B314	BEST TENNESSEE	[round, red on yellow & white]
B315	BEST YET	[rectangular, embossed]
B316	BEST YET	[die-cut leaf, black on green]
B317	BETH BIRCH	[round, black on yellow, picture of girl]
B318	BETTER BEAVER	[round, red on yellow, in center I T Cnfld]
B319	BETTER THEN GOLD	[rectangular, black on yellow, W. D. & Co.]
B320	**BETTER THAN GOLD**	**(a) [rectangular, gold on blue]**
		(b) [gold on black]
B321	**BETTER TIMES**	**[oval, embossed]**
B322	BETTER TIMES	[round, gold on black]
B323	B.G. TWIST	[round, red on yellow]
B325	B GRADE CC 4	[embossed]
B327	BICYCLE	[round, embossed]
B329	BID	[die-cut chain links]
B331	BIG	[cut-out diamond shape]
B333	BIG	[die-cut letters, embossed]
B335	BIG	[round, embossed]
B337	BIG AUGER	[round, black on yellow, picture of auger]
B339	**BIG B**	**[cut-out square]**

B340	BIG BALL	[round, black on yellow]
B341	**BIG BASS**	**(a) [round, picture of fish, P. Lorillard & Co. black on yellow]** (b) [green on white]
B342	BIG BAR	[round, embossed]
B343	BIG BB	[rectangular, embossed]
B344	BIG BEN	[oval, red on gold]
B345	BIG BEN	[round, red on yellow]
B347	**BIG BILL**	**[round, picture of man on horse, Taylor Bros. Inc.]**
B348	BIG BITE	[oval, red on yellow]
B349	**BIG BITE**	**[round, embossed]**
B350	BIG BOY	[round, black on yellow, picture of boy standing]
B351	BIG BOY	[rectangular, red on yellow]
B353	**BIG BUCK**	**(a) [round, white on blue, picture of buck, R.A. Patterson Tob. Co.]** (b) [round, yellow on brown]
B354	BIG BUCK	[round, yellow on black, picture of buck, Patterson]
B355	**BIG D**	**[die-cut D embossed]**
B357	**BIG D T**	**[round, embossed]**
B358	**BIG D TEN**	**[round, embossed]**
B359	**BIG C**	**[round, embossed]**
B361	**BIG C**	**(a) [round, large, black on red, Trade Mark]** (b) [round, small]
B363	**BIG CASINO**	**[diamond, embossed]**
B364	BIG CHARLOTTE	[rectangular, yellow on red]
B365	BIG CHIEF	[round, black on red & yellow, picture of Indian]
B366	BIG CHIP	[round, black on yellow, picture of lumberjack, Flynt]
B367	BIG CHIP	[round, black on red, picture of man chopping wood]
B369	**BIG CHUNK**	**[round, black on red]**
B371	BIG CUT	[rectangular, black on yellow & brown]
B373	BIG DEAL	[diamond, yellow on red]
B374	**BIG DEMAND**	**[rectangular, H. Clark & Sons, black on yellow]**
B375	BIG DIME	[rectangular, black on red]
B376	BIG DIME	[round, embossed, with stars]
B377	BIG DIME 10	[round, embossed]
B378	BIG DRIVE	[round, embossed]
B379	BIG DRIVE	[rectangular, red on black, A. Meigs & Co.]
B381	BIG E	[round, embossed]
B382	**BIG E**	**[die-cut E embossed]**
B383	**BIG 5**	**[die-cut 5 embossed, Emerson]**
B385	**BIG 5**	**[die-cut 5, embossed, Meigs]**
B386	**BIG 5**	**(a) [die-cut 5 embossed red, R & W]** (b) [silver] (c) [gold]
B387	**BIG 5**	**[round, large red & black on red]**
B388	BIG FIG	[round, embossed]
B389	BIG FIG	[round, yellow on black & green, picture of fig, Butler]
B390	BIG FOOT	[rectangular, black on yellow, picture of man & his foot]
B391	**THE BIG 5 CENTER**	**(a) [oval, embossed gold]** (b) [silver]
B393	BIG FOUR	[octagonal embossed]
B395	BIG FOUR	[hexagon black on yellow, H Bros & Co. Danville, Va.]
B397	**BIG FOUR TWIST**	**[round, Samson]**
B399	**BIG GUN**	**[die-cut, P.J. Sorg & Co. black on red]**
B401	**BIG GUN**	**[oval, black on red]**
B402	BIG GUN	[round, embossed picture of a gun]
B403	**BIG H**	**[round, embossed]**
B405	**BIG HORN**	**[die-cut embossed, Bailey Bros]**
B407	**BIG HORN**	**[diamond, embossed]**
B409	BIG IKE	[rectangular, embossed, J.K.L. & Co.]
B411	**BIG INDIAN**	**[diamond, black on red, tobacco]**
B412	BIG INJUN	[rectangular, diamond, black on yellow, picture of Indian]
B413	BIG J. 5¢ TWIST	[round, black on red]
B414	BIG JOHN	[round, black on yellow, picture of a man]
B415	BIG JOKER	[round, embossed, J.L.K. & Co.]
B417	**BIG KICK**	**[oval, black on red]**
B419	BIG KING	[round, black on red]
B420	BIG LEVER	[round, picture of lever]
B421	**BIG LEVER**	**[round, Taylor Bros. Chewing Tobacco]**
B422	BIG LICK	[oval, red on black]
B423	BIG LICK	[oval, embossed]
B425	BIG LINK	[die-cut, chain links embossed]
B426	BIG MULE	[square, embossed]
B427	**BIG MOGUL**	**[square, embossed]**
B428	BIG N	[die cut N embossed]
B429	BIG NICKEL	[oval, embossed]
B431	BIG NICK	[round, white on blue]
B433	BIG NIG	[rectangular, embossed]
B435	BIG 9	[die-cut 9 embossed]
B437	**BIG PLUG**	**[die-cut fireplug, gold on red]**
B439	BIG PLUG	[die-cut fireplug, black on red]
B440	BIG QQ PLUG	[oval, embossed]
B441	**BIG RING**	**[die-cut embossed ring, paper insert, picture of bell, Wilson & McCallay Tob. Co.]**
B442	BIG RUN	[rectangular, embossed]
B443	**BIG RUN**	**[round, embossed]**
B444	BIG RUN	[rectangular, picture of man on horse, Miller & Clifford]
B445	**BIG RUN**	**[die-cut embossed man on horse]**
B446	BIG ROCK	[round, red on white]
B447	BIG SAM	[rectangular, black on yellow picture of a bell]
B448	BIG SAM	[round, black on yellow]
B449	**BIG SCHOONER**	**(a) [rectangular, black on yellow, picture of ship]** (b) [black on red] (c) [black on orange]
B451	**BIG SCHOONER PLUG**	**[die-cut embossed]**
B453	**BIG SCORE**	**[rectangular, black on yellow, picture of a bat]**
B454	BIG SELLER	[rectangular, embossed, J.N.W. & Co., Danville, Va.]
B455	BIG SHE	[rectangular, embossed]
B456	BIG SHOT FIRED	[round, black on yellow]
B457	BIG 6	[die-cut 6 embossed]
B458	**BIG SNAP**	**[rectangular, embossed]**
B459	BIG STEAMER	[rectangular, black on red, picture of a ship]
B460	BIG STICK	[rectangular, black on white, picture of big stick]
B461	BIG STICK	[round, black on red, Newburgh Tobacco Co.]

B462 BIG STUMP [rectangular, embossed]
B463 BIG 10 [rectangular, large, cut-out]
B464 BIG 10 [rectangular, embossed, large, Cosgrave]
B465 BIG 10 CENTER [round, small, black on red, Venables]
B467 BIG TIMES 5 Cents [round, embossed]
B468 BIG 2 [round, large embossed, Traylor, Spencer & Co.]
B469 BIG 20 [rectangular, cut-out]
B470 BIG TWIST [round, white on red, Z.T.C. & Co.]
B471 THE BIG U-5-L NICK [oval, embossed]
B472 BIG WHIP [round, embossed, picture of whip]
B473 BIG WHISTLE [rectangular, multicolored]
B474 BIG WHISTLE (a) [die-cut, Bailey Bros. black on red]
(b) [die-cut embossed, Bailey Bros.]
B475 BIG WINSTON [rectangular, black on yellow, Dalton-Farrow Co.]
B477 BIG X [square, odd, embossed]
B479 BIGGEST & BEST [round, black & red on yellow]
B481 BIJOU [round, embossed]
B482 BILBO [rectangular, black on yellow]
B483 BILL BALEY [oval, picture of man]
B484 BILLY [round, picture of goat pulling a man in a cart]
B485 BILLY [rectangular, embossed]
B486 BILLY [round, black on red, picture of goat, J.N. Wyllie & Co.]
B487 BILLY [round, red on black, picture of goat, N. Wyllie & Co]
B488 BILLY BLACK [round, embossed]
B489 BILLY BUCK [oval, black on yellow]
B490 BILLY BUTTON [oval, embossed]
B491 BILLY POSSUM [rectangular, black, red & yellow, picture of possum]
B492 BINGO [round, blue on cream, picture of dog's head]
B493 BINGO SUNCURED [round, blue & red on cream, picture of dog's head]
B494 BIN [round, embossed]
B495 BIRD IN HAND [hexagon, black on yellow]
B497 BISHOP [rectangular, embossed]
B499 BITTING'S 1/2 LB PLUG [rectangular, embossed]
B501 BITTING & HAY [rectangular, shield embossed]
B503 BLACK ACE [square, embossed]
B505 BLACK BASS [oval, black on white]
B506 BLACK BASS [die-cut M, Finzer, black on white]
B507 BLACK BASS [oval, embossed]
B508 BLACK BASS [die-cut M, Musselman, black on white]
B509 BLACK BEAR [die-cut small bear embossed]
B511 BLACK BEAR [oval, black on yellow]
B512 BLACK BEAR [oval, black & brown on yellow, picture of bear]
B513 BLACK BERRY [round, multicolored, Suncured]
B514 BLACK BIRD [round, black on yellow, picture of bird facing left]
B515 BLACK BIRD (a) [round, black on yellow, picture of bird facing right]
(b) [round, black on red]
B516 BLACK BIRD (a) [round, black on gray, bird & man, P. Lorillard & Co.]
(b) [wording on side]
B517 BLACK CAKE (a) [round, black on red]
(b) [round, embossed]
B518 BLACK COCK (a) [diamond, brown on yellow, picture of fat chicken]
(b) [picture of skinny chicken]
B519 BLACK COIL TWIST [round, black on yellow]
B520 BLACK CROSS [round, multicolored, picture of black cross, F. Lasseher & Co]
B521 BLACK DIAMOND [round, black on red, G.T. Mfg. Co.]
B522 BLACK DIAMOND [diamond, gold on black, Greenville]
B523 BLACK DIAMOND [diamond, black on white, Greenville]
B524 BLACK DIAMOND [diamond, black on red, J.W. Daniel Tob. Co., Henry Co.]
B525 BLACK EAGLE [oval, black on red, Sun Cured, picture of eagle]
B526 BLACK EAGLE [oval, black on red, Genuine Sun Cured]
B527 BLACK EAGLE GENUINE [oval, black on red, Sun Cured]
B528 BLACK-EYE [oval, black on white, picture of a eye, Penn & Watson]
B529 BLACK TIGER [round, black on yellow]
B530 BLACK GEM [round, embossed]
B531 BLACK GRAPE [round, black on red]
B533 BLACK HAWK [round, black on white, picture of Indian]
B535 BLACK HORSE [rectangular, black on red, picture of horse]
B537 BLACK JACK [rectangular, embossed]
B539 BLACK JACK [round, embossed]
B541 BLACK JOE [round, black on yellow, picture of black man, L.B. & T]
B543 BLACK JOHN [oval, embossed]
B544 BLACKLEAF [rectangular, yellow on black]
B545 BLACK MAMMY [diamond, embossed, L.S. & Co.]
B547 BLACK MARIA (a) [rectangular, small, black & red on yellow]
(b) [rectangular, larger]
B551 BLACK MARIA (a) [oval, embossed, Taylor Bros., Winston, N.C.]
(b) [oval, embossed, T. Bros & Co, . Winston, N.C.]
B553 BLACK PATCH [rectangular, black on yellow & yellow on black]
B554 BLACK PATCH [rectangular, red on black]
B555 BLACK PETE [rectangular, embossed]
B557 BLACK PRINCE (a) [red die-cut embossed]
(b) [silver]
(c) [blue]
B559 BLACK PRINCE [round, embossed]
B561 BLACK PUSSY [round, black on red, picture of cat]
B562 BLACK RABBIT [round, black on yellow, picture of rabbit, Shield Shelburn T&T&Co.]
B563 BLACK SAM [rectangular, black on yellow, picture of man]
B564 BLACK SHEEP [round, black on yellow, picture of a sheep]
B565 BLACK SILK [rectangular, embossed]
B566 BLACK SKIN [diamond, black on yellow]
B567 BLACK STANDARD [oval, Scotten Tobacco Co. black on yellow]
B568 BLACK SWAN [round, black on yellow, picture of Swan]

B569	BLACK TIGER	[round, black on red, picture of Tiger head]
B570	BLACK TIGER	[round, black on yellow]
B571	BLACK TIGER 1878	[round, black on yellow, picture of Tiger head in center]
B572	**BLACK WATCH**	**[round, red on gold]**
B573	BLACK WAXIE	[diamond, white on red]
B574	BLACK B-W WAXEY	[rectangular, white on blue]
B575	BLAINE-LOGAN	[rectangular, two rounds, black on yellow, picture of two men]
B576	**J.G. BLAINE FOR PRESIDENT**	**[round, embossed, photo in center]**
B577	**J.G. BLAINE**	**[rectangular, black on red picture of man]**
B579	J.G. BLAINE	[rectangular, black on yellow, picture of man]
B581	**JAMES G. BLAINE**	**[die-cut shield large, picture of man, black on yellow]**
B582	**BLAND**	**[rectangular, Bland Tob Co. black on yellow & red]**
B583	BLAND	[round, black on red, Bland Tobacco Co., Petersburg, Va.]
B584	BLIND	[round, black on yellow, picture of Tiger]
B585	**BLIND BILLY NO. 1**	**[round, embossed]**
B586	BLIND BILLY	[round, embossed]
B587	BLIND TOM	(a) [rectangular, red on black] (b) [red on yellow]
B589	BLIZZARD	[rectangular, red on yellow]
B591	BLOCKADE	[rectangular, embossed]
B593	**BLOOD HOUND**	**[die-cut, black on red]**
B595	BLOOD'S PERFECTION	[round, black & red on yellow]
B596	BLUE BEARD	[rectangular, bow tie, embossed]
B597	BLUE BELL	[concave square, picture]
B598	BLUE BIRD	(a) [round, blue on yellow & red, picture of bird, Carmart & Brothers] (b) [blue on white & red]
B599	BLUE BOOK TOBACCO	[rectangular, black on yellow, picture of book, W.E.& Co.]
B600	BLUE EAGLE	[rectangular, red on black]
B601	**BLUE EAGLE**	**[rectangular, embossed]**
B602	BLUE EAGLE	[oval, embossed]
B603	BLUE EYE	[round, black on white & blue, picture of eye]
B604	BLUE GRASS	[rectangular, odd, black on red]
B605	THE BLUE AND THE GRAY	[round, embossed]
B606	BLUE HEN	[oval, white on blue]
B607	**BLUE JAY**	**[round, small, black on red]**
B608	**BLUE JAY**	**[round, embossed]**
B609	**BLUE JAY**	**[octagonal, blue on white, picture of bird]**
B610	BLUE LABEL	[round, white on blue, Homespun Twist]
B611	BLUE PETER	[oval, blue on white, picture of blue, white & red flag]
B612	BLUE RIBBON	[round, black & blue on red, picture of ribbon, Mfg. by Robt. Harris & Bro.]
B613	**BLUE ROCK**	**[die-cut bird & base, white on red & green, Scotten Dillon Co]**
B614	BLUE SEAL	[round, black on blue]
B615	BLUE STOCKING	[round, embossed]
B616	BLUE TAG SMOKING	[oval, yellow on blue, Tuckett, tag made in U.S.A.]
B617	BLUE WING	[rectangular, embossed, J. L. K. & Co.]
B618	BLUE WING	[round, embossed]
B619	**BUFF CITY**	**[round, embossed]**
B620	**BLUNDER BUSS**	**[die-cut rifle, black on red]**
B621	**BO PEEP**	**[round, embossed]**
B622	BOB HANCOCK	[round, black on red]
B623	BOB HERNDON	[rectangular, black on red]
B623	BOB LEE	[octagonal, blue on white, picture of man on a horse]
B625	BOB VANCE	[round, black on yellow, Dalton & Ellington]
B625	BOB WHITE	[round, brown on yellow, Traylor, Spencer & Co, picture of bird]
B626	BOB WHITE	[round, K.L. Ogburn & Co. black on red]
B627	**BOB WHITE**	**[die-cut bird, black on red]**
B628	**BOBS**	**[die-cut snow shoe, black on red]**
B629	BOBS	[round, black on red]
B631	BOBS	[round, embossed]
B633	**BOMB SHELL**	**[oval, black on red]**
B634	BOMB SHELL	[round, embossed]
B635	BOMB PROOF	[die-cut shield, Red, white & blue, Geo. O. Jones & Co.]
B637	BON BON	[round, red on yellow]
B639	BONNIE BLUEBELL	[round, black & blue on yellow, picture of girl]
B641	BONNIE BLUE FLAG	[rectangular, blue on white, picture of flag, Penn & Rison]
B643	BONNIE CAKE	[octagonal embossed, O.F.]
B645	BONNIE KATE	[rectangular, black on red, yellow, picture of a girl, mfg. by Irvin & Poston]
B647	BONNIE WOOD	[rectangular, embossed]
B649	BONNY JEAN	[round, black on red, picture of a girl, R.L. Candler & Co.]
B651	**BON TON**	**[rectangular, black on white]**
B653	**BON TON**	**[rectangular, embossed silver]**
B654	BON TON	[odd rectangular, black on green, D M & CO.]
B655	**BON TON**	**[round, embossed, Dick Middleton & Co.]**
B656	BON TON	[round, Musselman & Co.]
B657	BON TON	[round, black on red]
B658	**BOODLE**	**[rectangular, embossed]**
B659	**BOOKER'S KY. BURLEY**	**[die-cut leaf, black on yellow]**
B661	BOOM & OUT	
B663	BOOM	[die-cut a leaf, H L & Co., Winston]
B664	BOON	[rectangular, shield embossed]
B665	**BOON**	**(a) [rectangular, embossed silver]** (b) [blue]
B667	BOON'S OLD VA. WEED	[rectangular, embossed]
B669	BOOSTER NATURAL TWIST	[rectangular, black on red]
B670	**BOOSTER NATURAL TWIST**	**[rectangular, black on red]**
B671	**BOOT**	**[die-cut embossed, man]**
B673	**BOOT**	**[die-cut embossed, ladies]**
B675	BOOT	[round, embossed, picture of boot]
B677	**BOOT BLACK**	**[round, black on yellow, picture of a boy shining shoes, Ogburn Hill & Co.]**
B678	**BOOT JACK**	**[round, silver on blue, Jno. Finzer & Bro.]**
B679	**BOSS**	**[round, embossed]**

B681	BOSS	[rectangular black on red]
B682	BOSS	[oval, embossed]
B683	**BOSS BITE**	**[oval, embossed]**
B684	**BOSS BITE**	**[oval, embossed]**
B685	**BOSS LUMP**	**[round, embossed]**
B687	**BOSTON**	**(a) [round, small, black on yellow, block letters, P. Lorillard & Co.]** **(b) [black on yellow, script letters]**
B689	BOSTON BEST	[round, green on white]
B691	BOSTON CREAM	[rectangular, red on yellow]
B692	BOSTON GEM	[oval, embossed]
B693	BOSTON HUB	[round, picture of a hub, P. Lorillard & Co.]
B694	BOSTON MARKET	[round, red on yellow, A.B. Co.]
B695	BOSTON NAVY	[round, brown on green, trade D mark]
B696	BOSTON TWIST	[die-cut bell, black on yellow]
B697	**BOUQUET**	**[round, embossed, B.P. & Co.]**
B698	BOUNCER	[round, embossed]
B699	BOUND TO WIN	[diamond, shape, black on brown picture, Williamson Tob. Co.]
B700	**BOURBON TOBACCO**	**[round, brown on yellow, picture of a tobacco twist]**
B701	BOWERY BOY	[round, gold on red]
B702	**BOXER**	**[die-cut embossed man]**
B703	BOXER	[rectangular, black on yellow, picture of boxer]
B704	BOY BLUE	[oval, picture of boy]
B705	BOYKINS NATURAL LEAF	[rectangular, embossed]
B707	**BRACER**	**[oval, Scotten, Dillon Company, black on yellow]**
B709	BRADLEY'S TWIST	[diamond, white on blue]
B711	**BRAG**	**(a) [oval, small, black on yellow]** (b) [black on white]
B712	BRANCH	[diamond, black on red, Strater Bros.]
B713	**BRANDY WINE**	**[rectangular, picture of two bottles]**
B714	BRANDY WINE	[rectangular, small, black on red]
B715	BRASS TACK	[round, embossed]
B717	**BRAVO**	**[round, embossed]**
B718	BREAD AND BUTTER	[round, black on yellow, Butler]
B719	BREITWIESER'S U.S.	[round, embossed]
B721	**BRER RABBIT**	**(a) [round, black on yellow, picture of rabbit, Winston]** (b) [black on orange]
B723	BRICK FACTORY	[rectangular, black on yellow, picture of factory]
B725	**BRICK HOUSE**	**[round, picture of a house]**
B726	BRICK HOUSE	[rectangular, black on red]
B727	BRIDLE BIT	[rectangular, black on yellow, picture of a bit]
B728	BRIDE CAKE	[rectangular, embossed]
B729	**BRIDE CAKE**	**(a) [rectangular, banner, blue on white]** (b) [black on yellow]
B730	**BRIDGE TO LODI**	**[rectangular, P.H. Hanes & Co., black on red]**
B731	BRIER	[die-cut heart engraved W C McDonald, Montreal]
B735	BRIGHT IDA	[round, embossed]
B737	BRIGHT MARS	[round, yellow on red]
B739	**BRIGHT NAVY**	**[oval, Buchanan & Lyall, black on red]**
B741	BRIGHT 7 OZ. TWIST	[rectangular, black on red & yellow]
B742	**BRILLIANT**	**(a) [rectangular, embossed]** (b) [round, embossed]
B743	BRISTOL	[die-cut shape, bell embossed]
B744	BRITANNIA	[rectangular, multicolored, picture of lady]
B745	BRITISH COLONEL-B C	[round, yellow on red, large B C in center]
B746	BRITISH NAVY	[oval, picture, blue on yellow]
B747	**BRITISH NAVY**	**[oval, white on blue]**
B748	**BRITISH NAVY**	**[rectangular, black on green]**
B749	**BROAD AXE**	**[die-cut axe head, gold & red, Vaughn]**
B750	BROAD GAUGE	[round, embossed]
B751	BROAD LEAF	[die-cut leaf, black on yellow, Bluegrass Tobacco Co.]
B752	BROAD SWORD	[die-cut shape, large sword, black on gold]
B753	BROTHERHOOD	[rectangular, embossed]
B754	**BROWN**	**[rectangular, embossed]**
B755	**BROWN BISCUIT PLUG**	**[round, brown on white]**
B756	BROWN BROS.	[rectangular, shield embossed]
B757	**BROWN DICK**	**[round, black on red]**
B758	BROWN DICK	[oval, black on red
B759	BROWN JUG	[round, black on red]
B760	BROWN JUG	[octagonal brown on tan, picture of jug, L.V. & Co.]
B761	**BROWN P.**	**[rectangular, embossed]**
B762	**BROWN SLAB**	**[rectangular, Scotten, black on yellow]**
B763	BROWN'S BEST NATURAL LEAF	[die-cut leaf, black on yel low]
B765	BROWN'S COUNTRY	[oval, embossed]
B766	**BROWN'S COUNTRY**	**[oval, black on white]**
B767	BROWN'S TWIST	[rectangular, embossed]
B769	**BROWN'S MULE**	**(a) [die-cut mule, black on red]** (b) [larger tag, open legs]
B771	BROWN'S SUN CURED	[round, black & red on yellow]
B775	H. BROWN	[round, black on red]
B776	BROWNIE	[rectangular, black on yellow, picture of brownie, Eshelby Tob. Co.]
B777	BROWNIE	[die-cut of a man, embossed]
B779	BRUNETTE	[die-cut heart, W.C. McDonald]
B781	BUCCANEER	[rectangular, embossed]
B783	**BUDWEISER**	**[rectangular, black on red]**
B784	**BUCK DEER TWIST**	**[rectangular, Henry Co. black on yellow, picture of deer]**
B785	**BUCK GRAVELY**	**[rectangular, black on yellow]**
B786	BUCK HORN TWIST	[round, black on red, picture of deer, The Henry County Tob. Co.]
B787	BUCK SCOTT	[oval, embossed]
B788	BUCK SKIN	[oval, embossed]
B789	BUCKET	(a) [die-cut bucket, black on green] (b) [white on green]
B790	BUCK EYE	[die-cut embossed]
B791	BUCK EYE	[odd oval embossed]
B792	**BUCKEYE**	**[rectangular embossed]**
B793	BUCKEYE	[round, white on red]
B794	BUCKEYE TWIST	[rectangular, blue on white, Bates & Killinger Tob. Co.]
B795	**BUCKLE**	**[die-cut, embossed belt buckle]**
B796	BUCKSKIN	[oval, embossed]

B797 BUFFALO [round, black & brown on yellow, picture of buffalo, Geo. O. Jones & Co.]
B798 BUFFALO [die-cut buffalo, embossed, W.L.A. Co.]
B799 BUFFALO BILL [round, red on yellow, picture of a man]
B801 BUFFOON [round, black on yellow, picture of clown]
B803 BUGLE [die-cut, embossed bugle]
B805 BUGLE [diamond, G.S.S.A. Co. black on red]
B806 BUGLE BLAST [rectangular, brown on yellow, picture of bugle]
B807 BUGLER [round, blue on red & white, picture of horse]
B808 BUGLER [round, picture of man in uniform with bugle]
B809 BULL [die-cut bull embossed, Bull Durham]
B811 BULL [die-cut head embossed, K. & R.]
B812 BULL HORN [rectangular, curved, embossed, McAlpins]
B813 BULL OF THE WOODS [round, black on yellow, picture of bull, M.L. Ogburn & Co.]
B814 BULL OF THE WOODS [rectangular, Blackburn, Harvey & Leak]
B815 BULL OF THE WOODS [rectangular, Taylor Bros. Inc.]
B816 BULL OF THE WOODS [oval, embossed]
B817 BULL OF THE WOODS [rectangular, Leak Bros. & Hastan]
B818 BULL OF THE WOODS [rectangular, Whitaker-Harvey.Co.]
B819 BULL'S EYE [round, embossed]
B821 BULL'S HEAD [die-cut, embossed, K. & R.]
B823 BULL'S HEAD [rectangular, banner, black on yellow]
B825 BULL DOG [die-cut, embossed]
B827 BULL DOG [round, black on red & yellow, picture of dog, Double Thick, G. Penn Sons Tob. Co.]
B829 BULL DOG [oval, Brown]
B830 BULL DOG [round, black on white, picture of dog's head]
B831 BULL DOG TOBACCO [rectangular, black on yellow, British-Australian Tob. Co.]
B833 BULL DOG TWIST (a) [die-cut, small dog, Our New Tag, black on yellow]
(b) [larger dog]
B835 BULL DOG TWIST CHEWING [die-cut dog, Our New Tag, black on yellow]
B837 BULLION GRADE (a) [round, small, P. Lorillard & Co.]
(b) [larger tag]
B839 BULLION GRADE (a) [round, small, black on green, block letters, P. Lorillard Co.]
(b) [round, large]
B841 BULL TONGUE [die-cut long tongue, embossed]
B842 BUMBLE BEE [round, black on orange, picture of a bee]
B843 BUN [round, embossed]
B845 BUN [cross embossed]
B847 BUNTY ROOSTER [round, black on yellow, picture of a rooster, J.A. Hartley & Co.]
B849 BURGESS [round, embossed]
B851 BURLY [round, black on red]
B852 BURR OAK [die-cut embossed, black on red]
B853 BURR OAK [die-cut small, black on red]
B854 BURR OAK [die-cut, embossed]
B855 BUSINESS [oval, embossed]
B857 BUSTER [rectangular, embossed]
B858 BUSTER [rectangular, yellow on red]
B859 BUSTER [rectangular, white on red]
B861 BUSTER CHEW OR SMOKE [rectangular, white on red]
B862 BUSTLER [round, embossed]
B863 BUT OH HOW NICE [round, red & black on yellow]
B864 BUTCHER [die-cut knife, black on red]
B865 BUTLER [rectangular, embossed]
B867 BUTLER [round, small, black on red]
B869 BUTLER'S BEST [round, yellow on black]
B871 BUTLER'S BIG FIG [round, embossed]
B873 BUTLER'S CUT [rectangular, embossed]
B875 BUTLER'S DARK SMOKE [rectangular, Butler & Bosher]
B876 BUTTER'S FANCY CHEW [rectangular, Butler & Bosher]
B877 BUTLER'S GILT EDGE [die-cut, embossed leaf]
B879 BUTLER'S LIGHT SMOKE [rectangular, Butler & Bosher]
B881 BUTLER'S OLD PEACH (a) [die-cut peach, green on red]
(b) [black on yellow]
B883 BUTLER & BOSHER [round, embossed]
B884 BUTT CUTT [round, black on yellow]
B885 BUTTER [round, red on yellow]
B887 BUTTER CUP [odd square, brown on yellow, picture of flower, P. Lorillard & Co]
B889 BUTTERFLY [die-cut, embossed butterfly]
B891 BUTTERFLY [round, black & green on white, picture of a butterfly, P. Lorillard & Co.]
B893 BUTTON ROUND [die-cut, embossed, Wilson & McCallay Trade Mark]
B895 BUTTON [die-cut, embossed button, St. Charles, Mo.]
B896 BUZZ [die-cut saw blade, embossed]
B897 BUZZ SAW [die-cut, embossed saw blade]
B899 BY JINGO (a) [round, embossed, F. MacV. & Co.]
(b) [round, larger embossed]
B903 BY JINGO [rectangular, scalloped, W. & McC. Tob. Co. white on red]
B904 BY JINGO [rectangular, scalloped, white on red, A.T. Co.]
B905 BY JOE [round, embossed, Strater Bros.]
B907 BY JOE [round, Strater Bros., red on yellow]

C

C101 C (a) [die-cut, embossed, green]
(b) [red]
(c) [silver]
(d) [gold]
C103 C [round, small, embossed]
C105 C [die-cut, large]
C107 C [die-cut, black on red, Bailey Bros "C"]
C109 C [rectangular, black on red]
C110 C [die-cut large, red on yellow, Hess Elery & Co.]

C111	**CA**	**[rectangular, embossed]**
C112	C.A. RICHMOND CO	[round, with four points, black on red, Virginia]
C113	C A R	[die-cut letters, black on yellow, Caser & Wright Mfrs., Winston, N.C.]
C114	C A R	[die-cut letters, red]
C115	**C.B.**	**[round, embossed]**
C116	C & E	[round, black on yellow, picture of a head facing left]
C117	C & E	[oval, black on yellow, picture of a head facing left]
C119	**C E D**	**(a) [die-cut letters, Driver Brothers, black on white]**
		(b) [black on yellow]
C120	C.E. KENWORTH	[die-cut letter K, embossed]
C121	C.E.S.	[round, embossed]
C123	**C & H**	**[square, embossed]**
C124	C & H	[rectangular, embossed]
C125	C I B	[rectangular, large CIB, red on black, Charles Isaac Branan]
C126	**C K T Co**	**[die-cut cross, embossed]**
C127	C.O.D. 5¢	[diamond, embossed]
C128	**C.M. & Co. 6**	**[round, gold on black]**
C129	**C.M.K.**	**[round, gold on black]**
C131	C.P.	[odd shape, cut-out letters]
C133	C.S. & CO'S 5 S	[oval, embossed]
C135	C T B	[round, red on yellow]
C137	**C T C**	**[die-cut, embossed, three leaf]**
C139	**C W A**	**[cut-out, rectangular]**
C141	**C W A**	**[rectangular, embossed]**
C142	C.W.C.	[round, black on yellow]
C143	C. & W.	[round, yellow on blue, chew smoke]
C144	CABIN	[die-cut cabin, embossed, Hatcher & Stamps]
C145	**CABIN HOME**	**[rectangular, odd, embossed]**
C146	CABIN HOME	(a) [round, yellow on green & red, F.M. Bohannon]
		(b) [red on blue & yellow on red]
C147	**CABINET.**	**[round, small, black on red]**
C148	CABINET	[round, black on yellow]
C149	**CABINET**	**[oval, B.F. Gravely & Sons, black on red]**
C150	CABINET	[oval, black on red, W.D. & Co.]
C151	THE CABINET	[round, Pres. Cleveland & heads of men in his cabinet, Pegream & Penn]
C153	**CADILLAC**	**[round, Scotten Tobacco Co. black on gold & red]**
C154	CADILLAC	[round, Scotten Plug black on gold & red]
C155	**CAKE**	**[round, embossed]**
C157	CAKE BOX	[rectangular, embossed]
C159	**CALHOUN**	**(a) [oval, black on yellow, picture of man, D.H. Spencer & Son.]**
		(b) [oval, black on orange-yellow, older tag]
C161	**CALIFORNIA**	**[round, silver on black]**
C163	CALIFORNIA TWIST	[oval, embossed]
C165	CALLOWAY COUNTY TWIST	[rectangular, octagon, black on yellow]
C167	CALUMET	[round, red on white]
C169	**CAMEL**	**[die-cut large camel, brown on yellow]**
C171	CAMERON TOBACCO COMPANY	[round, embossed]
C172	CAMERON & BROS. WM.	[round, black & red on tan]
C173	CAMERON'S LION	[round, black on red, picture of a lion]
C174	CAN	[round, black on yellow]
C175	CANADA NAVY	[round, small, black on blue]
C176	CANADA NAVY	(a) [round, small, white on blue]
		(b) [large, white on blue]
C177	CANADA NAVY	[round, black on yellow]
C179	CANARY	[round, picture of bird, white on yellow]
C180	**CANARY**	**[round, yellow on green, picture of a bird]**
C181	**CANDY**	**[rectangular, red on white stripes]**
C182	CANEY FORK	[round, green on yellow]
C183	**CANNON**	**[die-cut embossed cannon]**
C184	CANNON BALL	[rectangular, red & yellow on black, picture of cannon]
C185	**CANNON BALL**	**[rectangular, Taylor Bros., black on yellow]**
C187	**CANNON BALL**	**[round, black on yellow]**
C189	CANNON BALL	[rectangular, black on red, picture, Ogburn]
C190	CANT B BEAT	[die-cut B, embossed]
C191	**CANTEEN**	**[round, white on blue]**
C193	CANTEEN	[round, embossed]
C194	CANTELOPE	[rectangular, black on yellow]
C195	**CANTELOPE**	**[rectangular, green on yellow, picture of cantelope in a dish]**
C196	CANTRAL	[round, embossed]
C197	CANUCK	[rectangular, black on yellow]
C199	CAP GREENVILLE	[round, embossed]
C201	CAPITAL TWIST	[oval, black on yellow, F.A. Davis & Co.]
C202	CAPITOL	[oval, black on yellow, A.H. Motley & Co., Reidsville, N.C.]
C203	**CAPITOL**	**[die-cut, embossed building]**
C204	CAPITOL	[oval, embossed]
C205	**CAPITOL**	**[round, embossed, E. & M.]**
C206	CAPITOL	[rectangular, embossed]
C207	CAPITOL CITY	[oval, embossed]
C208	CAPSTAN PLUG TOBACCO	[oval, gold on red, W.D. & H.D. Willis, Sydney]
C209	CAP'T. JACK	[rectangular, black on yellow, P.H. Hanes & Co.]
C210	CAPT. CLARK'S STEAM BOAT SMOKE	[rectangular, white on red]
C211	CAPTAIN KIDD	[oval, black on red]
C213	CAR TICKET	[rectangular, embossed]
C214	CAR WHEEL	[round, yellow on red, Gravely]
C215	CARDINAL	[rectangular, black & red on white, picture of bird, P.B. Gravely & Co.]
C216	CARDINAL 12 IN	[rectangular, black & red on yellow, picture of bird, P.B. Gravely & Co.]
C217	CARHART'S	[rectangular, red on yellow]
C218	CARLISLE	(a) [large round, yellow on red]
		(b) [small]
C219	CARL'S BEST	[round, black on green]
C220	CAROLINA BOSS	[odd oval, embossed]
C221	CAROLINA CHOICE	[embossed]
C220	CAROLINA MOUNTAIN PINK	[octagonal, black on yellow, Moody]
C223	CAROLINA SOUTH	[embossed]
C224	CAROLINE	[round, black on yellow]
C225	**CAROMEL**	**[rectangular, yellow on red]**
C226	CAROMEL	[round, gold on red]
C227	**CARTRIDGE**	**(a) [oval, P. Lorillard Co., brown & silver on green]**

		(b) [brown & silver on red]
C229	**CASH GROCER**	**[diamond, W.B.T. Co., black on red]**
C231	**CASH VALUE**	**[rectangular, embossed]**
C232	CASH VALUE	[rectangular, black on red]
C233	**CASTLE**	**[die-cut, embossed castle]**
C234	CAT	[die-cut shape, cat standing, embossed]
C235	CAT	[oval, black on yellow & brown, picture of a cat, Parker & Foster]
C236	CAT	[round, black on yellow, picture of a cat's head]
C237	CAT- BIRD	[rectangular, white, red & blue]
C238	**CAT HEAD**	**[round, embossed]**
C239	**CATALOGUE**	**[rectangular, black on yellow, picture of a cat on a log]**
C241	**CATAWBA GRADE**	**[round, small, P. Lorillard & Co., black on yellow]**
C243	CATCH PENNY	[round, embossed]
C245	CATCH ME WILLIE	[rectangular, black on yellow, picture of a boy running]
C247	**CATCHER**	**[oval, red on black & yellow, picture of a baseball catcher]**
C248	**CATCHER**	**[round, black on green, picture of a baseball catcher]**
C249	CATO	[round, embossed]
C250	CAVALIER	[small oval, black on red]
C251	CAVE	[round, brown on cream, picture of a horse]
C252	CAVE	[round, red & black on white, picture of a donkey]
C253	**CAVENDISH**	**[oval, small, Richmond Co., black on yellow]**
C255	CAVENDISH	[round, black on green, P. Lorillard & Co.]
C257	CECIL	[black on yellow, Clark & Co., Bedford, Va.]
C261	CEINGSTONE	[round, embossed]
C263	**CELEBRATED VIRGINIA**	**[rectangular, scroll, J.H. Maclin, black on yellow]**
C264	CENTER RUSH	[round, embossed]
C265	THE CENTRAL UNION	[round, The U.S. Tob. Co., blue on white & red]
C266	**CENTRAL UNION**	**[round, white on blue & red, The U.S. Tob. Co., Richmond, Va.]**
C269	CERES	[oval, black on red]
C271	**CHAIN**	**[die-cut, embossed chain links]**
C272	CHAMBER MAID	[rectangular, embossed]
C273	CHAMP	[round, red on white]
C275	CHAMP CARTER	[round, yellow on black, T.C. Williams Co., Richmond, Va.]
C277	CHAMPAGNE	[die-cut bottle, embossed]
C279	CHAMPAGNE CHAW	[rectangular, red on yellow]
C280	**CHAMPION A**	**[round, embossed]**
C281	CHAMPION	[rectangular, picture of two black boys, Penn Tobacco]
C282	CHAMPION	[small rectangular, black on red]
C283	CHAMPION	[rectangular, yellow on black, B.F. Gravely & Sons]
C284	**CHAMPION**	**[oval, black on red, R.A. Patterson Tob. Co.]**
C285	CHAMPION PLUG TOBACCO	[oval, gold on red]
C286	CHAMPION NATURAL	[oval, gold on red]

	FLAVOR	
C287	CHAMPION TOBACCO	[oval, embossed]
C288	**CHAPMAN'S 900 TOBACCO**	**[round, black on red]**
C289	CHARIOT	[oval, black on yellow]
C290	CHARLES ISAAC	[rectangular, black on red]
C291	CHARLIE ASHBY	[oval, black on yellow, picture of a man & dog, F.R. Penn & Co.]
C292	CHARLIE ASHBY	[octagonal, picture of a man & dog, F.R. Penn & Co.]
C293	CHARLIE PORTER	[oval, embossed]
C294	CHARLIE SHORT	[rectangular, black on yellow]
C295	**CHARM**	**[odd shape, embossed, T.C. & Co.]**
C297	CHARMER	[round, yellow on red]
C299	**CHAW DOG**	**[rectangular, embossed]**
C301	**CHAW DOG**	**[diamond, black on red]**
C303	**CHECK**	**[rectangular, black on white]**
C305	**CHECK**	**[rectangular, black on red]**
C307	**CHECK**	**[round, embossed]**
C309	**CHECK 7-20-8**	**(a) [die-cut large shield, embossed, D.M. & Co.]**
		(b) [die-cut small shield, embossed]
C311	CHECKMATE	[rectangular, embossed]
C312	CHERRY BLOSSOM	[round, red on yellow & green, picture of a tree, Registered]
C313	CHERRY BLOSSOM	[round, embossed, Farber]
C314	CHERRY FLAVOR	[round, black on yellow, multicolored picture of cherries]
C315	CHERRY LEAF	[round, yellow on black, multicolored picture, Sternberg & Sons]
C316	CHERRY RED	[round, red on silver, picture of a cherry, G. Penn Sons Tob. Co.]
C317	CHERRY RED	[round, red on white, picture of a cherry]
C319	CHERRY RIPE	[octagonal, embossed]
C321	**CHESS**	**[round, Scotten, Dillon Company, yellow on red]**
C337	CHESTNUT BURR	[round, yellow on brown]
C323	**CHESTNUTS**	**[rectangular, black on red]**
C339	CHEW AMERICAN BOY	[rectangular, black on white & red, blue]
C340	CHEW BIG LEVER TOBACCO	[round, multicolored pic ture, Taylor Bros]
C341	CHEW COTTON BELT	[octagonal black on red, picture of a horse & cart, Tobacco]
C342	CHEW GOLD SEAL	[diamond, black on gold]
C324	CHEW ME	[round, red on yellow, A.J. Town Co.]
C343	CHEW NOTARY SEAL	[round, black on red]
C344	CHEW TELEPHONE TOBACCO	[rectangular, embossed]
C325	**CHIC**	**[octagonal, yellow on red]**
C326	**CHICAGO**	**[round, small scalloped, black on orange, picture of a girl running]**
C327	CHICAGO	[round, black on yellow, cartoon girl with windblown hair]
C329	**CHICKEN**	**[oval, Crusoe Bros. Co., black on red]**
C330	**CHICKEN**	**(a) [die-cut small, chicken embossed]**
		(b) [die-cut large]
C331	CHICKEN	[round, red on yellow, Crusoe Bros. Co.]
C332	**CHIEF**	**[die-cut cross, embossed]**

C333	**CHIEFTAIN**	**[oval, black on red]**
C335	CHINESE CHARACTERS	[rectangular, embossed]
C346	**CHIP**	**(a) [hexagon, white on green, Hancock Bros. & Co.]** (b) [white in blue]
C347	CHIP	[round, embossed]
C349	**CHOCOLATE**	**[die-cut banner, red on yellow]**
C351	**CHOCOLATE CAKE**	**[oval, embossed]**
C353	CHOCOLATE CREAM	[die-cut shield, McAlpin]
C355	**CHOICE**	**[round, Bohannon, yellow & black on red]**
C357	**CHOICE**	**[round, embossed, R.J. Reynolds]**
C358	CHOICE	[round, embossed, cut-out cross in center, Gravely]
C359	**CHOICE**	**[round, embossed, Gravely]**
C361	CHOICE	[die-cut P, embossed]
C362	CHOICE	[rectangular, black on red & yellow, L. Ash]
C363	CHOICE	[round, yellow on red, Brown & Williamson]
C364	CHOICE	[rectangular, embossed]
C365	CHOICE CREAM	[round, red on yellow]
C367	**CHOPPS**	**[rectangular, black on yellow]**
C369	CHOW DOG	[diamond, black on red]
C370	CHRISTIAN	[round, embossed, picture of a crown]
C371	**CHRISTIAN CROSS**	**[die-cut, gold cross]**
C372	**CHRISTMAS TREE**	**[die-cut tree embossed]**
C373	**R. J. CHRISTIAN**	**[round, embossed]**
C375	CHROMO	[round, black on red, B & L]
C377	**CHUCK**	**[round, embossed]**
C378	CHUCK	[die-cut dice, black on white]
C379	**CHUCK A LUCK**	**[die-cut cage, embossed]**
C381	CHUNK	[round, embossed]
C382	CHUSO	[rectangular, black on blue]
C383	CHUSO	[triangular, black on yellow]
C384	CINCH	[rectangular, embossed]
C385	CIRCLE	[round, yellow on red]
C387	CIRCLE	[round, embossed]
C389	**CIRCLE TOBACCO**	**[die-cut ring, embossed]**
C391	CITY STOCK	[odd rectangular, embossed]
C393	CITY TALK	[rectangular, embossed]
C395	CLAPBOARD	[rectangular, embossed]
C397	CLARET	[oval, black on red]
C398	**CLARET**	**[oval, black on red]**
C399	**CLAW HAMMER**	**[die-cut shape, black on red, Strater]**
C401	CLAW HAMMER	[die-cut shape, black on red]
C403	CLAW HAMMER	[round, black on yellow, picture of hammer]
C405	CLAY	[round, embossed]
C407	**CLEAN SWEEP**	**[diamond, embossed, picture of broom]**
C409	**CLEAN SWEEP**	**[round, embossed]**
C410	CLEAN SWEEP	[rectangular, embossed, picture of a broom]
C411	**CLEAR QUILL**	**[die-cut embossed, I.C. & M., Chicago]**
C412	CLEAR THE TRACK	[oval, black on yellow, picture, Stafford]
C413	CLEAR TRACK	[rectangular, white & black on red, picture of a train]
C414	CLEAVER	[rectangular, black on yellow, picture of a cleaver, Liipfert & Jones]
C415	**CLEOPATRA**	**[rectangular, embossed]**
C416	CLEOPATRA	[round, embossed, picture of a woman's head]
C417	CLEOPATRA	[oval, embossed]
C418	CLEVELAND BAKERY	[die-cut large B & C, embossed]
C419	CLEVELAND-HENDRICKS	[rectangular, two rounds, black on red, picture of two men]
C421	CLEVELAND-HENDRICKS	[rectangular, two rounds, black on yellow, picture of two men]
C425	CLEVELAND ROLLING MILL CO. 10 1/2	[round, embossed]
C429	CLIFTON	[rectangular, banner embossed]
C431	CLIMAX	[oval, embossed]
C433	**CLIMAX GRADE**	**(a) [large round, black on gold, P. Lorillard & Co.]** (b) [small round black on red]
C435	CLIMAX GRADE	[small round, in script, black on green]
C437	**CLIMAX PLUG**	**[round, black on red, Lorillard]**
C439	CLINCH	[rectangular, embossed]
C441	CLINCHER	[diamond, G.S.S. & Co., black on red]
C443	**CLING PEACH**	**[octagonal, black on yellow & orange, picture of peaches]**
C445	CLIP	[oval, blue on white]
C446	CLIP	[round, black on red]
C447	**CLIPPER**	**(a) [rectangular, large, white on red]** (b) [small]
C448	CLIPPER	[round, embossed]
C449	CLIPPER	[oval, embossed]
C451	**CLIPPER L & M**	**[die-cut oval, embossed]**
C453	**CLIPPER L & M**	**[oval, embossed]**
C455	**CLIPPER L & M**	**[round, ring embossed]**
C457	**CLIPPER L & M**	**[round, embossed small hole]**
C459	CLIPPER NAVY	[rectangular, red & blue on white]
C460	CLOCK	[round, black on white, picture of clock face]
C461	CLOCK	[round, embossed clock face]
C462	CLOSE CALL	[rectangular, black on yellow, picture of man after melons, Hancock Bros.]
C463	CLOSE FIGURES	[round, black on orange, picture of a man & woman]
C465	**CLOSE SHAVE**	**[round, picture, red on yellow]**
C467	CLOTH OF GOLD	[rectangular, odd, yellow on red]
C469	**CLOVER LEAF**	**[die-cut embossed, three-leaf clover]**
C471	**CLOVER LEAF**	**[die-cut embossed, four-leaf clover]**
C472	**CLOVER LEAF**	**[round, three leaf, green on white]**
C473	**CLOVER BLOOM**	**(a) [round, green on orange, picture of a clover leaf & bloom]** (b) [green on yellow]
C475	**CLOVER BLOSSOM**	**[round, black on white & red, picture of a blossom]**
C477	**CLOWN**	**[round, picture of a clown, P. Lorillard & Co.]**
C478	CLUB	[round, embossed, picture of a house]
C479	**CLUB**	**[die-cut clover leaf, embossed]**
C480	CLUB	[die-cut clover leaf, black on yellow]
C481	**CLUBS**	**[rectangular, black on red, picture of card clubs]**
C483	CLUBS	[round, embossed, picture of card clubs]

C485	CLUNET, BROWN & CO.	[die-cut crescent man in the moon embossed]
C486	**C. McL & CO.**	**[rectangular, embossed]**
C487	COANS CLEAR CUT	[diamond, embossed]
C488	**COCA**	**[rectangular, embossed]**
C489	COCHRAN J.L.	[round, embossed]
C491	COCHRAN J.L.	[diamond, embossed]
C492	COCHRAN'S TWIST	[oval, embossed]
C493	COCK ROBIN	[rectangular, black on yellow, picture of a bird]
C494	COCK ROBIN	[rectangular, black on yellow, picture of a multicolored bird]
C495	**COCK SPUR**	**[round, white on red, Bailey Bros.]**
C496	COCKSPUR	[round, black on yellow, picture]
C497	**COCKTAIL PLUG**	**[round, embossed]**
C499	COFFEE POT	[rectangular, black on yellow, picture of a coffee pot, H. & D. Good & Strang]
C500	COFFEE POT	[rectangular, black on red, picture of a coffee pot, H & R Good & Strang]
C501	**COGNAC COCKTAILS**	**[die-cut glass, P. Lorillard & Co., black on yellow]**
C503	COGNAC COCKTAILS	[rectangular, multicolored, picture of glass]
C504	**COLD WAVE PLUG**	**[square, Tinsley Tobacco Co., black on white]**
C505	**COLONIAL TWIST**	**[diamond, E. Shelby Tob. Co., black on yellow]**
C507	COLUMBIA	[oval, flag of Columbia]
C508	COMBINATION	[diamond, embossed, L]
C509	COME AGAIN	(a) [round, large, red on gold, Blackburn & Dalton] (b) [small]
C511	**COMET**	**[oval, embossed]**
C512	**COMET**	**[die-cut shape, star & tail, black on red]**
C513	**COMET**	**[round, embossed]**
C515	**COMET**	**[round, smaller embossed]**
C516	COMFITS	[oval, embossed]
C517	COMFORT	[round, embossed]
C519	COMING RACE	[rectangular, black on yellow, Stanley & Finley]
C520	**COMMERCIAL TRAVELER**	**[round, black on red]**
C521	COMMERCIAL TRAVELER	[Daniel Scotten & Co.]
C522	COMMON SENSE	[rectangular, black on yellow]
C523	COMMON SENSE	[oval, embossed]
C524	COMMON WEALTH	[diamond, shape, black on red]
C525	COMMON WEALTH	[round, black on red]
C526	**COMPASS**	**[round, embossed, picture of compass]**
C527	COMPASS	[round, embossed]
C529	COMPLETE	[rectangular, black on orange, F.A. Davis & Co]
C530	COMRADE PLUG	[round, red, white & blue]
C531	**CONCORD GEM**	**[round, black on red]**
C533	CONE	[oval, embossed]
C534	CONFIDENCE	[diamond, embossed]
C535	**CONGRESS**	**[round, small, L. & B. Tob. Co., black on red]**
C537	CONGRESS OLD NAVY	[round, black on yellow]
C538	**CONNECTICUT**	**[rectangular, Wingfield & Taylor, black on gold]**
C539	CONQUEROR	[octagonal, embossed]
C540	CONQUEROR	[round, embossed]
C541	**CONQUEST**	**[hexagon, red on yellow]**
C543	CONSTITUTION GRADY	[round, white on black & brown, Bitting & Hay]

C545	**CONTINENTAL**	**[oval, Daniel Scotten & Co., black on red, yellow, blue]**
C546	CORAL	[round, black on red, W.H.I. Hayes]
C547	CORA MOORE	[rectangular, black on yellow, picture of a girl, Blackburn, Dalton & Co.]
C548	**CORK SCREW**	**[round, black on yellow]**
C549	CORK SCREW	[oval, black on yellow, picture of a cork screw, tag made in U.S.A.]
C551	CORDUROY	[rectangular, embossed]
C553	**CORKER**	**[oval, black on white, red & green]**
C555	CORKER	[round, embossed]
C557	**CORN BREAD**	**[rectangular, blue on white]**
C558	CORN BUG	[rectangular, red on black]
C559	CORN CRACKER	[oval, embossed]
C560	CORN JUG	[rectangular, red on black]
C561	**CORN JUICE**	**[die-cut ear of corn, embossed]**
C563	**CORN JUICE**	**[die-cut bottle, white on red]**
C565	CORN JUICE	[oval, embossed, picture of ear of corn]
C567	CORN JUICE	[rectangular, red on black]
C569	CORN JUICE	[round, embossed]
C571	**CORN PLANTER**	**[rectangular, black on red]**
C573	**CORNER STONE**	**(a) [small rectangular, embossed]** (b) [larger rectangular]
C575	CORNER STONE	[square, embossed]
C576	**CORONET**	**[die-cut crown, embossed]**
C577	COSBY'S CHALLENGE	[rectangular, embossed]
C578	COSBY'S HAMBONE	[octagonal black on yellow, picture of a black man with bone]
C579	**COSTA RICA**	**[oval, flag of Costa Rica]**
C581	COTTAGE HOME	[die-cut B, black on red]
C583	COTTON AIDER	[oval, embossed]
C585	COTTON BALL	[round, black on red, picture of a ball of cotton]
C587	**COTTON BOLL**	**(a) [round, small, black on silver]** (b) [black on gold] (c) [black on brown]
C589	COTTON BELT	[octagonal, black on red, picture of negroes in cart]
C590	COTTON EYED JOE	[round, embossed]
C591	COTTON PLANT	[rectangular, black on red]
C592	**COUNTRY JUDGE**	**[oval, black on yellow]**
C593	COUNTRY LAD	[oval, black on yellow, picture of a boy, Cumberland Tobacco Works]
C594	COUNTRY MADE	[round, red on yellow, picture of a girl's head]
C595	COW BELL	[die-cut bell, black on red]
C596	**COW BELL TWIST**	**[square, black on yellow]**
C597	**COW BOY**	**[rectangular, embossed]**
C598	**CRACK A JACK**	**[round, Liipfert, Scales & Co., black on red]**
C600	CRAB	[round, embossed]
C601	CRACKER BOY	[rectangular, scroll embossed]
C603	CRACKER JACK	[rectangular, odd embossed]
C605	CRAFTS-MY TOM	[oval, black on red, picture of man]
C607	**CRANE**	**[oval, picture of a bird, black on yellow, white & green]**
C609	**CRAPPS**	**[square, embossed]**
C610	CRAPS	[rectangular, picture of two pair of dice, 7 & 11]
C611	CRAPS	[rectangular, embossed, picture of two dice]

C613 CREAM OF N.C. [rectangular, black on tan, picture of a girl milking cow]
C614 CREAM OF THE SOUTH [round, black on cream]
C615 CREAM NUT [oval, black on yellow]
C617 CREAM PUFF [round, black & brown on yellow, picture of a cream puff]
C619 CREOLE [round, embossed]
C621 CREOLE [round, embossed, C.W. Allen]
C623 CREOLE GIRL [rectangular, bow shape, embossed]
C625 CREOLE GRADE [round, red on white]
C627 CRESCENT (a) [small die-cut crescent, embossed]
(b) [larger]
C629 CRESCENT [round, embossed, S.R. & Co.]
C631 CRESCENT [die-cut crescent, black on red]
C633 CRESCENT GREENVILLE [diamond, black on yellow]
C635 CREST OP [round, black on red & yellow, picture of a lion]
C637 CREWS BIG 4 [round, white on red]
C639 CREWS-44 [round, black on brown, picture of a bullet]
C641 CREWS NATURAL LEAF [round, black on green, picture of a leaf]
C642 CREWS TOLU [round, red on yellow]
C643 CREWS TRIUMPH [round, black on blue]
C644 CRITIC [banner embossed]
C645 CRICKET [rectangular, embossed]
C646 CRICKET [rectangular, odd embossed, W.H.L.]
C647 CROCODILE [die-cut crocodile embossed, K. & R.]
C649 CROCODILE [round, yellow on black, picture of crocodile, T.C. Williams Co.]
C651 CROSS BOW [die-cut cross bow, embossed]
C653 CROSS [die-cut cross]
C654 CROSS CUT [oval, embossed, P & M]
C655 CROSS CUT SAW [die-cut saw, embossed]
C657 CROSS MARK [round, red on yellow, picture of X in center]
C658 CROSS MARK [round, orange on black, picture of X in center]
C659 CROSS MARK [round, orange on black, Stanley & Finley]
C661 CROSS SWORDS [rectangular, embossed, Ogburn Hill & Co.]
C663 CROWN [die-cut embossed crown]
C665 CROWN [diamond, embossed crown, C.K. Allen]
C666 CROWN [round, embossed]
C667 CROWN [die-cut crown, black on gold]
C669 CROWN JEWEL [octagonal, picture of crown, black on red]
C670 CROWN JEWEL [diamond, embossed, J. N. W. & Co. Danville, Va.]
C671 CROWN OF DIAMONDS [round, embossed]
C673 CROWNED HEADS [rectangular, bow tie shape, black on yellow]
C675 CRUSADER [round, embossed]
C677 CRUSHER [rectangular, black on red]
C679 CRUSOE BROS. CO [oval, picture of rooster, black on red]
C681 CUBA FREE [round, black on red, Henry County Tobacco Co.]
C682 CUBAN [oval, black on red, W.H.I. Hayes]
C683 CUBAN (a) [diamond, black on red]
(b) [diamond, black on yellow]
C685 CUP [die-cut large, embossed cup]
C687 CUP [die-cut small, embossed cup]
C689 CUP [die-cut large cup embossed, Greenville]
C690 CUPID [oval, white on blue & black, picture of Cupid, P. Lorillard & Co.]
C691 CURB STONE [round, yellow on black & red]
C692 CURRENCY [die-cut snow shoe, embossed]
C693 CURRY [round, black on yellow & brown, picture of a horse head, Tunis & Norwood]
C694 CUSHINGS [rectangular, embossed]
C695 CUT RATE [rectangular, odd, red & yellow on cream]
C696 CUT SHORT [oval, embossed]
C697 CUT SHORT [rectangular, brown on yellow, picture of a man & donkey]
C698 CUTE [round, black on yellow, picture of girl]
C699 CUTTER [round, red on yellow]
C701 CUTTER [small oval, embossed]
C702 CUTTER [large oval, embossed]
C703 THE CUTTER [rectangular, picture of a ship, black on yellow]
C705 CYCLONE [oval, black on yellow]
C707 CYCLONE [round, embossed]

D

D101 D [cut-out]
D103 D [round, small embossed]
D105 D & D [die-cut criss-cross, embossed, D & D]
D107 D & D [round, embossed]
D108 D 49 [round, blue & red on white]
D109 D.L. [rectangular, embossed]
D111 D-W [die-cut black on yellow, Danforth and Willard]
D113 D.E.W. [rectangular, scalloped, embossed]
D114 D H D [diamond, black on yellow]
D115 D.L.G. [die-cut letters]
D116 D.M. & CO. [square, black on red]
D117 D.O.L. [die-cut letters]
D119 D-S [rectangular, black on white]
D121 D.S. CO. [die-cut star small, embossed]
D123 D.S. CO. [round, small embossed]
D125 D.T.C. [diamond, black on red]
D127 DAD'S CHOICE [round, black on silver]
D128 DAD'S HOME SPUN TWIST (a) [round, large letters, black on red]
(b) [round, small letters, black on red]
D129 DADDY'S DOLLAR [round, embossed]
D131 DAILY BREAD [odd rectangular, embossed]
D133 DAILY CALL [oval, embossed]
D135 DAISY [round, cut-out embossed, H. Wirgman]
D137 DAISY [round, embossed, S. A. Ogburn]
D139 DAISY [round, cut-out large]
D141 DAISY C.V. [rectangular, embossed]

D143	DAISY KING	[round, white on red, D.F. King, Leaksville, N.C.]
D145	DAISY TWIST	[round, gold on red, Burley Tobacco Co.]
D146	DALTON FARROW & Co.	[octagonal, embossed]
D147	DAN PATCH	[rectangular, picture of a horse]
D149	DANBURY	[rectangular, odd embossed]
D151	**DANDY**	**[round, embossed]**
D152	DANDY 5	[octagonal, embossed]
D153	DANDY LION	[round, green on black & yellow]
D154	DANDY SMOKE	[round, red on yellow]
D155	**DANIEL BOONE**	**[round, embossed arrows in center]**
D157	**DANIEL BOONE**	**[round, embossed lines in center]**
D161	**DANIEL SCOTTEN & Co.**	**[oval, small, black on red, yellow & green]**
D162	DANIEL'S FIG	[hexagon black on yellow, Henry Co.]
D163	**DARBY & JOAN**	**(a) [gold square embossed, C.W. Allen Plug]**
		(b) [silver]
		(c) [blue]
D167	**DARK**	**[rectangular, picture of horse, black on red]**
D169	**DARK CUT**	**[diamond, Henry Co. red on black]**
D170	DARK CUT	[die-cut diamond, black on red, B & D]
D171	DARK HORSE THE	[round, black on red, picture of a horse & rider]
D175	DARK TWIST	[round, white on blue, Brown & Williamson]
D177	DARLING FANNY PANCAKE	[round, black on yellow, pic ture of a girl]
D179	**DAUNTLESS**	**[round, black on red, picture of a man in armor]**
D181	DAUSMAN	[round, embossed, picture of a horse's head]
D183	DAUSMAN	[rectangular, banner embossed]
D184	DAUSMAN	[round, embossed]
D185	**DAY O WORK**	**[square, black on red]**
D186	**DAY'S WORK**	**(a) [diamond, Strater Brothers, black on red]**
		(b) [with branch]
D187	**DAZZLER**	**[round, gold on black, Walker Bros.]**
D188	DAZZLER	[rectangular, embossed]
D189	THE DEACON'S PLUG	[die-cut shield, embossed]
D190	DEAR BOY	[round, black on yellow, picture of a boy's head]
D191	DEATON'S	[round, red on yellow]
D192	DEAL	[rectangular, white, brown & orange, picture of a square]
D193	DEBT PAPER	[die-cut banner, embossed]
D194	DEBT PAYER	[diamond, embossed]
D195	**DEER HEAD**	**[die-cut small, embossed head]**
D196	DEER HEAD	[die-cut large, embossed head]
D197	**DEER SKIN**	**[diamond, black on yellow]**
D199	DEFIANCE	[rectangular, black on yellow, Lorillard]
D201	DEFIANCE	[rectangular, embossed]
D203	DEFIANCE PLUG	[rectangular red, white & blue, Lorillard]
D204	**DELAWARE TWIST**	**[round, F. & H. Tob Co., white on red]**
D205	DELIGHT	[round, black on red]
D206	DELTA	[round, black on white, black triangle in center]
D207	DELTA	[oval, black on red]
D208	DEMOCRAT	[rectangular, red on black]
D209	DEN MAR	[rectangular, black on red]
D210	**DERBY**	**[oval, black in yellow, picture of a cap, red, white & blue]**
D211	DERBY	[die-cut hat, black on red, P. Ritchie Co., trademark]
D213	DEVIL	[round, red on white]
D214	DEW DROP	[die-cut embossed, W.W.T.]
D215	**DEW DROP**	**(a) [oval, embossed, red]**
		(b) [silver]
		(c) [gold]
D217	DEWEY TWIST	(a) [round, small white on red, Moore]
		(b) [round, large white on red, Moore]
D218	DEWEY TWIST	[round, black on yellow]
D219	DEXTER	[oval, embossed]
D221	DIADEM	[round, black on yellow, Johnson Humphrey]
D222	DIAMOND	[rectangular, black on yellow, B.F. Gravely & Sons]
D223	**DIAMOND**	**[diamond, embossed]**
D224	DIAMOND	[rectangular, red diamond on white, D.J.F. & Co.]
D225	**DIAMOND**	**[round, cut-out diamond]**
D226	DIAMOND	[rectangular, black on red]
D227	DIAMOND	[round, black on yellow, D in center, Sun Cured]
D229	**DIAMOND CROSS**	**[diamond, black on silver]**
D231	**DIAMOND CROWN**	**[oval, embossed]**
D232	**DIAMOND IN THE ROUGH**	**[round, F.M. Bohannon, black on red, white, blue]**
D233	DIAMOND TOBACCO	[diamond, black on red, H.H. Reynolds]
D234	DIANA	[rectangular, gold on red]
D235	DIANORA	[round, black on yellow, picture of a girl, Pace]
D236	DICE	[rectangular, embossed, two dice, 5 & 5, red & blue]
D237	**DICE**	**[die-cut shape, black on white dots, 6, 3, & 5]**
D239	**DICE**	**[round, black on white, picture of dice]**
D241	DICE	[round, black on yellow, picture of dice]
D243	**DICK MIDDLETON & Co.**	**[rectangular, embossed, red]**
D245	**DICK TURPIN**	**[rectangular, picture of a man on a horse, black on red]**
D246	DICK'S BUCKET TWIST	[round, black on yellow, picture of a bucket]
D247	**DICK'S PET**	**[oval, black on yellow, picture of a horse running]**
D248	DICK'S HIGH GRADE A A	[octagonal, embossed]
D249	DIDO	[octagonal, black on red, picture of a girl, Brodnax & Co.]
D250	DIEM 10¢ NAVY	[round, black on cream]
D251	DIETZ & HARRISON	[die-cut bear]
D253	**DILL'S BEST**	**[oval, picture of girl, green & red]**
D255	DIME	[round, embossed, looks like a dime]
D257	DIME	[round, silver on black, looks like a dime]
D258	DIME SMOKE	[round, black on yellow, J. Wright Co.]
D259	**DIPLOMA**	**[oval, embossed]**
D260	DIPLOMAT	[rectangular, black on red & yellow, Natural Leaf, Ryan-Hampton Tob. Co.]

D261	**DIPPER**	**[die-cut small dipper, black on red]**
D263	**DIPPER**	**[die-cut large dipper, black on red]**
D265	DIPPER	[die-cut embossed shape, large]
D267	DIRECT DISPATCH	[rectangular, black on yellow, picture of man]
D269	**DIXIE**	**[rectangular, embossed]**
D271	DIXIE	[oval, white on yellow & red, picture of a leaf]
D273	DIXIE	[round, black on yellow, picture of a black man, Ogburn Hill]
D275	DIXIE KID	[round, embossed]
D277	**DIXIE PLUG**	**[round, black on red]**
D279	DIXIE QUEEN	[round, blue on yellow, picture of a girl]
D280	DIXIE SUN CURED	[round, black on red]
D281	**DIXIE TWIST**	**[round, white on red, Moores]**
D283	DIXIE'S CHOICE	[odd oval, embossed]
D285	DIXON'S IDEA	[hexagon embossed, W.E. & Co.]
D286	DOCTOR'S PRESCRIPTION	[rectangular, black on yel low]
D287	DODGER	[rectangular, black on yellow, White-Hurley Tob. Co.]
D289	**DOG**	**[die-cut, embossed dog]**
D291	DOG LEG TWIST	[rectangular, embossed]
D293	DOG'S HEAD SMOKING	[round, multicolored, picture of a dog's head]
D295	DOLLAR MARK	[round, yellow on red, picture of a dollar sign in center]
D297	DOMINION NAVY	[rectangular, odd, embossed]
D299	DOMINO	[rectangular, black on white, twelve dots]
D300	DON PEDRO	[rectangular, embossed]
D301	**DONKEY HEAD**	**[die-cut, embossed head]**
D302	DONKEY HEAD	[round, black on red, picture of a donkey head]
D303	DOT	[round, black on red]
D304	DOT	[round, gold on red]
D305	**DOT**	**(a) [die-cut round small, red]**
		(b) [green]
		(c) [yellow]
D306	**DOT**	**(a) [die-cut round large, red]**
		(b) [green]
		(c) [yellow]
D307	**DOT**	**[round, red with gold cross]**
D308	**DOT**	**[oval, Little, black on green]**
D309	DOUBLE CROSS	[round, red on yellow, picture of two crosses]
D310	DOUBLE EAGLE	[round, picture of spread wing eagle, Sutters]
D311	DOUBLE SIX	[rectangular, white on black, Domino]
D313	DOUBLE SIX CHEWING	[rectangular, black on white, picture of two dice, sixes]
D314	DOUBLE TEAM	[rectangular, black on yellow, picture of two horses & a wagon, F. & R.]
D315	DOUBLE THICK	[rectangular, black on red, Rucker & Whitten]
D317	**DOUBLER**	**[round, embossed]**
D319	**DOUGH BOY**	**[round, picture of a man, black on yellow]**
D321	**DOUGH NUT**	**[diamond, black on red]**
D323	DOUGH NUT	[rectangular, black on yellow & tan]
D325	**DOUGLAS'S BEST**	**[rectangular, embossed]**
D327	DOVE	[die-cut dove on a branch, embossed]
D329	DOVE	[round, black on red, picture of a dove]
D330	DR. PUFF	[oval, black on red, picture of a horse]
D331	**DRAGOON**	**(a) [rectangular, blue on white]**
		(b) [black on white]
D333	DRAW KNIFE	[rectangular, red on yellow & black, picture of a knife]
D334	DRIED PEACH	[round, black on yellow, picture of a peach]
D335	DRIED PEACH	[round, black on yellow]
D336	DRIED PEACH	[die-cut peach, red on yellow]
D337	THE DRIVING RING	[round, scalloped, black on red]
D338	THE DRIVING PLUG	[rectangular, black on yellow]
D341	DRUID HILL	[rectangular, black on red]
D343	DRUM	[die-cut small drum, picture of eagle, red, white & blue]
D344	DRUM	[die-cut large drum, picture of eagle, red, white & blue]
D345	DRUM MAJOR	[rectangular, embossed]
D347	**DRUMMER**	**[rectangular, white on red, picture of a drummer boy, Osburn, Hill & Co.]**
D349	DRUMMOND	[small rectangular, embossed]
D350	**DRUMMOND**	**[large rectangular, embossed]**
D351	**DRUMMOND TOB CO.**	**[die-cut horse shoe, red on yellow]**
D353	**DRUMSTICK**	**[die-cut shape, Strater, black on yellow & red]**
D354	DRY DOCK	[round, yellow on black, L.W. Davis]
D355	DUBLIN TWIST	[rectangular, tan on yellow]
D357	**DUCHE PLUGS**	**[oval, black on red]**
D359	**DUCHESS**	**[rectangular, white on red]**
D361	DUCHESS EAGLE	[round, black & white on red, picture of bird]
D362	DUCK	[round, picture of a duck, multicolored]
D363	DUCK BILL	[die-cut duck, embossed]
D364	DUCK LEG	[die-cut duck, black on cream, Casey & Wright]
D365	DUDLEY'S JEWEL	[octagonal, embossed]
D367	DUKE OF ATHOL	[oval, black on red]
D369	DUKE'S CHOICE	[hexagon, embossed]
D370	DUMB BELLS	[die-cut shape, Dumb Bells]
D371	DURATION	[round, black on yellow]
D373	**THE DURHAM**	**[oval, picture of bull, black on red]**
D375	DURHAM TWIST	[die-cut shape, embossed]
D377	DUXBURY	[round, black on yellow]
D378	DYNAMITE	[rectangular, black on red]
D379	DYNAMITE	[rectangular, black on yellow, picture, Leak Bros. & Hasten]

E

E101	**E**	**[die-cut]**
E103	**E**	**[die-cut, embossed]**
E105	E A D	[die-cut red on yellow, Mfgd. for E.A. Davis & Co.
E107	E B & Co.	[round, embossed]

E109	**E C**	**[square, cut-out letters]**
E111	E.H.G.	[oval, embossed]
E113	**E.H.C.**	**[oval, embossed]**
E117	**E J**	**[oval, embossed]**
E118	E.J.S.	[round, black on red, white & blue]
E119	**E & M**	**[round, embossed]**
E120	E-Z	[round, white on red, Easy Twist]
E121	**EAGLE**	**[die-cut, embossed small eagle]**
E123	**EAGLE**	**[die-cut, embossed large eagle]**
E125	**EAGLE**	**[oval, picture of eagle sitting on a branch, gold on black]**
E126	EAGLE	[round, embossed picture of eagle]
E127	EAGLE	[round, black on yellow, picture of eagle]
E129	EAGLE CHEW	[rectangular, black on yellow & brown, picture of eagle, P.B. Gravely & Co.]
E131	EAGLE HEAD	[round, black on red, picture of bird's head]
E133	**EAGLE GRADE**	**[octagonal, G.B., black on red]**
E134	EAGLE TWIST	[round, black on red]
E135	EAGLE'S PLUG	[round, black on yellow, picture of an eagle]
E136	**EARLY BIRD**	**[round, black & red on yellow, picture of a bird, P.H. Hanes & Co.]**
E137	**EARLY BIRD**	**[round, embossed black & red on yellow, picture of a bird, P.H. Hanes]**
E139	**EARLY BIRD**	**[round, embossed & black on yellow & orange]**
E140	EARLY BIRD	[round, red on yellow, picture of a bird]
E141	**EARTH QUAKE**	**(a) [red round, embossed]** (b) [silver] (c) [gold]
E143	**THE EARTH TWIST**	**[round, yellow on red, J & B Tob. Co.]**
E145	EAST SIDE	[round, red on yellow]
E146	**ELBERT PAYNE & CO.**	**[oval, embossed]**
E147	EBY & Co.	[rectangular, embossed]
E148	**ECHO**	**[die-cut E, embossed]**
E149	**ECHO**	**[round, green on gold, trademark]**
E150	**ECLIPSE NO. 1**	**[die-cut, embossed apple]**
E151	ECLIPSE	[round, embossed, Weller Repetti]
E153	ECLIPSE	[rectangular, banner embossed]
E155	**ECLIPSE**	**[round, embossed]**
E156	EDGAR D.	[round, black on red]
E157	ED LOCKETT	[round, black on orange]
E159	**EDEN**	**[rectangular, black on yellow & red]**
E162	EGG SHELL TWIST	[round, white on red, picture of an egg breaking]
E163	EGBERT	[oval, black on red]
E164	EIGHT HOUR WORK DAY	[die-cut heart shape, Carhart]
E165	**8 HOURS**	**[die-cut 8, embossed]**
E167	8 OZ	[oval, small, white on green, R.J. Reynolds]
E168	**86**	**[rectangular, embossed]**
E169	**86**	**[round, embossed]**
E170	86	[square, embossed]
E171	**88**	**[die-cut 8-8, embossed]**
E172	**E. LEIDY**	**[die-cut arrow, embossed]**
E173	**ELBERTA**	**[round, Sparger Bros. Co. red on green, multicolored picture]**

E175	ELBERTA	[round, multicolored picture, Mfgd. by Sparger & Co.]
E177	**ELEGANT CHEW**	**[rectangular, embossed, Beck]**
E179	**ELEPHANT WHITE**	**[die-cut elephant, embossed]**
E181	**ELEPHANT**	**[large die-cut elephant]**
E182	**ELGIN**	**[oval, embossed]**
E183	ELI	[round, black on red]
E184	**ELI**	**[round, white on black & yellow]**
E185	ELI	[odd rectangular, embossed]
E186	ELK	[round, embossed elk head]
E187	ELKO	[round, black on red, J.G. Dill, Richmond, Va.]
E188	ELKS	[rectangular, yellow on red]
E189	ELKANNA	[rectangular, round embossed]
E190	ELLEN FISCHER	[rectangular, embossed]
E191	ELLEN FISHER	[rectangular, black on yellow, picture of a girl]
E192	ELLIE	[square, black on red, J.W.E.]
E193	ELLIS BEST	[rectangular, black on yellow, picture, 12 inch 3s]
E194	ELVA	[rectangular, embossed]
E195	**EMBLEM**	**[oval, embossed picture, world with an arrow]**
E197	EMERSON'S BIG 5	[die-cut 5, embossed]
E199	EMIGRANT'S DELIGHT	[rectangular, multicolored, picture of a wagon & people]
E201	EMMA H	(a) [oval, black on yellow] (b) [black on red]
E202	EMMA LEE	[round, yellow on black, Thomas & Meadams, Danville, Va.]
E203	EMMA VIRGINIA	[rectangular, red on yellow]
E205	EMPIRE	[die-cut snowshoe, black on red]
E207	EMPRESS	[rectangular, black on yellow, The R.A. Patterson Tob. Co.]
E209	ENCHANTED	[rectangular, embossed]
E211	**ENSIGN**	**[round, picture of American flag, black on white]**
E213	**ENTERPRIZE**	**[diamond, embossed]**
E215	**E. RICE GREENVILLE**	**[rectangular, embossed]**
E217	ESCORT	[odd rectangular, embossed]
E219	**ESPERANZA**	**[oval, black on green]**
E221	**EUCHRE**	**[rectangular, embossed]**
E223	**EU-CHU-MEE**	**[hexagon, picture of an Oriental man, black on orange]**
E224	EU-CHU-MEE	[diamond, embossed]
E225	**EUREAK**	**[rectangular, Major & Mackey, black on red]**
E227	EUREAK	[round, embossed]
E229	**EUREAK GRADE**	**[square, embossed]**
E231	EVAN'S BEST	[rectangular, embossed]
E233	**EVEN CHANGE 5 CENT**	**[round, embossed]**
E237	**EVERSON WILSON & CO.**	**[die-cut, embossed pocket knife]**
E239	**EVERSON WILSON & CO**	**[die-cut heart, embossed]**
E241	EVERYDAY PLUG TOBACCO.	[round, red & black on yel low]
E243	**EVERYDAY SMOKE**	**[arch banner, white on red]**
E245	**EVERYDAY SMOKE**	**[round, small, N & W, red on gold]**
E247	EVERYDAY SMOKE	[round, large, N & W, red on gold]
E249	EXCELLENT	[round, embossed]
E250	**EXCELSIOR**	**[round, scalloped, black on red]**

E251	EXPANSION	[rectangular, embossed]
E252	EXPANSION	[oval, black on green & orange, The United States Tobacco Co.]
E253	EXPANSION	[rectangular, yellow on red, U.S. Tob. Co.]
E254	**EXPERT**	**[die-cut cross, Pace Plug, black on red]**
E255	EXPERT	[die-cut scroll, black on red]
E257	EXPORT	[rectangular, black on red]
E258	**EXPRESS**	**[round, embossed, Seal Rock Tob Co.]**
E259	**EXTRA**	(a) **[round, black on yellow, Fine Tobacco, J.W. Burton]** (b) [rectangular]
E260	EXTRA	[round, embossed J.R. & Co.]
E261	EXTRA FIG	[round, black on yellow, Hcb.]
E262	**EXTRA FINE**	**[diamond, Mfg. by Job Deshazo, white on red]**
E263	**EXTRA FINE**	**[oval, black on white, R.C. Payne]**
E264	EXTRA GOLDEN	[diamond, embossed]
E265	EXTRA GOLDEN	[half round embossed, Jas. R. Millner]
E266	EXTRA GOLDEN	[oval, embossed]
E267	EXTRA 3 PLY	[octagonal, embossed, Gravely]
E268	EXTRA 9 INCH	[rectangular, black on yellow, B.F. Gravely & Sons]
E269	EXTRA 9 IN 5'S	[rectangular, black on yellow, E.B. Gravely & Co.]
E271	**EZELL'S CHOICE**	**[round, red on yellow]**

F

F101	**F & Co.**	**[round, embossed]**
F103	**F & D**	**[rectangular, embossed]**
F104	**F D**	**[rectangular, cut-out & embossed Muscatine]**
F105	F & E	[round, embossed]
F107	F A D	[cut-out letters, black on yellow, F.A. Davis & Co.]
F109	F A S & Co CHEW	[round, green & tan on yellow]
F110	F.B.C.	[die-cut star, black on red, green, orange & yellow, R & K, Henderson, Ky.]
F112	F.F.D.	[die-cut letters red on yellow, mfgd. for F.A. Davis & Co. Baltimore MD.]
F113	F.F.V.	[rectangular, embossed]
F114	**F O B**	**[rectangular, cut-out]**
F115	F.O.W.	[round, embossed]
F116	F.P.	[rectangular, embossed]
F117	F.P.U.	[rectangular, red on black]
F118	**FACE**	**[die-cut man in the moon, Clunet, Brown & Co.]**
F119	**FACE**	**[die-cut embossed boy's face]**
F120	FAIR DEAL	[rectangular, black on red]
F121	FAIRMOUNT	[die-cut banner, black on blue]
F122	FAIR PLAY	[round, red & black on yellow, Larus & Bro. Co. Richmond, Va. U.S.A.]
F123	**FAIR PLAY**	**[round, red & black on yellow]**
F124	FAIR PLAY	[round, embossed, Larus & Bro]
F125	FAIR PLAY	[round, red on yellow, Larus & Bro Co.]
F126	FAIR REBEL	[rectangular, odd, black on yellow, picture of girl, Penn Bros. Co.]
F127	**FAMOUS PLUG**	**[round, black on red]**
F128	**FAMOUS PLUG**	**[round, black on red, Mueller]**
F129	FAMOUS PLUG	[round, embossed]
F130	**FANCY**	**[round, red on yellow]**
F131	**FANCY BOY**	**[oval, large, brown & black on yellow Taylor Bros.]**
F132	FANCY BOY	[oval, large, black & brown on white Taylor Bros.]
F133	**FANCY BOY**	**[oval, small, black & brown on white Taylor Bros.]**
F135	FANCY BOY	[oval, small, black & brown on red & yellow Taylor Bros.]
F137	**FANCY BOY**	**[oval, small black & brown on red & yellow]**
F138	**FANDANGO**	**[die-cut guitar, black on red, J.A. Andrews & Co.]**
F139	**FANNIE MAY**	**[rectangular, banner embossed]**
F141	FANNY EDEL	[rectangular, black on yellow, wording across top, picture of girl & dog]
F142	FANNY EDEL	[rectangular, black on yellow, wording on left, picture of girl & dog]
F143	FANNY FERN	[rectangular, black on yellow, picture of girl]
F145	FARBER'S COMET	[odd rectangular, embossed]
F147	FAR WEST	[rectangular, black on red, picture of buffalo]
F148	FARMER BOY	[round, black on yellow, picture]
F149	FARMER GIRL	[round, black on gold, The E. R. Pen Tob Co.]
F150	FARMER GIRL	[round, embossed, F. R. Penn & Co.] ***? check name**
F151	**FARMERS 6 E CHOICE**	**[round, embossed]**
F153	FARMER'S CHOICE	[rectangular, black on red, picture of boy, S.A. Ogburn]
F155	FARMER'S CHOICE	[oval, embossed, W.P.P. & Co.]
F157	FARMER'S CHOICE	[rectangular, embossed]
F159	**FARMER'S DELIGHT**	**[round, black on red, Robt. Harris & Bro.]**
F160	FARMER'S FRIEND	[round, black on red, picture of plow]
F161	**FARMER'S FRIEND**	**[round, embossed, picture of plow]**
F163	FARMER'S FRIEND	[oval, embossed, Ridgeway, Va.]
F165	FARMER'S JOY	(a) [round, large, black on yellow, J & A.G. Stafford] (b) [small]
F166	FARMER'S JOY	[round, black on yellow, picture of horse & plow]
F167	**FARMER'S PLUG PLAIN**	**[rectangular, embossed]**
F169	FARMER'S PRIDE	[round, gold on black]
F171	FARMER'S SOLACE	[round, embossed]
F173	FARMER'S TWIST	[rectangular, embossed]
F175	**FASHION**	**[die-cut F, black on red, John Finzer & bros.]**
F177	**FASHION**	**[die-cut F embossed, J.F. & bros]**
F179	FASHION	[oval, red on green]
F180	FAST FRIEND	[diamond, embossed]
F181	FAST FRIEND	[diamond, black on yellow]
F182	**FAT BACK**	**[rectangular, multicolored, picture of a pig, Liipfert Scales & Co.]**

F183	FAT BOY	[rectangular, picture of baby boy]
F185	FAT POSSUM	[round, embossed, Bitting & Hay]
F186	FATTY FELIX	[round, multicolored, picture of boy and dog]
F187	**FAVORFTE**	**[round, black on yellow & red, F.M. Bohannons]**
F189	**FAVORITE**	**[octagonal, gold on black, F.R.R. & Co.]**
F190	FAVORITE	[rectangular, black on yellow & red]
F191	FAST MAIL	[round, embossed]
F192	FEDERATION SMOKE	[diamond, black on yellow, picture of two hands]
F193	FERRIS WHEEL	[round, black on yellow, picture of ferris wheel, Toledo Tob Wks.Co.]
F194	FIDDLE	[die-cut fiddle, Bachmanns]
F195	**FIDELITY**	**(a) [rectangular, blue on white, picture of a dog]** (b) [green on white]
F196	FIELD'S CHOP AXE	[oval, black on red]
F197	FIELD'S RED HEART	[die-cut heart]
F198	FIG	[round, embossed, W.T. Hancock]
F199	FIG	[octagonal embossed, W.A. Brown]
F200	FIG	[hexagon black on yellow, Hickok, Richmond, Va.]
F201	FIG	[round, yellow on green, Edel]
F202	FIG	[hexagon black on yellow, Daniel, Henry Co. Va.]
F203	**FIGS**	**[rectangular, embossed]**
F207	FILBERT	[round, embossed]
F209	**FINE**	**[die-cut X embossed]**
F210	FINE	[rectangular, odd embossed, G.T.L.]
F211	FINE	[diamond, black on gold & red]
F212	FINE 8 0Z.	[diamond, red on gold]
F213	FINE NATURAL LEAF	[rectangular, embossed, Brown & Williamson]
F214	FINE NATURAL LEAF	[banner red on yellow & black & red trim, Brown & Williamson]
F215	FINE NINE INCH	[rectangular, red on yellow, B.F. Gravely & Sons]
F216	FINE 3 PLY	[round, embossed, Gravely]
F217	FINE 3 PLY	[round, black on red, Brown & Williamson]
F219	**FINE VIRGINIA LEAF**	**[die-cut leaf rectangular, embossed]**
F221	FINK'S AAAA	[round, gold on red]
F223	**FINK'S GOLDEN**	**[oval, black on gold]**
F225	FINK'S TWIST	[round, black on yellow, picture of twist tobacco]
F226	**FINZER**	**[round, small, black on red, Finzer]**
F227	**FINZER**	**[rectangular, black on red]**
F231	FINZER'S	[die-cut S black on red]
F232	FINZER'S, BLACK BASS	[die-cut M black on yellow]
F233	FINZER'S HINDOO	[rectangular, black on red, picture of horse]
F234	**FINZER'S OLD HONESTY**	**(a) [round, multicolored, picture of dog's head, small H]** (b) [large H]
F235	FIRED TWIST	[round, red on yellow, Hopkinsville]
F237	FIRST CALL	[round, black, yellow & red, picture of rooster with trumpet]
F238	FIRST DEAL	[round, red on yellow]
F239	FIRST FRUIT	[round, picture of a strawberry, B.W. Tob Bo.]
F241	**FIRST GRADE**	**[round, embossed, C.W. Allens]**
F243	FIRST OPTION	[rectangular, embossed]
F245	FIRST STEP	[octagonal black on red]
F247	FISH	[rectangular, odd, black on red, picture of a fish]
F249	FISH	[die-cut, a small fish, embossed]
F251	**FISH**	**[die-cut, a large fish, embossed]**
F253	FISH	[round, embossed picture of fish]
F255	**FISH HOOK**	**[die-cut large hook, embossed]**
F256	FITZ LEE	[oval, yellow on black, Dixie Tob. Co. Bedford City Va.]
F257	**FITZHUGH LEE**	**[rectangular, gold on black]**
F259	**5**	**[die-cut large 5]**
F261	**5**	**[oval, cut -out 5]**
F262	5	[round, embossed, Tinsley Tobacco Co.]
F265	**5¢ C.O.D.**	**[diamond, embossed]**
F266	5¢ HARD TIMES	[round, embossed, Cures]
F267	5 CENT JACK	[round, red on black]
F268	5¢ PIECE	[round, embossed, Ogburn. S.A.]
F269	**5 CENT TOBACCO**	**[round, embossed, Fred Daut & Co.]**
F271	**5 SQUARE DEAL**	**[odd square embossed]**
F272	5 STAR	[round, red on yellow, Suhers]
F273	**FIVE BROTHERS**	**[round, black on red]**
F275	FIVE POINTS	[square, embossed]
F277	FIVE TIMES	[oval, red on yellow]
F279	**FLAG**	**[die-cut embossed, American flag]**
F280	FLAG	[round, red, white & blue on white, picture of American Flag]
F281	FLAG	[round, multicolored, picture of girl & flag, "We All Unite Here," mfg. by King Bros]
F282	FLAG	[die-cut shape, American flag on a pole, red, white & blue]
F283	**FLAG STAFF**	**[round, embossed]**
F285	**FLAP JACK**	**[round, embossed, hole in center]**
F287	**FLAP JACK**	**[odd rectangular, embossed]**
F289	FLAP JACK	[rectangular, embossed]
F291	**FLAT IRON**	**[die-cut iron, yellow on black & orange]**
F292	FLAT IRON	[round, black on yellow & red, picture of flat iron]
F293	FLIRT	[black on cream, small cartoon tag, picture of a lady]
F294	FLY ROD	[octagonal, picture of fish jumping to catch fly]
F295	FLOME	[oval, embossed]
F296	**FLORA BRAND**	**[oval, black on red, Buchanan & Lyall]**
F297	**FLORA BRAND**	**[oval, black on red, Buchanan & Lyall]**
F298	FLORA BRAND	[oval, embossed, Buchanan]
F301	**FLORA McFLIMSY**	**[hexagon, black on green]**
F302	**FLORA McFLIMSY**	**[round, scalloped, black on red]**
F303	FLOURNOY TOB. Co.	[round, black on yellow, picture of rabbit]
F305	FOOT PRINTS	[die-cut foot, black on red]
F307	FOR YOU	[round, embossed]
F309	**FORE**	**[round, white on red & green stripe]**

F311	**FORSEE'S RAINBOW**	**[rectangular, bow, black on red, best chew made]**
F313	**FORGET ME NOT**	**[hexagon, black on yellow, P. Lorillard & Co.]**
F314	FORGET ME NOT	[hexagon, black on yellow, block letters, P. Lorillard Co.]
F315	**FORGET ME NOT**	**[octagonal, gold on blue]**
F317	**FORKED DEER**	**[round, gold on dark brown, trade mark Tobacco]**
F318	**FORT ALAMO TWIST**	**[rectangular, embossed, A.H. Motley & Co.]**
F319	FORT DEFIANCE	[round, black on red]
F320	FORTUNE'S PRIDE	[round, red on yellow, mfg. By J. & W. Fordsville, Ky.]
F321	FOUND IT	[oval, embossed]
F322	FOUNTAIN NATURAL LEAF	[round, black on red, Lovell & Buffington Tob Co.]
F323	FOUNTAIN	[die-cut embossed fountain]
F324	**4**	**[die-cut 4]**
F325	**48**	**[round, black on yellow, Taylor Bros]**
F327	48	[octagonal, red on white]
F328	**48**	**[round, red on white]**
F329	49 SMOKING	[round, black on yellow, Nolan Bros.]
F330	4 D	[square, embossed]
F331	**4'S NATURAL LEAF**	**[round, embossed]**
F333	4 W P	[banner embossed]
F334	FOX	[oval, picture of a fox, yellow on red]
F336	**FOX**	**[oval, picture of a fox, red on black]**
F337	FOX	[rectangular, red on yellow, picture of fox]
F338	FOX HORN TWIST	(a) [round, black on blue, picture of horn] (b) [black on green] (c) [white on red]
F339	**FOX HORN TWIST**	**[oval, red on white]**
F341	**FOX RIDGE TWIST**	**[rectangular, octagon, black on white, picture of red fox]**
F343	FOXY KID	[round, black on yellow]
F345	F.R. PENN & Co. TOBACCO	[round, black on red & yel low]
F346	F.R. PENN & Co. TOBACCO	[round, embossed]
F347	F.R. PENN, REIDSVILLE, N.C.	[round, black on yellow, pic ture of man]
F349	FRAGRANT	[round, embossed]
F351	FRANCIS DELIGHT	[round, red & white on yellow]
F352	FRANK LEE	[rectangular, round ends, embossed]
F353	FREDERICK	[oval, white on blue]
F354	FREE COINAGE	[round, black on tan, picture of Liberty head, B.F. Gravely & Co.]
F355	FREE COINAGE	[round, embossed]
F357	FREE GRASS	[die-cut shape, scroll embossed]
F359	**FREE LUNCH**	**[rectangular, odd, embossed]**
F361	FREE LUNCH	[round, embossed]
F362	FREE SILVER	[round, embossed]
F363	**FREE SILVER**	**[round, embossed, two stars in center]**
F364	FREE SILVER	[rectangular, black on white, L.B. & T.]
F365	**FREE TRADE**	**[rectangular, odd embossed]**
F366	FREE TRADE	[round, embossed]
F367	**FRIED CAKE**	**[round, blue on white, red, white & blue, picture of cake]**
F371	FROG EYE	[round, multicolored, picture of a eye, Liipfert & Sons]
F373	**FROLIC**	**[oval, embossed]**
F375	**FRONTIER GRADE**	**[rectangular, embossed, A.T. & Co.]**
F377	**FRUIT CAKE**	**[rectangular, black on yellow]**
F379	FRUIT CAKE	[rectangular, embossed]
F380	FRUIT CAKE	[die-cut odd block, black on yellow]
F381	FRUIT NUGGET	[round, multicolored picture of fruit, S.F. Hess & Co]
F382	**FRUIT OF THE VINE**	**[diamond, black on red]**
F383	**FRUIT OF THE VINE**	**[round, small, black on red]**
F384	FRUIT OF THE VINE	[round, black on yellow]
F385	**FULL BLOOM**	**(a) [rectangular, large, black on yellow, red flower, L. Ash]** (b) [rectangular, small]
F386	**FULL BLOOM**	**[rectangular, banner, black on yellow]**
F387	FULL FARE	[round, embossed, picture of eagle]
F388	FULL MOON	[round, blue on white, picture of moon face]
F389	**FULL PAY**	**(a) [small die-cut money bag, black on red]** (b) [black on pink] (c) [black on yellow]
F390	FULL PAY	[larger die-cut money bag, red on yellow]
F391	FULL SAIL	[round, red on black]
F392	FULL VALUE	[round, embossed, flower in center]
F393	**FULL WEIGHT**	**[rectangular, embossed]**
F395	FULLER & Co.	[round, embossed]
F397	**FUROR**	**[rectangular, odd black on red]**
F399	**FUTURE PLUG 5¢**	(a) [round, red on black & yellow] **(b) [red on black & green]**
F401	FUTURITY	[round, oval gold on red, 100% pure N.Y. red burley]
F402	FUTURITY	[flat oval gold on red, 100% pure N.Y. red burley]
F403	FUTURITY	[small oval gold on red, 100% pure N.Y. red burley]

G

G101	G	[die-cut embossed]
G103	**G**	**[square, embossed G in a diamond]**
G104	G	[round, black on red G in a diamond]
G105	**GA-AR**	**[cross embossed]**
G107	**G.B.**	**[round, embossed]**
G109	G A R	[die-cut cross embossed
G111	**G & G**	**[cross embossed]**
G113	G & G	[rectangular, curved, embossed]
G115	G & M	[round, black on yellow]
G116	G.I.G. TWIST	[round, black on orange, The E.O. Eshelby Tobacco Co]
G117	G L C	[round, white on red]
G118	**G O P**	**[rectangular, embossed, silver on black, good old plug]**

G119	G P E	[rectangular, black on red]
G121	G T L	[rectangular, red on white]
G122	G.T.W.	[octagonal, black on red, Owensboro, Light Pressed]
G123	G.T.W.	[rectangular, black on yellow]
G125	**G.T.W.**	**[rectangular, embossed, black on yellow]**
G127	**G.W.A.**	**[rectangular, embossed]**
G129	G W T	[round, embossed]
G131	G. WHIZ	[diamond, gold on dark green]
G132	GALA TEA	[hexagon black on yellow, Camerons]
G133	GALE'S NAVY	[round, embossed]
G135	**GALLAHER'S**	**[rectangular, banner embossed]**
G136	GALLAHER'S ARMY & NAVY	[rectangular, banner em bossed]
G137	GAME	[rectangular, white on red]
G139	GAME BIRD	[oval, embossed]
G141	GAME COCK TOBACCO	[round, multicolored picture of rooster]
G143	**GAME G. B. GRADE**	**[octagonal, gold on dark green]**
G144	**GANT HOOK**	**[die-cut embossed]**
G145	GARFIELD	[round, embossed]
G147	**GARLAND PLUG**	**[odd round embossed]**
G149	GARRETT'S 10¢ PLUG	[rectangular, black on white, picture of a man]
G151	GARTER BUCKLE	[round, black on yellow, multicolored picture]
G153	GATE CITY	[round, black on red]
G154	GAUNTLET	[oval, blue on yellow, Taylor Bros & Co. Winston N.C.]
G156	GAY BIRD	[octagonal, picture of bird, Irvin & Poston]
G157	**GAY BIRD**	**[rectangular, black on yellow, picture of a bird, Irvin & Poston]**
G158	GAY BIRD	[rectangular, black on yellow, picture of a red bird, Irvin & Poston]
G159	**GAYLORD**	**[rectangular, black on red]**
G161	GEM	[round, embossed, lion face embossed]
G162	GEM	[round, embossed]
G163	GEM	[round, black on red, Concord]
G164	GEM	[round, black on red, A.H. Motley Co., Reidsville, N.C.]
G165	GEM	(a) [round, white on red] (b) [gold on red]
G166	GEM	[rectangular, hexagon black on yellow]
G167	GEM LIGHT PRESSED	[round, black on red]
G168	GEM ROSENFELD'S	[rectangular, odd]
G169	**GEN'L FORREST**	**[round, black on yellow, picture of a man]**
G170	**GEN'L MORGAN**	**[half round black on red, U.S. Tob. Co.]**
G171	GENOA	[oval, black on red]
G172	GENUINE LUXURY	[rectangular, banner, black on red]
G173	GENUINE LUXURY TOBACCO	[rectangular, banner, black on yellow, British Tob. Co.]
G174	GENUINE RED CROSS TOBACCO	[die-cut cross, yellow on red]
G175	**GENUINE VIRGINIA BEAUTY**	[oval, embossed]
G176	GEORGE	[rectangular, curved, embossed]
G177	GEORGE WASHINGTON	[round, multicolored, pic ture of Washington]
G181	GEORGE WASHINGTON	[oval, black on cream, pic ture of Washington]
G182	GEORGIA BUCK	[diamond, embossed]
G183	**GEORGIA CRACKER**	**[round, black on red, F.R. Penn Tob Co.]**
G185	GEORGIA CRACKER	[oval, black on red, Watt, Penn & Co.]
G187	GEORGIA CRACKER	[oval, black on red, F.R. Penn Co.]
G189	GEORGIA MELON	[black on yellow & red, picture of melon, Turner Powell & Co]
G190	GEORGIA MELON	[round, black on yellow & green, picture of a melon]
G191	GEORGIA SENATOR	[round, black on yellow, picture of bust of a man]
G193	GEORGIE	[rectangular, embossed]
G195	GET THERE	[round, Mitchell & Dunlop, Boxwood, Henry Co.]
G197	GET THERE EARLY	[rectangular, red on yellow]
G199	GETTYSBURG	[round, multicolored, picture of two soldiers]
G201	**GIANT**	**[oval, black on red, W.J. Gould & Co.]**
G203	**GIANT**	**[round, black on red, W.J. Gould & Co.]**
G205	**GIG TWIST**	**[round, black on orange, The E.D. Eshelby Tob. Co.]**
G207	**GILDED AGE**	**[round, embossed]**
G209	GILMORE BROS.	[rectangular, yellow on red]
G211	GILT	[round, embossed]
G213	GILT EDGE	[oval, black on orange, P.H. Hanes, H]
G215	**GILT EDGE**	**[die-cut leaf, gold on brown]**
G216	**GILT EDGE**	**[round, embossed, G in center]**
G217	**GINGER BREAD**	**[round, black on yellow, picture of bread, G. Penn Sons & Co.]**
G219	GINGER CAKE	[round, embossed]
G221	**GINO**	**[round, embossed, Hawkins, Morris & Co.]**
G223	**GIPSY**	**[round, embossed]**
G224	GIPSY	[rectangular, black on yellow, T.C. Williams Co.]
G225	GIPSY GIRL	[round, black on red, Stone]
G227	**GIT ON**	**[oval, embossed]**
G229	GIT ON	[oval, red on black]
G231	**GLADIATOR**	**[oval, embossed]**
G232	**GLADIATOR**	**[die-cut man with spear & shield]**
G233	**GLOBE**	**[round, small, black on red]**
G234	GLOBE	(a) [large round black on red, West Winfree Tobacco Co.] (b) [small round black on red, West Winfree Tobacco Co.]
G235	GLOBE	[round, small embossed]
G236	**GLOBE**	**[round, long points embossed, Chew Fine Cut, patent appl'd for]**
G237	**GLORY**	**[rectangular, blue on red, white & blue]**
G239	**GLORY**	**[round, yellow on black, McNamara]**
G241	**GLORY**	**[round, embossed small, McNamara]**
G243	**GLORY**	**[round, embossed large, McNamara]**
G245	GLORY PLUG	[rectangular, embossed]

G246	GLOBE TOBACCO CO.	[round, black on red, Detroit, Mich.]
G247	**GNU PLUG**	**[oval, embossed]**
G248	GO QUICK	[oval, embossed]
G249	**GO QUICK**	**[round, embossed, with wreath]**
G251	**GO QUICK**	**[round, embossed]**
G253	GOLD BAND	[rectangular, black & gold on red, picture of banner]
G255	GOLD BAR	[rectangular, yellow on red]
G256	GOLD BASIS	[round, red on yellow]
G257	GOLD BASIS	[round, gold on red]
G258	GOLD BOND	[rectangular, yellow & red on white]
G259	GOLD BRICK	[rectangular, black on yellow, Cannelton Tob. Co.]
G260	GOLD BRICK	[rectangular, black on yellow, Mitchell & Dunlop, Henry Co.]
G261	**GOLD BUG**	**[rectangular, gold on green, picture of bug]**
G263	**GOLD CHUNK**	**[oval, embossed]**
G265	**GOLD CLUB**	**[die-cut bow tie shape, black on red W & McO Tob. Co.]**
G266	GOLD COIN	[round, red on yellow, Bradley-Banks Co.]
G267	**GOLD COIN**	**[round, embossed Liberty head type]**
G269	**GOLD CROSS**	**[round, gold cross on red]**
G271	GOLD CROSS	[round, multicolored, picture of gold cross, F. Lasseher & Co.]
G273	GOLD DOLLAR	[round, embossed, copy of gold dollar]
G275	**GOLD DOT**	**[round, small embossed]**
G277	**GOLD DOT**	**[round, comes in many colors]**
G278	GOLD DUST	[round, gold on green]
G279	**GOLD DUST**	**[rectangular, with embossed horse]**
G281	GOLD FIG	[round, gold on brown, picture of fig]
G283	GOLD FINCH	[round, black on yellow, picture of bird]
G284	GOLD FIND	[round, embossed]
G285	GOLD FISH	[die-cut fish, embossed]
G287	GOLD HAWK	[round, embossed, picture of hawk]
G289	**GOLD LEAF**	**[rectangular, embossed]**
G291	GOLD LEAF	[die-cut leaf]
G292	**GOLD LEAF**	**[round, black on red, D.F. King., Leaksville N.C.]**
G293	**GOLD LUMP**	**[oval, embossed]**
G295	GOLD MEDAL	[round, black on white & yellow]
G296	**GOLD MINE**	**[die-cut leaf, gold on silver]**
G297	GOLD NUGGET	[odd shape, embossed, W.E. & Co.]
G299	**GOLD RING**	**[die cut ring embossed]**
G301	GOLD ROCK	[rectangular, gold on black]
G302	GOLD ROCK	[rectangular, bar, yellow on green]
G303	**GOLD ROLL**	**[rectangular, black on gold, Lottier]**
G305	**GOLD BOLL**	**[round, embossed, Lottier]**
G307	GOLD ROLL	[round, black on gold, Lottier]
G309	**GOLD ROPE**	**(a) [rectangular, gold on red, Wilson and McCallay Tob.Co.]**
		(b) [Wilson and McCally Tob. Co., The American Tob Co. Suc.]
G311	GOLD ROPE	[rectangular, gold on red, Continental Tob. Co.]
G312	GOLD SEAL	[rectangular, black on red]
G313	**GOLD SHIELD**	**[die-cut embossed shield, McAlpin]**
G315	GOLD SHORE	[rectangular, embossed]
G317	**GOLD TAG**	**[round, embossed]**
G318	GOLD WEDGE	[die-cut wedge shape, embossed]
G319	GOLD WEDGE	[die-cut rectangular, wedge, red on yellow]
G321	GOLDEN	[round, black on yellow]
G322	**GOLDEN AGE**	**[die-cut S embossed]**
G323	GOLDEN APPLE	[round, yellow on blue & red, picture of apple, R.A. Patterson Tobacco Co.]
G325	GOLDEN BELL	[die-cut bell, embossed]
G327	GOLDEN CAGE	[round, yellow on green, picture of cage, Gravely]
G329	**GOLDEN CALF**	**[round, embossed]**
G330	GOLDEN CHAIN	[round, black on yellow & red, picture of a girl, F. Hanes]
G331	**GOLDEN CHAIN**	**[oval, embossed, picture of chain]**
G333	**GOLDEN CHARM**	**[rectangular, embossed, Gravely]**
G334	GOLDEN COIN	[round, embossed, L. & B.]
G335	GOLDEN COIN	[round, black on yellow, Luchs & Bro.]
G336	GOLDEN CROWN	[rectangular, embossed]
G337	**GOLDEN CROWN**	**[round, embossed, B.F.P. & Co.]**
G339	GOLDEN EAGLE	[oval, gold on black, picture of an eagle]
G341	GOLDEN EAGLE	[round, embossed]
G343	GOLDEN EGG	[die-cut S embossed]
G344	GOLDEN FLEECE	[octagonal black on yellow, Boykin Seddon & Co.]
G345	GOLDEN GEM	[round, embossed]
G346	GOLDEN GEM	[round, black on red, Egerton Bros, Reidsville, N.C.]
G347	**GOLDEN GRAIN**	**[round, red on yellow & black]**
G349	GOLDEN LAND	[rectangular, embossed, Gravely]
G351	GOLDEN LINK	[round, black & red on yellow, picture of torch]
G353	**GOLDEN PHEASANT**	**[round, embossed]**
G355	**GOLDEN PRIZE**	**[round, embossed]**
G357	**GOLDEN PRIZE**	**[round, black on white]**
G359	GOLDEN PRIZE	[round, small, black on red]
G361	GOLDEN PROSPECT	[rectangular, embossed, Gravely]
G363	GOLDEN RING	[round, embossed, C.R. & Co.]
G364	**GOLDEN ROBIN**	**[oval, embossed]**
G365	GOLDEN ROD	[oval, black on yellow]
G366	GOLDEN ROD	[oval, black on red, C. T. M. C.]
G367	**GOLDEN RULE**	**(a) [large die-cut rule, black on yellow]**
		(b) [small]
G368	**GOLDEN RULE**	**[die-cut rule, embossed]**
G369	GOLDEN RULE	[oval, black on red]
G370	GOLDEN SCEPTRE	[rectangular, black on yellow, picture of staff]
G371	**GOLDEN SEAL**	**(a) [round, scalloped embossed, large]**
		(b) [round, scalloped embossed, small]
G375	**GOLDEN SEAL**	**[odd rectangular, embossed, D.B. & Co]**
G376	GOLDEN SHEAF	[round, black on red]
G377	**GOLDEN SLIPPER**	**(a) [rectangular, black on yellow & red, Whitaker-Harvey Co]**
		(b) [black on yellow & green]
G379	**GOLDEN SLIPPER**	**[die-cut slipper, black on**

		yellow]
G381	GOLDEN SLIPPER	[die-cut embossed]
G383	GOLDEN SOUTH	[rectangular, round ends, embossed]
G385	**GOLDEN SPARKS**	**[rectangular, black on white, picture of a spoke wheel]**
G387	GOLDEN SPARKS	[rectangular, red on black]
G388	GOLDEN SPARKS	[rectangular, shield, embossed]
G389	GOLDEN STAFF	[round, black on cream, picture of staff]
G391	GOLDEN STAIRS	[rectangular, embossed]
G392	GOLDEN TAG	[round, embossed]
G393	GOLDEN TWINS	[round, black on red, Lorillard]
G394	**GOLDEN TWIST**	**[oval, cut-out embossed]**
G395	**GOOD**	**[rectangular, banner embossed]**
G396	GOOD AS GOLD	[round, black on yellow, picture of a lady]
G397	**GOOD AS WHEAT PLUG**	**[die-cut bundle of wheat, black on yellow, Scotten Tob Co]**
G399	GOODBODY'S	[round, embossed]
G401	**GOOD BOY**	**[rectangular, embossed]**
G403	GOOD CHECK	[diamond, embossed]
G405	**GOOD CHEW**	**[rectangular, picture of a man, black on yellow, L. ASH]**
G407	GOOD CHEW	[diamond, embossed]
G409	**GOOD ENOUGH**	**[round, embossed]**
G410	**GOOD ENOUGH**	**[round, embossed fancy]**
G411	GOOD FRIEND	[rectangular odd, embossed]
G413	**GOOD HOPE**	**[rectangular, black on red]**
G415	**GOOD LUCK**	**[round, small, embossed]**
G417	**GOOD LUCK SMOKE**	**[round, black on red & green four leaf clover]**
G419	**GOOD MEAT**	**(a) [oval, black on red, picture of a pig]**
		(b) [blue on white]
G421	**GOOD MONEY**	**[round, black on red]**
G423	GOOD MORNING	[die-cut G embossed]
G424	GOOD NEWS	[rectangular, black on brown, The F.R. Penn Tob Co.]
G425	GOOD OLD PLUG	[round, black on red, W.B.T. Co.]
G427	GOOD PLUG	[rectangular, embossed]
G429	**GOOD SMOKE**	**(a) [small, round, embossed center Lorillard, gold on red]**
		(b) [large, round, embossed center Lorillard, gold on red]
G431	GOOD TOBACCO	[oval, black on red]
G433	GOOD TO EAT	[octagonal red on yellow & black]
G434	**GOOD TRADE**	**[rectangular, odd, black on red, W.B.T. Co.]**
G435	GOOD TWIST	[round, green & tan on white, Penn Tob Co.]
G437	GOOD VALUE	[rectangular, embossed]
G439	GOOD VALUE	[rectangular, larger, embossed]
G441	GOOD WILL	[diamond, embossed]
G442	GOOD X	[oval, embossed]
G443	GOOSE EGG	[oval, black on yellow, D.S. & Co. Detroit]
G444	GORDON	[octagonal brown on yellow, Penn, picture of a man]
G445	GORILLA	[rectangular, black on yellow, picture of gorilla]
G447	GOSSIP	[round, embossed]
G448	GOULD'S DIAMOND	[rectangular, black on red]
G449	GOURD	[die-cut dipper, black on yellow]
G450	**THE GOVERNOR**	**[small round, black on white, picture of a man]**
G451	GRAB 10 CENTS	[round, embossed]
G452	GRACE DARLING	[oval, black on red, Clark & Co., Bedford, Va.]
G453	GRADY	[round, picture of a man]
G454	GRADY CONSTITUTION	[round, black on yellow, picture, Bitting & Hay]
G455	**GRAND**	**[rectangular, black on green, picture of a Lady, J.H. Spencer]**
G457	GRAND ARMY PLUG	[round, black on tan, picture of men marching]
G459	GRAND CHARM	[rectangular, black on green, picture]
G461	GRAND LADY OUT	[rectangular, black on yellow]
G463	**GRAND OPENING**	**[small round scalloped, black on yellow, man playing a banjo]**
G465	GRAND OPENING	[rectangular, banner black & red on yellow]
G466	GRAND REPUBLIC	[round, black on red]
G467	GRANGER	[oval embossed]
G468	GRANGER	[oval, embossed, rope edge]
G469	GRANGER	[octagonal embossed]
G471	**GRANGER**	**(a) [round, with tab, white on red, W. & D.T. CO.]**
		(b) [on tab additional wording, "added this piece"]
G474	GRANGER	[round, black on red]
G477	**GRANGER TWIST**	**(a) [round, with tab, white on red]**
		(b) [on tab additional wording, "added this piece"]
G481	**GRAPE**	**[round, black on yellow, picture of grapes, R.A. Patterson Tob Co.]**
G483	**GRAPE JUICE**	**[round, embossed]**
G485	GRAPELEAF A B	[die-cut leaf embossed]
G487	GRAPE SHOT	[round, embossed]
G489	GRAPE TWIST	[round, embossed]
G491	**GRAPE VINE TWIST**	**[round, red on yellow, picture of grapes & vine multicolored]**
G492	GRAPE WINE TWIST	[round, black on yellow, picture of grapes]
G493	**GRAVELY & MILLER'S 9 INCH**	**[rectangular, black on light green, picture of two men]**
G494	**GRAVELY'S**	**[rectangular, yellow on red, J.T. & H. Clay]**
G495	**GRAVELY'S CHOICE**	**[round, embossed]**
G497	**GRAVELY'S SIX INCH**	**(a) [small, rectangular, orange on black]**
		(b) [large, rectangular, orange on black]
G498	GRAVELY'S I X L	[round, embossed]
G499	GRAY EAGLE	[rectangular, shield embossed]
G500	GRAY HORSE	[rectangular, black on yellow, picture of a horse]
G501	GRAYLINE F H & Co.	(a) [die-cut fish black on red]
		(b) [black on yellow]
G502	GREAT SCOTT	[round, embossed]
G503	**GREAT WESTERN**	**[round, embossed]**
G504	GREECE	[oval, black on white, flag of Greece, white & blue]
G505	GREEK SLAVE	[oval, black on cream, picture of in center, P.H. Hanes & Cos.]

G506	GREEN LIGHT	[round, light green on green]
G507	**GREEN RIVER**	**[round, black on orange, smoke & chew, East Tenn. Tob. Co.]**
G508	GREEN RIVER EIGHTS 5¢	[octagonal black on red, G.R.T. Co.]
G509	GREEN SEAL	[round, embossed]
G510	GREEN SHAMROCK	[round, green on white, picture of shamrock]
G511	**GREEN TURTLE**	**[small round, black on brown, P. Lorillard Co.]**
G513	**GREEN TURTLE**	**[round, black on white & green turtle P. Lorillard Co.]**
G515	**GREEN TURTLE**	**[rectangular, black on white & green turtle P. Lorillard Co]**
G516	GREENVILLE	[diamond, black on red, S.E. Rice, Greenville, Ky.]
G517	GREENVILLE	[round, green on yellow, P. & H. Paris, Tenn]
G518	GREENVILLE R & B's	[rectangular, red on yellow]
G519	GREENVILLE GTW	[round, embossed]
G521	GREENVILLE P.D.	[rectangular, embossed]
G523	GREENVILLE 5¢	[round, embossed]
G531	GRIDIRON	[round, embossed, picture of a grate]
G532	GRIFFIN	[round, embossed]
G533	**GRIP**	**[rectangular, embossed, two hands shaking]**
G535	GRIP PLUG	[oval, embossed]
G537	GRIT	[D.S. & Co.]
G538	**GRIZZLY**	**[round, embossed, picture of bear head]**
G539	**GRIZZLY**	**[round, embossed, picture of bear]**
G540	**G. CLEVELAND FOR PRESIDENT**	**[round, embossed, center photo]**
G541	**GROVER CLEVELAND**	**[round, center photo, red white & blue]**
G543	**GROVER CLEVELAND**	**[rectangular, black on red, picture of man]**
G544	GROVER CLEVELAND	[round, black on yellow, picture of man]
G545	GROWLER	[oval, embossed]
G546	GROWLER	[oval, black on red]
G547	**G.T.L. 3 PLY'4 S**	**[rectangular, banner embossed]**
G548	GUARDIAN	[hexagon embossed]
G549	GUCKENHEIMER'S	[rectangular, black on yellow]
G551	**GUIDE**	**[die-cut embossed hand, finger pointing]**
G553	GUINTER'S BEST	[round, embossed]
G555	GUINTER'S TWIST	[rectangular, odd embossed]
G557	GUN	[die-cut a rifle embossed]
G558	GUN PLUG	[oval, embossed]
G559	GUSSIE GRADY	[round, yellow on black, F.R. Penn & Co.]
G561	**GUSTO**	**[oval, embossed]**
G563	GUSTO	[oval, black on red]
G565	GYPSY	[round, embossed]

H

H100	H	[die-cut H embossed]
H101	H.B.H. & CO.	[die-cut H embossed]
H102	H B L	[die-cut letters black on red, Hoogin Bros & Lunn]
H103	H.D.C.	[oval, embossed]
H104	**H.E.C.**	**[rectangular, red on black & red, picture of a leaf, trademark]**
H105	**H & E**	**[rectangular, embossed]**
H106	H & H	(a) [rectangular, black on yellow, Jno. J. Bagley & Co.] (b) [star on left of H, star on right of H]
H107	H & H TOBACCO	[die-cut leaf, black on tan]
H108	H H G	[die-cut letters, red, yellow & green]
H109	H.H.K.	[die-cut letters, black on red, H.H. Kohlsaat]
H110	H & L	[rectangular, red on black]
H111	**H.M. & Co.**	**[round, embossed]**
H112	**H P**	**[diamond, embossed]**
H113	**H P**	**[cut-out rectangular]**
H114	**H.W.T. Co.**	**[round, black on red]**
H117	**1/2 ACRE**	**[round, embossed]**
H119	1/2 BUSHEL TOBACCO	[rectangular, scalloped, black on red]
H121	1/2 POUND B	[rectangular, embossed]
H123	1/2 POUND V	[rectangular, embossed]
H125	**HALF DIME**	**[diamond, black on red]**
H127	HALF & HALF	[die-cut half round, half red & half white]
H129	HALF SHELL	[oval, black on yellow, picture of a hand & oyster]
H131	HALF SOLE	[die-cut a half shoe sole, black on yellow]
H132	HAM 5¢	[round, white on blue, sweet twist]
H133	HAM MEAT	[round, embossed, picture of ham]
H135	HAMBONE	[rectangular, odd, picture, black on yellow, Cosby]
H137	**HAMILTON'S**	**[die-cut X embossed]**
H139	**HAMMER**	**[small die-cut embossed]**
H141	HAMMER	[rectangular, embossed, picture of a claw hammer]
H143	**CLAW HAMMER**	**[die-cut, black on red, Strater]**
H145	**HANCOCK**	**[round, embossed, hole in center, Richmond. Va.]**
H146	HANCOCK'S BEST	[die-cut shield, black on red, W.T. Hancock]
H147	HANCOCK W.T.	[die-cut shield embossed]
H149	**HAND**	**[die-cut embossed, left hand]**
H151	HAND	[rectangular, yellow on black, picture of hand]
H153	HAND BAG	[round, embossed]
H155	**HAND IN HAND**	**[die-cut embossed, shaking hands]**
H157	HAND SHAKE	[rectangular, embossed, picture of a hand shake, J.D. & Co.]
H159	HAND SPIKE	[rectangular, embossed, picture of hand & spike, Strater Bros]
H160	HAND MADE	[die-cut hand, black on red]
H161	**HANDMADE**	**[rectangular, picture of hand holding a plug, Globe Tob. Co]**
H163	**HANES**	**[rectangular, bow tie shape, embossed]**
H165	HANES B.F.	[rectangular, shield, embossed]
H166	**HANES B.F. WINSTON N.C.**	**[octagonal embossed]**
H167	HANES FROG	[die-cut frog, embossed]
H169	HANES OWL	[round, black on multi-colors, picture of owl]
H171	**HANES P.H. & Co.**	**[octagonal embossed]**
H173	**HANES P.H. & CO.'S NATURAL LEAF**	**(a) [rectangular, black on orange leaf & yellow]** (b) [with R.J.R. Tob. Co. Suc'sr]

H175	HANNIBAL	[rectangular, embossed]
H177	HAPPY GREETING	[rectangular, odd ends, embossed]
H179	HAPPY HOLLOW TWIST	[round, black on red]
H183	HAPPY JOE	[round, black on yellow, picture of black man standing, Vandyke & Henley]
H184	HAPPY JOE	[round, black on yellow, picture of black man's face, Rome, Ga.]
H185	HAPPY NIG	[oval, black on yellow, picture of a black man laughing]
H185	HARD PAN	[square, embossed]
H187	HARD TO BEAT	[rectangular, banner type, embossed]
H189	HARDEN	[round, yellow on red]
H190	HARDGROVE	[round, black on yellow, three stars]
H191	HARFORD	[round, embossed, B.F. Gravely & Son]
H192	HARGRAVES H DIAMOND	[diamond, black on red]
H193	**HARP**	**[die-cut embossed]**
H195	**HARP**	**[small die-cut embossed, C.W. Allen]**
H197	**HARP**	**[larger different die-cut embossed, C.W. Allen]**
H199	**HARPER'S HOME SPUN**	**[round, black on red]**
H205	**HARRISON**	**[round, picture of man in center, red, white & blue edge]**
H207	HARRY LEE	[rectangular, embossed]
H209	**HARRY WEISSINGER**	**[diamond, black on red, Tobacco Co.]**
H211	**HARVEST**	**[oval, black on red, Buchanan & Lyall, Bright Navy]**
H212	HARVEY	[rectangular, odd, embossed]
H213	HARVEY'S	[die-cut leaf, red on yellow]
H215	**HARVEY'S NAT. LEAF**	**(a) [die-cut, black on yellow, man pictured]**
		(b) [die-cut black on yellow & orange, man pictured]
H217	HASTEN'S	[die-cut leaf, black on yellow]
H219	**HAT**	**[die-cut hat, embossed]**
H221	HAT	[rectangular, black on yellow, picture of tall hat]
H223	HATCHER & STAMPS	[die-cut shape, log cabin embossed]
H225	**HATCHET**	**[die-cut, red on black]**
H226	HATCHET	[die-cut large embossed]
H227	**HATCHET**	**[die-cut small embossed]**
H228	HATCHET	[die-cut black on red, Mf'd By Rob't. Harris & Co.]
H229	HATTIE MOORE	[round, embossed]
H230	HATTIE WALKER	[die-cut H, embossed]
H231	HAVELOCK	[oval, small, black on yellow]
H232	HAVELOCK	[octagonal black on orange, Tobacco, British Aust, Tobacco Co.]
H233	HAVELOCK	[rectangular, black on red, Aromatic Tobacco, British Aust, Tob Co.]
H235	HAWAIIAN FLAG	[oval, picture of flag]
H236	HAWK	[oval, gold on black, picture of hawk]
H237	**HAWK EYE**	**[rectangular, embossed]**
H239	HAWKINS MORRIS Co.	[round, embossed]
H241	HAZEL NUT	[round, multicolored, picture of tree]
H242	HAZEL KIRKE	[rectangular, odd, embossed]
H243	**HAZEL & KIRKE**	**[square, black on red]**
H245	**H. BRUNN**	**[rectangular, embossed, 1323 Wash. Ave, Minneapolis]**
H246	**HEADS I WIN**	**[die-cut & round, black on yellow & brown picture of a dog]**
H247	HEADLIGHT	[oval, black on gold & red, Clark, Thurmon & Co., Redford City, Va.]
H249	**HEART**	**[die-cut black on red small letters]**
H251	**HEART**	**[die-cut black on red fancy letters]**
H253	HEART	[die-cut embossed, Wilson Everson & Co]
H254	**HEART**	**[die-cut plane, red]**
H255	**HEART'S CONTENT**	**[die-cut embossed heart]**
H256	**HEARTS CONTENT**	**[die-cut heart, yellow on red]**
H257	**HEBE**	**[round, embossed]**
H258	HEBERT	[rectangular, embossed]
H259	HEEL TAP	[round, embossed]
H260	HEEL TAP	[rectangular, black on red]
H261	HEEL TAP	[round, black on red, Samson]
H262	HEEL TAP	[round, white on red]
H263	HELEN WILSON	[round, black on yellow, picture of a girl, C.A. Raine & Co.]
H264	HELLO	[round, black on red]
H265	HELMET	[die-cut black on red, fireman's helmet]
H266	**HEN**	**[round, red on yellow, picture of a hen]**
H267	HEN FRUIT	[round, black & green on white, picture of hen & fruit, W.E. & Co.]
H268	HEN SPARROW	[rectangular, round center, yellow on silver, picture of a bird]
H269	HENRIETTA	[rectangular, black on yellow, picture of a girl]
H270	HENRIETTA	[round, black on yellow, picture of a girl]
H271	**HENRY A. HERSEY**	**[die-cut embossed H]**
H273	HENRY Co.	[round, black on yellow, Traylor, Spencer & Co. Tobacco.]
H275	HENRY COUNTRY	[rectangular, black on red]
H277	HENRY COUNTRY	[rectangular, embossed]
H278	HENRY GRADY	[diamond, yellow on red]
H279	**HENTSCHEL'S**	**(a) [rectangular, black on red, picture of a horse]**
		(b) [black on yellow]
		(c) [black on green]
H281	HERBERT	[rectangular, black on yellow]
H283	HERE'S LUCK	[oval, black on red]
H285	HERO	[round, red on yellow]
H286	HERO	[rectangular, black on red, picture of a dog, T.M. Co.]
H287	**HERO**	**[round, embossed, Smith & Chandler]**
H288	HERO	[round, embossed]
H289	**HERO**	**[oval, embossed, Penn & Rison]**
H290	**HERO**	**[rectangular, black on red, Moreland Bros & Crane]**
H291	HESS	[rectangular, embossed, Elert & Co.]
H293	HESTER	[oval, yellow on red]
H295	HI FOX	[round, yellow on red]

H297	**HI-YELLOW**	**[rectangular, yellow on black]**
H299	**HIAWATHA**	**(a) [die-cut Indian, black on red]** **(b) [black on orange]** **(c) [black on yellow]** **(d) [blue on white]** **(e) [blue on yellow]**
H300	HIAWATHA	(a) [die-cut Indian black on yellow, Scotten] (b) [red on yellow]
H301	**HIAWATHA**	**(a) [rectangular, bow black on red, D.S. & Co. Detroit]** **(b) [gold on black]**
H302	HIAWATHA	[rectangular, black on yellow, picture of a Indian]
H303	HICKOK'S	[hexagon black on yellow, Richmond, Va.]
H304	HICKOK'S NO. 1	[round, embossed]
H305	HICKOK'S NO. 1	[rectangular, embossed]
H306	**HICKORY**	**[round, white on blue picture of a nut]**
H307	HICKORY	[round, embossed, white on blue, picture of a nut]
H308	HICKORY NUT	[round, black & brown on yellow]
H309	HIGH ART	[round, black on yellow, Fluhrer Bros, Boonville Ind.]
H311	**HIGH COURT**	**[rectangular, black on red]**
H312	HIGH FIVE 10¢	[round, embossed, C.R.T. Co.]
H313	HIGH GRADE	[rectangular, black on red]
H314	HIGH GRADE	[round, black on brown, plug tobacco]
H315	HIGH GRADE	[rectangular, black on yellow, J.W. Daniel Tob Co., Henry Co.]
H317	**HIGH KICKER**	**[rectangular, picture of man and lady, Dalton Farrow & Co.]**
H319	HIGH LIFE	[round, black on cream, picture of man sitting back]
H321	**HIGH SPEED**	**[round, large, black on white, picture of car]**
H323	HIGH SPEED	[round, small, black on white, picture of different car]
H325	HIGH STANDARD	[oval, black on yellow, picture of man, R.J. Reynolds]
H327	HIGH TEA	[oval, embossed]
H329	HIGH TIDE	[oval, white on red, Wilson & McCallay]
H331	**HIGH TOP SHOE**	**[die-cut embossed ladies shoe]**
H332	HIGH TRUMP	[rectangular, embossed, cut-out heart in center]
H333	**HI-YELLOW**	**[rectangular, yellow on black]**
H335	**HILL BILLY**	**[square, orange on black, picture of goat]**
H337	**HILLSIDE**	**[rectangular, gold on black & red]**
H339	HINDOO	[rectangular, round small, black on red picture of moose]
H341	**HIT**	**[round, red on yellow, Taylor Bros]**
H343	HIT	[oval, embossed]
H345	**HOBBY**	**[rectangular, odd shape, embossed]**
H347	H-O	[round, embossed]
H348	HOCH-PAYNE NATURAL LEAF	[die-cut leaf black on green]
H349	HODGIN BROS & LUNN	[oval, embossed, Winston, N.C.]
H350	HOE CAKE	[rectangular, embossed]
H351	HOE CAKE	[rectangular, black on yellow]
H353	HOE MAID	[round, black on yellow & red, picture of girl hoeing, T. P. & Co.]
H355	**HOE PLUG**	**[die-cut of hoe embossed]**
H357	HOG AND HOMINY	[round, embossed, picture of hog in center]
H359	HOG & HOMINY	[round, embossed, word Hog in center]
H361	HOG MEAT	[die-cut hog standing, black on yellow]
H363	**HOLD FAST**	**[round, gold on red]**
H365	**HOLD FAST**	**[round, red on gold]**
H367	**HOLD FAST**	**[round, embossed]**
H369	HOLD FAST	[rectangular, embossed]
H370	**HOLLAND TOBACCO CO. BURTON**	**[rectangular, embossed]**
H371	HOLLAND'S PINE APPLE	[die-cut pine apple]
H372	HOLLARD'S R & R SEAL	[round, embossed]
H373	HOLLY	[rectangular, embossed]
H374	HOLLY	[oval, black on red, Dominion Tob Co. Montreal]
H375	HOLLY	[round, green on white, picture of red berries & leaves]
H376	HOLLY FORK	[oval, black on white, green wreath]
H377	HOME AGAIN	[round, embossed, J.B. Pace]
H378	HOME COMFORT	[round, embossed]
H379	HOME GUARD	[die-cut shield embossed]
H380	HOME MADE	[rectangular, embossed]
H381	HOME MADE TWIST	[round, embossed]
H382	HOME PLATE	[round, picture of man sliding into home plate]
H383	HOME RULE	[rectangular, embossed]
H384	HOME RULE	[rectangular, hexagon shape, embossed]
H385	HOME RULE	[rectangular, black on red, picture of roller]
H386	HOME RUN	[round, baseball, black on yellow & white]
H387	HOME RUN	[die-cut banner, black on yellow, B & Co.]
H388	HOME SPUN	[round, black on red, Fluhrer Bros.]
H389	HOME SPUN	[round, black on red, Flenge]
H390	**HOME SPUN**	**(a) [small round black on red, Hampton]** **(b) [larger round]**
H391	HOME SPUN	[round, black on yellow, Preston]
H392	HOME SPUN	[round, black on red, Newburgh]
H393	**HOME SPUN**	**[round, red on yellow, K.Y. Bests]**
H394	HOME SPUN	[oval, white on red, Harper]
H395	HOME SPUN	(a) [oval, black on yellow, outside prongs] (b) [inside prongs]
H396	HOME SPUN TWIST	[round, black on yellow, Robards]
H397	HOME SPUN TWIST	[round, red on yellow, Mattingly]
H398	HOME SPUN TWIST	[round, white on blue, Farmers Co-Op Assn.]
H399	HOME STRETCH	[round, embossed]
H400	HOME TWIST	[round, yellow on blue, Peerless Tob. Co.]
H401	HONEST BEN	[round, black on yellow, picture of a dog head, G.W. Smitit]
H402	HONEST BILL	[round, embossed]

H403 HONEST BOY [round, embossed]
H404 HONEST BOY T 5¢ H [round, embossed]
H405 HONEST CHEW [diamond, embossed]
H406 HONEST FACT [round, embossed]
H407 HONEST GEORGE [oval, black on yellow, picture of George Washington]
H409 HONEST JIM [rectangular, embossed]
H411 HONEST JOHN [rectangular, gold on dark blue, D.H. Spencer & Sons]
H413 HONEST JOHN PLUG [round, black on white]
H415 HONEST JOHN TWIST [round, black on yellow]
H417 HONEST LABOR [round, black on yellow & red, picture of an Arm & Hammer]
H419 HONEST MONEY [round, black on red]
H421 HONEST RACER [oval, embossed]
H422 HONEST 7 [die-cut B, black on green]
H423 HONEST TOIL [round, white on red]
H425 HONEST TOIL [round, black on yellow, picture of working-man with a pick]
H427 HONEST TOM [round, yellow on red]
H429 HONEST WEIGHT [rectangular, black on red, picture of scale]
H431 HONEY (a) [die-cut B embossed, red]
(b) [yellow]
H435 HONEY [die-cut B, black on yellow]
H437 HONEY BEE [round, red on yellow & brown, picture of a bee]
H439 HONEY BOY [round, white on red & green]
H441 HONEY COMB TWIST [round, red on yellow, picture of a bee hive]
H442 HONEY CUT [square, black on red, D.H. Spencer & Sons]
H443 HONEY CUT [rectangular, square ends, black on yellow]
H444 HONEY CUT [rectangular, round ends, black on yellow]
H445 HONEY DEW [round, small embossed]
H446 HONEY DEW [round, large embossed]
H447 HONEY DEW [rectangular, embossed]
H449 HONEY DEW [rectangular, yellow on white]
H451 HONEY DEW TWIST [round, black on yellow]
H453 HONEY DIP TWIST (a) [rectangular, light blue on white]
(b) [dark blue on white]
H454 HONEY KRUST [round, black on white, Borlow-Moore, Tobacco Co.]
H455 HONEY NUT [round, red on orange & yellow]
H457 HONEY PLANT [round, embossed]
H459 HONEY SUCKLE [die-cut heart black on red, W.C. MacDonald, Montreal]
H461 HONEY SUCKLE [oval, embossed]
H463 HONEY SUCKLE TWIST [round, black on yellow]
H464 HONEY TWIST [oval, embossed, L.S. & Co.]
H465 HONOR [banner embossed]
H467 HONOR BRIGHT [round, embossed]
H469 HOOK [die-cut large embossed shape]
H471 HOOK [die-cut small shape red]
H473 HOOK [rectangular, black on red, picture of fish hook]
H475 HOOK [round, black on red]
H476 HOOSIER [round, black on green]
H477 HOOSIER TWIST [round, white on blue, E.O.E. Shelby Tob Co.]
H478 HOPE NAVY [round, red on white & blue, picture of a sailor]
H479 HOPS [rectangular, black on red]
H481 HORN OF PLENTY [round, multicolored, picture of cornucopia]
H483 HORN OF PLENTY [rectangular, embossed, picture of horn]
H485 HORNET [round, red & brown on yellow, Miller & Clifford]
H487 HORNET'S NEST [oval, red & black on yellow, Miller & Clifford]
H488 HORSE [die-cut, running horse embossed]
H489 HORSE [die-cut small horse, embossed]
H491 HORSE [die-cut large, tan & yellow]
H492 HORSE APPLE [octagonal black on yellow & red, Adams, Powell & Co.]
H493 HORSE APPLE [octagonal black & orange on yellow, Irvin & Poston]
H494 HORSE BIT [die-cut a horse bit]
H495 HORSE HEAD (a) [small, die-cut embossed shape facing right]
(b) [medium]
(c) [large]
H496 HORSE HEAD [round, embossed, picture of horse head, Dausman]
H497 HORSE HEAD [round, embossed head, orange & tan on yellow]
H499 HORSE SHOE (a) [small die-cut shape]
(b) [medium]
(c) [large]
H501 HORSE SHOE [die-cut shape, gold on green]
H502 HORSE SHOE (a) [die-cut shape, brown on tan]
(b) [brown on yellow]
H503 HORSE SHOE [round, black on green]
H505 HOT CAKE [oval, embossed]
H507 HOT BALL [round, red on yellow]
H509 HOT DROPS [die-cut cross, black on white & red dots]
H511 HOT SCOTCH [round, embossed]
H512 HOT SHOT [round, embossed]
H513 HOT SHOT [hexagon embossed]
H515 HOT STUFF (a) [rectangular, black on gold, picture of a pepper, B.F. Hanes]
(b) [red on yellow]
H516 HOT TAMALE [round, black on yellow, Dalton Farrow & Co.]
H517 HOURS [die-cut 8 embossed]
H518 HOUND [die-cut dog running, black on yellow, B.W.]
H519 H.S. EXPRESS [octagonal embossed]
H520 H.S. EXPRESS [round, embossed]
H521 HUB (a) [die-cut shape, black on red Trade Mark]
(b) [black on yellow]
(c) [black on white]
(d) [blue on white]
(e) [blue on yellow]
H522 HUB [oval, embossed]
H523 HUB (a) [small die-cut embossed]
(b) [large]
H525 HUB SCOTTEN'S [round, black on red & orange, picture of a hub]
H527 HUCKLEBERRY [rectangular, embossed]
H529 H. HUDSON [round, embossed]
H531 HUMMING BIRD [octagonal black on yellow & red, blue bird, Bitting & Hay]
H533 HUMMING BIRD [oval, picture of a bird, brown & orange on

		white]
H535	HUMMING BIRD	[round, embossed, picture of bird]
H537	HUMMING BIRD	[rectangular, black & yellow on red, picture of bird]
H539	HUNNICUTT'S CHAMPION	[rectangular, embossed]
H541	HUSTLER	[rectangular, banner type, embossed]
H543	HUNTER'S CHOICE	[round, embossed]
H545	HUNTER'S TWIST	[rectangular, embossed]

I

I100	I C M	[rectangular, odd, embossed]
I101	I.C.U.	[round, red on green, S.F. Laumfield & Co. Mt. Airy]
I102	**I-I-C**	**[round, scalloped, black on yellow, picture of a man on a horse]**
I103	I O U	[rectangular, cut-out letters]
I104	I T C	[diamond, black on yellow]
I105	I W B	[die-cut letters, red, yellow & green]
I106	I X L	[die-cut letters black on red, R.J. Reynolds & Co.]
I107	I.X.L.	[rectangular, black on yellow, British-Australian Tobacco Sydney]
I109	**IBEX NAVY**	**[oval, white on red, Scotten, Dillon Company]**
I110	I-CHU-U	[diamond, embossed]
I111	IDA MOTLEY	[round, black on yellow, picture of girl, peach flavor]
I112	IDA MAY	[round, black on yellow, picture of girl]
I113	IDEAL	[die-cut cross embossed]
I115	**IDEAL**	**[die-cut girls head, black on yellow]**
I117	ILL-MO	[round, white on red]
I121	IMPERIAL	[banner embossed]
I123	**IMPERIAL**	**[round, embossed, W. & McC.]**
I125	IMPERIAL	[rectangular, black & yellow on red, Strater Bros]
I127	IMPERIAL RUBY-BIRD'S EYE	[rectangular, gold on red, T.C. Williams Co.]
I128	IN STYLE	[rectangular, embossed]
I129	IN THE SWIM	[round, blue on white, picture of man in water]
I130	INDEPENDENT	[rectangular, bar, red on black]
I131	**INDEPENDENT**	**[die-cut cross black on red]**
I133	**INDEPENDENT NAVY**	**[round, white on red, picture of a man]**
I135	**INDEX**	**[die-cut hand pointing embossed, Strater Bros]**
I136	INDEX	[rectangular, embossed]
I137	INDEX	[round, embossed]
I139	INDIAN	[round, embossed, picture of Indian head]
I141	INDIAN CLUB	[rectangular, black on red picture of Indian standing]
I142	INDIAN GAME TOBACCO	[round, red on yellow & brown, picture of a rooster]
I143	**INDIAN MAID**	**[rectangular, black on red picture of an arrow]**
I145	INDIAN PLUG	(a) [round, small black on red, picture of Indian head]
		(b) [round, small black on yellow]
		(c) [round, small black on white]
I147	**INDIAN PLUG**	**(a) [round, large black on red, picture of Indian head]**
		(d) [round, large black on yellow]
I149	**INDIAN PENNY 1885**	**[round, ring penny in center]**
I151	INDIANA PLUG	[round, embossed]
I152	INDIANA TWIST	[octagonal yellow on red]
I153	INDUSTRY	[rectangular, embossed]
I155	**INTER STATE**	**[die-cut black on red]**
I156	**INVINCIBLE**	**[oval, black on green]**
I157	IREDELL'S BEAUTY	[round, picture of girl, Iredell-Statesville]
I158	IRISH MIKE	[oval, embossed]
I159	IRISH TWIST	[round, embossed]
I160	IRON CLAD	[octagonal red on black, picture of ship, Penns.]
I162	IRON SIDES	[oval, black on red]
I163	ISLAND OF CUBA	[round, small, black on yellow]
I164	ISLAND OF CUBA	[round, black on blue]
I165	**ISLAND OF CUBA**	**(a) [oval, black on yellow, Luchs & Bro.]**
		(b) [black on red]
I167	IT'S NAUGHTY	[small octagonal, black on green, picture of girl, cartoon type]
I168	**ITALY FLAG**	**[oval, picture of flag]**
I169	IVANHOE	[die-cut embossed]
I170	IVANHOE	[round, red on yellow, Cameron Tobacco]
I171	IVANHOE	[octagonal black on yellow, A.H. Motley Co.]
I172	IVY	[round, red on yellow, picture of green ivy, tag made in U.S.A.]
I173	IXL 3 PLY	[round, embossed, Gravely]

J

J101	J	[round, cut-out J]
J103	**J A T & Co.**	**[oval, embossed]**
J104	J A E	[die-cut letters black on red, Ellie, Trademark]
J105	J.A.G.	[round, yellow on black]
J106	**J B**	**[large rectangular, cut-out]**
J107	**J B**	**[small rectangular cut-out]**
J108	**J B**	**[die-cut shield, cut-out]**
J109	J.B.	[round, blue on white, Trademark]
J110	**J.B. HOME SPUN**	**[round, black on red]**
J111	J. BUGG TWIST	[round, black on yellow picture of a bug]
J112	J.C.L.	[round, black on yellow, chew, Homes & Miller Salisbury, N.C.]
J113	**J D**	**[rectangular, cut-out]**
J115	J.D. & C.	[round, embossed]
J117	**J D & Co.**	**[rectangular, embossed, clasped hands]**
J119	**J.F.S. WE R 7**	**[oval, black on red]**
J121	J H	[die-cut letters J H, black on red, Hilton Real Stuff]
J122	J H	[round, black on red, Sweet Hilton Twist]
J123	J.H.H.	[rectangular, embossed]
J124	J I C	[rectangular, embossed]
J125	**J.J.**	**[round, red on yellow & green, Wilson &**

		McCallay Tobacco Co.]
J127	J J	[rectangular, embossed]
J129	J.J.	[rectangular, cut-out letters]
J130	J K M	[die-cut letters, red]
J131	**J.L.**	**[rectangular, cut-out]**
J133	J L KING & Co	[oval, black on yellow]
J135	J. & M.	[oval, green & yellow on red]
J137	**J N**	**[rectangular, cut-out]**
J138	J N W	[die-cut letters]
J139	J.O.	[diamond, black on green, P. Lorillard & Co.]
J141	J.O.L.	[round, black on orange, Homes & Miller, chew, Salisbury, N.C.]
J143	J.O.C.	[round, embossed]
J144	J R N	[die-cut letters black on red, Mfg by J.R. Noel Tob. Co.]
J145	J.S.N. & Co.	[rectangular, embossed, New York]
J147	**J T**	**[rectangular, cut-out small]**
J149	J T	[rectangular, cut-out larger]
J151	J T	[round, embossed]
J153	**J. & W.**	**[rectangular, embossed]**
J155	J T A & Co.	[oval, embossed]
J157	**J.T. & H. CLAY**	**[rectangular, white on red Gravely]**
J159	**J T J**	**[round, black on red]**
J160	JACK	[die-cut a mule head embossed]
J161	**JACK**	**[round, embossed, donkey head]**
J163	**JACK**	**[round, black on white]**
J165	JACK	[round, black on red]
J167	JACK	[rectangular, small embossed]
J169	**JACK GRAVELY**	**(a) [round, black on yellow]** (b) [round, black on white] (c) [round, black on green]
J171	JACK HAVERLY	[square, black on red, D.M. & Co.]
J173	**JACK KNIFE**	**[die-cut embossed knife]**
J174	JACK NAVY	[diamond, black on yellow, picture of a jackass]
J175	JACK OF DIAMONDS	[rectangular, multicolored, picture of Jack of Diamonds card]
J176	JACK POT	[rectangular, embossed, S.L. Co.]
J177	**JACK RABBIT**	**[rectangular, black on red, picture of a rabbit]**
J178	JACK RABBIT	[round, picture of a rabbit]
J179	JACK ROSE	[round, embossed]
J181	JACK SHEPPARD	[round, gold on green]
J183	**JACK SCREW**	**[round, black on red, Taylor Bros]**
J185	JACK SPRATT	[octagonal brown on yellow, T.C. Williams Co. Continental Tob.Suc]
J186	**JACK SPRATT**	**[octagonal brown on yellow, T.C. Williams Co. Richmond, Va.]**
J187	JACK THE GIANT KILLER	[rectangular, black on red]
J189	**JACK THE RIPPER**	**[octagonal, black on yellow, picture of black man]**
J191	JACKO'S BEST	[round, embossed]
J193	JACK'S BEST	[round, black on yellow]
J195	JACKSON'S JACK	[rectangular, red on yellow]
J197	**JACKSON'S BEST**	**[hexagon embossed]**
J199	JACKSON'S TWIST	[banner embossed]
J201	JACKSON'S JULEP	[rectangular, embossed]
J203	**JACOBS LADDER**	**[rectangular, black on yellow, picture of man & ladder, Leak Bros. & Hasten]**
J204	JAGUAR	[round, black on red, picture of cat]
J205	JAMES RIVER TWIST	[triangular, black on yellow]
J206	JANUS	[gold on black, Richmond, Va.]
J207	JAVELIN	[rectangular, black on yellow, picture of spear, S.A. Ogburn]
J209	JAW BONE	[die-cut embossed]
J211	JAYBIRD	[round, embossed, picture of bird]
J212	J.B. HOMESPUN	[round, black on red]
J213	**J.B. PAGE**	**[rectangular, banner embossed]**
J214	J. BIG	[small round red on white]
J215	JEFF DAVIS	[rectangular, picture of man with goatee, on flag, Arnold & McCord]
J216	JEFF DAVIS	[rectangular, picture of man on flag, Full Beard, Atlanta, Ga.]
J217	JEFF DAVIS	[round, black on yellow, picture of man]
J218	JEFFERSON GREENVILLE	[rectangular, black on yel low]
J219	JEFFERSON SUN CURED	[octagonal orange on black, Gravely]
J220	JEFFERSON'S 4 S CHOICE	[round, embossed]
J221	**JELLY CAKE**	**[diamond, white on dark blue]**
J222	JELLY ROLL	[round, white on red, Bradley-Preston Tobacco Co.]
J223	JELLY ROLL	[oval, white on red]
J224	JENNY	[rectangular, embossed]
J225	JENNIE LIND	[octagonal black on yellow, Mfd. by Dodson Bros]
J226	**JENNIE LIND**	**[octagonal black on yellow, Dodson Bros & Stockton]**
J227	**JERSEY**	**[round, engraved, picture of a cow]**
J228	JERSEY	[round, embossed, picture of cow]
J229	JERSEY	[round, black on yellow, multicolored picture of cow, R.A. Patterson Tobacco Co.]
J230	JERSEY COW	[rectangular, red on yellow, picture of cow, S. A. Ogburn]
J231	JERSEY CREAM	[round, black on red, picture of a cow]
J232	JERSEY LILY	[round, embossed]
J233	**JESSIE**	**[round, black on red, picture of girl's head]**
J235	JESTOBACKER	[rectangular, black on yellow]
J236	JET	[round, black on red]
J237	**JEWEL**	**[rectangular, odd embossed]**
J239	JEWEL	[round, embossed]
J241	J I B	[rectangular, black on yellow, picture of ship]
J242	JIM CROW	[rectangular, black on red, picture of a crow facing left]
J243	JIM QUICK	[round, embossed]
J244	**JINGLE**	**[rectangular, black on red]**
J245	JITNEY TWIST	[round, black on yellow]
J246	JO	[die-cut heart shape, yellow on red]

J247	JOCKEY CAP	[die-cut a cap, multicolored]
J249	JOCKEY CAP	[die-cut embossed a cap]
J251	JOCKO NAVY	[round, embossed]
J253	JOE B.	[round, black on yellow]
J255	JOE BAILEY	[rectangular, odd red on yellow]
J257	JOE BOWERS	[rectangular, half oval crest, black on yellow]
J258	JOHN BULL SMOKING	[round, black on yellow, picture of a bull's head]
J259	**JOHN GILPIN**	**[round, embossed]**
J261	**JOHN L.**	**[rectangular, black on yellow, picture of boxer]**
J263	JOHN WESLEY	[round, black on red, picture of man]
J265	JOHN'S OX	[die-cut letters O X embossed]
J267	JOHN'S 7	[die-cut a 7 embossed]
J269	JOHN'S	[oval, red on yellow]
J270	JOHNNY CAKE	[octagonal black on red, wood]
J271	JOHNNY CAKE	[round, black on red, picture of boy eating]
J272	JOHNNY CAKE	[octagonal embossed]
J273	JOHNNY CAKE CHEWING	[round, embossed]
J274	JOHNNY CAKE SMOKE	[round, red on yellow]
J275	**JOHNNY GET THE GUN**	**[round, black on green picture of man]**
J276	JOHNSTON	[rectangular, embossed]
J278	JOLIETTE	[round, black on yellow J.U.G. Tob. Co.]
J279	**JOLLY TAR**	**(a) [round, black on yellow & orange, Finzer & Bros. American Tob Co.]**
		(b) [JNO. Finzer & Bros, Con. Tob. Co. Successor]
J283	JOLLY TAR	[round, black on yellow & orange, Finzer in dot]
J285	JONES 5 CENT PIECE	[round, silver on black, picture of coin]
J287	JOSIE	[oval, embossed]
J288	**JOY**	**[round, cut-out letters]**
J289	JOY	[round, black on orange]
J290	**J.P. MILLNER**	**[oval, embossed, Brosville, Va.]**
J291	**J.R. PACE**	**[rectangular, banner embossed]**
J292	**J.R. PACE**	**[rectangular, embossed]**
J293	**JUBILEE TWIST**	**[round, black on red]**
J294	**JUBILEE**	**[rectangular, embossed]**
J295	JUDGE	[round, scalloped, embossed, Empire Tob. Co.]
J296	JUDGE	[round, embossed, Empire Tob. Co.]
J297	JUG	[die-cut jug large embossed]
J298	**JUG**	**[die-cut jug small embossed]**
J299	JUICY CHEW	[black on yellow & red, J.B. Taylor Tob Co., Leaksville, N.C.]
J300	JUICY PLUM	[round, black on red]
J301	JULEP	[oval, black on yellow]
J302	**JULIET**	**[round, scalloped black on gray-green]**
J303	JULIET	[round, black on green]
J305	JULY	[round, multicolored, picture of fireworks]
J306	JULY	[round, yellow on red, Gravely]
J307	JULY TOBACCO	[rectangular, black on yellow, W.M. Stultz]
J308	**JUMBO**	**[round, scalloped, black on yellow, picture of elephant]**
J309	**JUMBO**	**[die-cut embossed elephant, facing right]**
J311	**JUMBO**	**[die-cut embossed elephant, facing left]**
J313	JUMBO	[die-cut embossed elephant, Tinsley]
J314	**JUMBO**	**[oval, embossed]**
J315	JUMBO PLUG	[rectangular, black on yellow]
J317	JUMBO SMOKING	[round, white on blue, D.C. & Co]
J319	**JUMBO TWIST**	**[round, black on white, picture of a elephant]**
J320	JUMPER	[rectangular, odd embossed]
J321	JUNE APPLE	[die-cut apple, yellow on red]
J323	**JUNE TOBACCO**	**[rectangular, yellow on black, W.M. Stultz]**
J325	JUNE BUG	[round, yellow on red & yellow, picture of a bug, F.M. Bohannon]
J326	JUNE BUG	[oval, embossed]
J327	JUNIOR NAVY	(a) [oval, black on red]
		(b) [oval, red on yellow]
J329	**JUST OUT**	**[hexagon black on green, picture of a chick]**
J331	**JUST OUT**	**[round, scalloped black on yellow, picture of bee]**
J333	**JUST OUT**	**[round, scalloped black on yellow, picture of bee, P.H. Hanes Co.]**
J335	JUST OUT	[round, embossed]
J337	JUST RIPE	[round, black on red]
J339	JUST SUITS	[rectangular, embossed]
J341	JUST THE THING	[rectangular, banner black & red on yellow]

K

K101	**K**	**[die-cut K embossed]**
K103	**K**	**[die-cut K plane]**
K105	K K K	[round, green on yellow]
K106	**K O**	**[octagonal, black on yellow, Bailey Bro.]**
K107	**K P**	**[rectangular, cut-out K P]**
K109	K.P.	[round, black on red]
K111	K. & R.	[oval, embossed]
K113	K.S.	[die-cut shaking hands, red on black]
K115	**K.T.**	**[round, white on red]**
K117	K.T.	[round, embossed]
K118	K of L	[odd rectangular, embossed, Made by M & J]
K119	**K. of L. PLUG**	**[die-cut, embossed king on a throne]**
K120	KAISER	[rectangular, embossed]
K121	KANT STOP	[rectangular, red on yellow]
K125	**KATE GRAVELY**	**(a) [rectangular, black on yellow, picture of bird]**
		(b) [square, black on yellow, picture of bird]
K126	KATIE	[round, black on red]
K127	KELLY'S IDEAL	[round, orange on green & yellow]
K129	KELLY'S PLUG	[round, black on red, picture of plug tobacco]
K130	KENO	[banner embossed]
K131	KENO	[round, embossed]
K132	KENO	[die-cut K, black on red, Middleton & Co.]
K133	**KENO**	**[die-cut K, black on yellow, M & Co.]**
K134	KENO	[round, black on red]
K135	KENT	[oval, black on yellow]
K136	KENTUCKY	[rectangular, red & gold on

		yellow, Penn]
K137	KENTUCKY BOY	[round, black on green, picture of a boy]
K139	KENTUCKY BURLEY	[oval, black on yellow, Selected Natural Leaf]
K141	KENTUCKY CARDINAL	[round, black on yellow, picture of a bird]
K142	**KENTUCKY CARDINAL**	**[round, black on yellow & red, picture of bird]**
K143	KENTUCKY DERBY	[round, black on red picture of horses racing, G.R.T. Co.]
K144	KENTUCKY DERBY	[rectangular, black on yellow, picture of horses racing]
K145	KENTUCKY DEW	[rectangular, embossed]
K146	KENTUCKY DEW	[oval, red on yellow]
K147	KENTUCKY DEW SWEET TWIST	[round, black on yellow]
K149	KENTUCKY DIAMOND	[diamond, black on red, H.B. Scott Tobacco Co.]
K150	KENTUCKY HOME SPUN	[oval, black on red, Ky. Tob. Co.]
K151	KENTUCKY KERNEL	[round, black on red, picture of a man, G.R.T. Co.]
K152	KENTUCKY KING	[round, gold on red, full weight]
K153	KENTUCKY LEADER	[round, picture of horse head in center]
K154	**KENTUCKY LEAF**	**[round, silver on blue]**
K155	KENTUCKY PLUG	[round, embossed]
K156	**KENTUCKY TWIST**	**[rectangular, octagon embossed]**
K157	KENTUCKY SEAL	[oval, black on red]
K158	KENTUCKY SMILE	[round, black on yellow]
K159	KETCHON	[round, embossed]
K160	KEY	[rectangular, black on yellow, picture of key]
K161	KEY	[die-cut small, shape key, embossed]
K162	KEY	[die-cut skeleton key]
K163	KEY	[die-cut key, Michigan Tobacco Co.]
K165	KEY & CO	[round, black on red, picture of key]
K164	KEY HOLE	[rectangular, die-cut key hole, embossed]
K167	KEY HOLE	[rectangular, cut-out keyhole, black on red]
K169	KEY TO SUCCESS	[rectangular, black on red, picture]
K170	KEY TO SUCCESS	[die-cut shape, large key embossed]
K171	KEY WEST	[rectangular, embossed]
K172	KEY NOTE	[die-cut shape, key, black on yellow, Penn & Sons Tob Co.]
K173	**KEY NOTE**	**[round, embossed]**
K174	KEY RING	[round, embossed]
K175	**KEY STONE**	**[die-cut keystone, embossed]**
K176	KEY STONE	[die-cut keystone embossed, WSK & Co.]
K177	**KEY STONE**	**[die-cut embossed, Sneeringer]**
K179	**KEY STONE**	**[die-cut horizontal, black on red]**
K180	KEY STONE	[square, black on yellow, picture of keystone, Wills]
K181	**KEY STONE**	**[rectangular, black on yellow, ?—land Bros & Crane]**
K182	KEY STONE	[rectangular, black on yellow]
K183	KEY STONE	[die-cut vertical, black on yellow]
K184	KEY STONE	[rectangular, embossed, picture of key, Penn & Rison]
K185	C. E. KEYWORTH	[die-cut K embossed]
K187	**KICKER**	**[octagonal embossed]**
K189	**KICKAPOO**	**(a) [small round black on orange, picture of Indian]**
		(b) [large round]
K197	KICKAPOO	[rectangular, embossed]
K201	**KIDS**	**[rectangular, black on yellow, picture of two babies, Gravely & Miller]**
K203	KIDS	[round, embossed]
K205	KIM-BO	[rectangular, black & yellow on red]
K207	**KING**	**[oval, embossed]**
K209	KING	[square, embossed]
K211	KING, A. & B. TWIST	[round, embossed]
K212	**KING BEE**	**[square, black on red]**
K213	KING BEE	[rectangular, black on yellow, picture of bee, Koonts & Hartley]
K215	KING BEE	[rectangular, black on yellow, picture of bee]
K219	**KING BOLT**	**(a) [die-cut bolt, gold on dark blue, The W & McC. Tob. Co.**
		(b) [silver on dark blue]
		(c) [gold on black]
K220	KING BROS	[round, embossed]
K221	KING BROS BEST	[rectangular, Sun Dried]
K222	KING BROS OLD VA.	[round, embossed]
K223	**KING CORN**	**[round, black on orange & yellow, picture of ear of corn]**
K225	KING CORN	[rectangular, yellow on black, picture of ear of corn]
K227	KING FISHER	[banner embossed]
K229	KING GEORGE'S NAVY	[oval, black on red]
K231	KING JACK	[octagonal white on red, Wetmores]
K233	KING OF ALL	[round, yellow on red, Penn & Rison]
K235	**KING OF ALL**	**[round, embossed, A.H. Motley & Co.]**
K237	KING OF ALL	[rectangular, black & yellow on orange]
K239	KING OF DIAMONDS	[rectangular, King of Diamonds playing card]
K241	**KING PIN**	**[rectangular, gold on blue, Liggett & Myers]**
K243	**THE KING**	**[round, blue on white, picture of a king]**
K245	KISMET	[round, embossed]
K247	**KISMET**	**[rectangular, black on red]**
K249	**KISMET**	**[round, small, black on red]**
K251	**KITE**	**(a) [die-cut kite hexagon embossed, red on white]**
		(b) [die-cut kite narrow hexagon embossed, red on white
		(c) [die-cut kite hexagon, red on white]
K252	KITE EXTRA FINE	[die-cut, black on yellow]
K253	KITE SUN CURED	[die-cut black & yellow on yellow & black]
K255	KITTY MAY	[round, black on yellow, picture of a girl, Penns]
K256	KNAFFL'S	[rectangular, scalloped, red on yellow picture of elephant, R. Knaffl]
K257	**KNAPSACK**	**[die-cut embossed]**
K259	**KNIGHT**	**[die-cut embossed knight on a horse]**
K261	KNIGHT'S DELIGHT	[round, embossed]
K263	KNOX ALL	[rectangular, red on yellow,

		picture of man]
K264	KNOX ALL	[round, red on yellow, E.T. & Co.]
K266	KO DAK	[die-cut letters, red]
K267	KY. BEST'S HOME SPUN	[round, red on yellow]
K268	KY-LO	[oval, black on red, Scotten]

L

L101	L	[die-cut letter]
L103	L.D.	[oval, white on red]
L105	**L F**	**[rectangular, cut-out]**
L106	**L G C**	**[round, white on red]**
L107	L.H. NAVY	[round, yellow on red]
L108	**L & M**	**[rectangular, cut-out]**
L109	**L & M**	**[rectangular, embossed]**
L111	**L & M NATURAL LEAF**	**[large oval, black on red]**
L113	L.P.	[rectangular, cut-out letters]
L115	L & R 9s	[round, black on red]
L117	L & R 10s	[round, black on yellow]
L119	**L.S. 5¢**	**[oval, embossed]**
L121	L.T.	[rectangular, embossed]
L123	**L W**	**[rectangular, cut-out]**
L125	**L W**	**[square, white on blue diamond in center]**
L126	L.W.A. & SONS	[rectangular, embossed]
L127	LABOR	[die-cut letters]
L128	LABOR DAY	[round, red on yellow]
L129	LABOR KING	[diamond, black on red]
L130	LABOR KING	[die-cut a bell, black on red]
L131	LABOR'S CHOICE	[die-cut shield black on red]
L132	LADY ELGIN	[octagonal black on yellow, picture of girl]
L133	LADY ELGIN	[octagonal black on red, picture of a girl, Penn & Rison]
L134	**LADY SLIPPER'S**	**[die-cut lady's leg engraved]**
L135	LAFAYETTE	[round, yellow on yellow, picture of man]
L136	LAGER	[rectangular, picture, Raines]
L137	LAGER, 10¢ TAGS, 5¢ CASH	[rectangular, red on yellow]
L139	LALLA ROOKH	[round, black on red]
L140	LAME DUCK	[round, black on yellow, picture of red duck, Virginia-Carolina Tob Co.]
L141	**LARGEST CUT X 10. CENTS**	**[round, embossed]**
L143	**LATEST**	**[rectangular, red on white, B.F. Gravely & Sons]**
L145	LATEST TWIST	[round, red on yellow]
L146	LAUREL BELLE	[round, red on white, picture of a girl, J.N. Wyllie & Co.]
L147	LAURUS & BRO. CO.	[round, black on red]
L148	LAWYER	[rectangular, bar, gold on red]
L149	L & B LUXURY TOBACCO	[round, black & yellow on tan]
L150	L & BRO PLUG	[round, brown on cream]
L151	LEACH'S DELIGHT	[rectangular, embossed]
L152	**LEADER**	**[rectangular, large embossed]**
L153	**LEADER**	**[rectangular, round embossed]**
L154	LEADER	[rectangular, octagon, Arbuckle & Co.]
L155	**LEADER**	**[die-cut shield embossed]**
L157	LEADER	[round, black on yellow, W.R. White & Sons]
L159	LEADER	[round, yellow on black, Harrelson & Crump]
L160	LEADER	[odd embossed, Philadelphia]
L161	LEADER	[oval, black on yellow]
L162	LEADER	[round, red on yellow]
L163	**LEAF**	**[die-cut embossed]**
L165	**LEATHERWOOD**	**[round, embossed, B.F.P. & Co]**
L166	LEATHERWOOD	[round, embossed, Gravely]
L167	LEE	[oval, red on white]
L169	LEE & MORTON	[rectangular, odd, embossed]
L171	**LEG**	**[die-cut leg, red]**
L173	**LEGAL TENDER**	**[rectangular, black on red, H. Bros & Co.]**
L175	**LEGAL TENDER**	**(a) [round, blue on yellow $ red dot in center]**
		(b) black on yellow
L176	LEGAL TENDER	[oval, red on white, I.C. & M.]
L177	LEGAL TENDER	[rectangular, round, black on red, F.R. Penn & Co.]
L179	LEGAL TENDER	[round, embossed]
L181	LEGAL TENDER	[rectangular, embossed]
L183	LEGAL TENDER	[round, black on yellow, Gravely]
L185	**LEGGETTS**	**[round, embossed]**
L187	E. LEIDY	[die-cut an arrow embossed]
L189	**LEO**	**[round, embossed]**
L191	LEON PLUG	[diamond, yellow on red, Lottier]
L192	LEON PLUG	[round, yellow on green, Lottier]
L193	LEONARD'S BEST	[oval, embossed]
L195	LEONORA	[rectangular, black on red]
L197	**LET GO**	**[round, white on black & red, W. & McC. Tob. Co.]**
L199	LET HER GO	[oval, embossed]
L200	LEVEL BEST	[diamond, black on yellow, Zeph Gravely]
L201	LEVI	[rectangular, bar embossed]
L202	LEWIS POUNDS	[die-cut letter W, black on red]
L203	LEW'S BEST	[banner embossed]
L205	LEW'S PLUG	[rectangular, embossed]
L206	LEXINGTON TWIST	[round, red on cream, picture of a horse head]
L207	**LIBERTY**	**[rectangular, black on red picture of a bell]**
L208	**LIBERTY**	**[oval, white on blue, P. Lorillard & Co.]**
L209	**LIBERTY**	**[die-cut banner embossed]**
L211	**LIBERTY**	**[round, embossed]**
L212	LIBERTY BELL	[round, red on yellow, picture of bell]
L213	**LIBERTY BELL**	**(a) [square, small, white on black & tan, picture of bell]**
		(b) [large]
L214	**LIBERTY BELL**	**(a) [square, large, embossed, white on black & tan, picture of bell]**
		(b) [small]
L215	LIBERTY BELL	[rectangular, black on red, picture of bell]
L218	**LIBERTY HEAD**	**[round, black on white, picture of only Liberty's head]**
L219	LIC QUID	[rectangular, embossed]
L221	LIFE	[oval, red on yellow]
L223	LIFE GUARD	[rectangular, black on yellow, picture of man]
L225	**LIGGETT & MYERS**	**[round, embossed hole in center]**
L227	LIGGETT & MYERS	[round, embossed]
L229	**LIGHT**	**[round, black on red, pic-**

Code	Name	Description
		ture of shield & star in center]
L231	LIGHT PRESSED	[octagonal black on red, Strater]
L235	LILAC TASTE	[round, embossed]
L237	**LILLIE DALE**	**[rectangular, black on tan, picture of girl, Penn & Rison]**
L239	LILLY MAY	[octagonal red on yellow & black, picture of girl, Mfd. by Irvin & Poston]
L240	LILY	[die-cut a lily, embossed]
L241	**LIMBERTWIG**	**[rectangular, black on yellow, picture of fruit, Miller & Clifford]**
L243	LIMBERTWIG	[round, black on yellow, picture of fruit, Turner Powell & Co.]
L244	LIMBERTWIG	[round, picture of fruit, Adams-Powell & Co.]
L245	LIMITED	[round, embossed]
L247	LIMITED	[round, red & green on yellow, Liipfert & Co]
L249	**LINKS**	**[rectangular, black on yellow, picture of chain]**
L250	LION	[die-cut shield, picture of a lion]
L251	**LION**	**[hexagon red on white, picture of a lion]**
L253	**LION**	**[rectangular, black on green picture of a lion facing left]**
L254	LION	[round, black on red, picture of lion, Cameron]
L255	LION	[round, black on red, picture of a lion]
L256	**LITTLE**	**[round, black on yellow, picture of a rock, Conner & Gilbert]**
L257	**LITTLE ALICE**	**[round, black on red]**
L258	LITTLE ANGELS	[rectangular, odd embossed]
L259	LITTLE BANTAM	[oval, multicolored, picture of a chicken, Hadley & Smith]
L260	LITTLE BILLY	[round, embossed]
L261	LITTLE BOB	[round, red on cream]
L262	LITTLE DICK	[round, black on red]
L263	LITTLE DORRIT	[hexagon embossed]
L265	**LITTLE DOT**	**[oval, black on green]**
L266	**LITTLE EDNA**	**[octagonal embossed, F.M. & Co.]**
L267	LITTLE EDWIN	[diamond, black on red, picture of boy on a horse, Penn & Watson]
L268	LITTLE ETHELL TOB	[round, black on yellow, picture of a girl, G. Penn. Sons & Co.]
L269	**LITTLE FELLOW**	**[round, black on yellow, picture of a baby]**
L271	LITTLE FELLOWS	[rectangular, embossed]
L272	**LITTLE FLORENCE**	**[round, embossed]**
L273	LITTLE FRED	[oval, black on yellow, picture of boy, Rankin Bros.]
L274	**LITTLE GEMS**	**[round, embossed]**
L275	LITTLE GIANT	[rectangular, banner embossed]
L279	LITTLE GRADY	(a) [round, black on red, picture of baby face] (b) [round, black on yellow]
L281	LITTLE HARDY	[round, embossed]
L282	LITTLE HATTIE	[rectangular, banner embossed]
L285	LITTLE HENRY	[rectangular, embossed]
L286	LITTLE JIM	[round, black on red]
L287	**LITTLE JOE**	**(a) [rectangular, black on yellow, Ogburn, Hill & Co.]**
		(b) [rectangular, red on yellow]
L291	**LITTLE JOE**	**[rectangular, black on yellow, Webb & Crawford]**
L293	LITTLE JOKER	[rectangular, black on red, picture of Joker]
L294	LITTLE KATE	[round, embossed, picture]
L295	LITTLE KATIE	[round, large, picture of a girl]
L296	LITTLE LAURIE	[octagonal black on red, picture of little girl]
L297	LITTLE LIDA	[oval, embossed]
L299	LITTLE MAJOR	[round, black on red, picture of boy]
L300	LITTLE MATTIE	[rectangular, round banner, embossed]
L301	LITTLE MAUD	[diamond, embossed]
L302	LITTLE MAY	[octagonal embossed]
L303	LITTLE MAY	[octagonal black on blue]
L304	**LITTLE MOSES**	**[rectangular, black on yellow, Stanley & Finley]**
L305	LITTLE NECK	[rectangular, odd embossed]
L306	LITTLE NICK	[rectangular, red on yellow]
L307	LITTLE NINA	[round, black & red on yellow]
L308	LITTLE PAUL	[round, embossed]
L309	LITTLE PAUL	[rectangular, embossed]
L310	LITTLE ROCK	[rectangular, picture of rock, Conner & Gilberts]
L311	LITTLE ROCK	[round, black & brown on yellow, picture of rock]
L313	LITTLE ROSE BUD	[round, red on yellow, picture of rosebud]
L314	LITTLE SUSIE	[oval, embossed]
L315	LITTLE SWAN	[round, embossed]
L317	**LITTLE TROTTER**	**[round, embossed]**
L318	LITTLE TROTTER	[round, black on red]
L319	LITTLE WORTH	[octagonal black on yellow]
L321	LIVE INDIAN	(a) [rectangular, black on yellow, picture of a Indian] (b) [black on orange]
L323	LIVER REGULATOR	[oval, embossed]
L325	LIZARD TAIL	[rectangular, embossed]
L327	LLOYD TWIST	[die-cut letter L, embossed]
L329	LLOYD'S PLUG	[die-cut letter L, embossed]
L330	LOCK	[die-cut embossed shape, a padlock]
L331	LOCK OUT	[oval, embossed]
L333	LOCKETT	[oval, black on red, picture of a lock]
L334	LOG BARN	[square, octagonal embossed]
L335	LOG BARN	[rectangular, octagon, black on yellow, picture of barn]
L337	**LOG CABIN**	**[round, black on yellow, Roth Tobacco Company, picture of cabin]**
L339	LOG CABIN	[round, black on yellow, picture of cabin, Fluhrer Bros, Boonvill Ind.]
L341	**LOG CAMP**	**[round, black on yellow, Roth Tobacco Co., picture of a tent]**
L343	**LONDON CLUB 8**	**[round, embossed]**
L345	LONE FISHERMAN	[octagonal black on yellow, picture of man fishing, Iredell Tob. Co.]
L347	LONE JACK	[octagonal black on yellow, picture of black man]
L348	LONE STAR	[rectangular, white on blue]
L349	LONE STAR	[rectangular, shield, black on red]
L351	**LONG BILL**	**[die-cut black on yellow & red, bird head]**
L353	**LONG COTTON**	[octagonal black on yel-

		low, picture of a black man picking cotton]
L355	LONG FELLOW	[oval, black on yellow, picture of race-horse]
L356	LONG HORN	[round, black on silver]
L357	**LONG HORN**	**[die-cut horn, black on red]**
L358	LONG LEAF	[die-cut leaf, gold on red]
L359	**LONG POLE**	**(a) [rectangular, black on red, R. & W.]** **(b) [rectangular, red on black, R. & W.]** **(c) [rectangular, yellow on black, R. & W.]**
L365	LONG POLE	[rectangular, embossed]
L367	**LONG RANGE**	**[rectangular, black on yellow, picture of man & rifle, Walker Bros.]**
L368	LONE STAR	[rectangular, octagon, white on blue]
L369	LONE STAR	[rectangular, embossed]
L370	LONUS	[round, black on yellow, Petersburg, Virginia, Made In America]
L371	**LOOK OUT**	**[rectangular, black on yellow, picture of light house]**
L372	LORILLARD	[rectangular, embossed]
L373	**LORILLARD**	**(a) [large round embossed, with star]** **(b) [small round embossed, with star]**
L374	LORILLARD	[round, small embossed]
L376	LORILLARD 1ST	[round, black on red]
L379	**LORILLARD 1ST**	**(a) [small round embossed]** **(b) [large round embossed]**
L383	**P. LORILLARD & Co.**	**[die-cut arrow, black on yellow]**
L385	LORILLARD NATURAL LEAF	[round, black on yellow]
L387	**LOST CHORD**	**[round, black on yellow, picture of girl & mandolin, Watt Penn & Co.]**
L388	LOTTIER'S	[rectangular, embossed]
L390	LOTTIER'S TRUCK	[round, black on red, picture of hand truck]
L391	**LOTTIER'S TRUCK**	**[rectangular, gold on black, picture of hand truck]**
L393	LOTUS	[oval, embossed]
L394	LOU DILLON	[rectangular, banner gold on red]
L395	**LOUIS MEHNER**	**[rectangular, embossed]**
L396	LOUISVILLE TWIST	[die-cut crescent, white on blue]
L397	LOUISA	[oval, black on yellow, R.J.Hancock]
L399	LOUISA SUNCURED	[oval, black on yellow, R.J. Hancock]
L401	**LOVE**	**[rectangular, small, embossed]**
L403	LOVE	[rectangular, large embossed]
L404	LOVE	[die-cut heart embossed]
L405	LOVELL'S BEST	[round, black on yellow]
L407	**LOVELL'S EXTRA**	(a) [round, red on yellow] **(b) [round, black on yellow]**
L411	LOVELL'S PLUG	[round, brown on yellow]
L413	LOVER'S LEAP	[rectangular, embossed, L. Buhler & Co]
L415	LUCILE	[die-cut large, black on yellow, picture of girl]
L417	**LUCILE**	**[die-cut small, black on yellow, picture of a girl]**
L419	LUCK	[round, picture, black on yellow]
L421	LUCK	[oval, embossed, S.A. Ogburn, Winston, N.C.]
L422	LUCK	[round, green on yellow, picture of four-leaf clover, Ogburn]
L423	**LUCKY**	**(a) [oval, embossed red]** **(b) [silver]**
L425	LUCKY COLORS	[rectangular, Kerner Bros. Winston, N.C.]
L427	LUCKY DICK	[oval, black on yellow, picture of horse]
L428	LUCKY DUCK	[round, black on yellow, picture of duck]
L429	**LUCKY 8**	**[large die-cut 8 embossed]**
L431	**THE LUCKY**	**[die-cut 8 embossed]**
L433	THE LUCKY	[die-cut solid 8 embossed]
L435	**LUCKY HIT**	**[oval, black on red]**
L437	**LUCKY JOE**	**[round, black on red & yellow F.M. Bohannon]**
L438	LUCKY 6'S	[round, embossed]
L439	**LUCKY STRIKE**	**[square, black on red & green, R.A. Patterson Tobacco Co.]**
L441	**LUCKY STRIKE**	**[round, black on red, R.A. Patterson Tobacco Co.]**
L444	LUCY ASHTON	[rectangular, odd, gold on black]
L445	**LUCY ASHTON TOBACCO**	**[rectangular, gold on black, G. Penn Sons & Co.]**
L446	**LUCY HINTON**	**(a) [octagonal black on gray, T.C. Williams Co.]** (b) [T. C. Williams Co., Continental Tob. Co. Suc.]
L447	LUCY LONG	[rectangular, banner embossed]
L448	LUCY LYLE	[round, black on red, picture of a girl]
L449	**LUCY NEAL**	**[round, black on red picture of a girl]**
L450	LUCY NEAL	[octagonal black on red, picture of a girl]
L452	LUCY PEYTON	[oval, black on red, P.B.G. & Co.]
L453	**LULA BELLE**	**[rectangular, black on red L.W. Ashby & Sons]**
L454	LUNAR	[round, embossed, A.M. & Co.]
L455	**LUNAR**	**[octagonal embossed, A.N. & Co.]**
L456	LUSCIOUS	[rectangular, embossed]
L458	LUTE GORDON	[round, yellow on red]
L460	LUXURY	[round, black on red, picture of fruit bowl]
L461	LUXURY TOBACCO	[round, yellow on dark blue, L & B. Co.]
L462	**A.M. LYON & CO**	**[round, embossed circle in center]**
L463	LYON & GOODSON	[large oval embossed, Danville, Va.]
L465	LYRE	[die-cut shape, harp-type instrument]
L467	LYRE	[round, black on red, picture of harp]

M

M101	**M**	**[die-cut letter, embossed]**
M103	**M**	**[diamond, small, black on yellow]**
M105	M.A.	[odd diamond black on red]
M107	M & B	[rectangular, embossed]
M109	M B F	[rectangular, odd, embossed]
M111	M C	[round, black on red, fancy M on C]
M112	M.C.	[rectangular, cut-out]
M113	M C	[round, embossed]

M114	M C	[rectangular, black on red]
M115	M C R	[round, embossed]
M117	**M D**	**[rectangular, cut-out]**
M118	M.F.R.S.	[rectangular, banner embossed]
M119	M G	[rectangular, brown on white]
M120	M and L	[round, black on red, trademark]
M121	M P & Co	[oval, embossed]
M122	M.T. Co.	[die-cut a bell embossed]
M123	M.T.	[round, white on red]
M124	**M.Y.**	**[rectangular, black on yellow]**
M125	MABEL LEE	[round, black on yellow, picture of girl, Mfg. by Irvin & Poston]
M126	**MABEL LEE**	**[rectangular, embossed]**
M127	MABEL LEE	[rectangular, black on yellow, picture of a girl, Mfd. by Irvin & Poston]
M128	MABEL LEE	[rectangular, black on yellow, different picture of a girl, Mfg. by Irvin & Poston]
M129	**MACKINAW**	**[round, embossed]**
M131	J.H. MACLIN'S CELEBRATED	[scroll, black on yellow]
M133	**MAC ZIM**	**[rectangular, black on yellow]**
M135	MADEIRA	[round, multicolored, picture of wine bottle]
M137	MAGGIE JONES	[rectangular, black on yellow, picture of girl]
M139	MAGGIE MITCHELL	[round, black on yellow, G. Penn Sons Co.]
M140	MAGGIE SPENCER	[rectangular, black on red]
M141	MAGGIE SPENCER	[round, embossed]
M142	**MAGIC**	**[diamond, black on red, Jack-in-the-Box]**
M143	MAGIC	[rectangular, embossed]
M144	MAGINTY TWIST	[oval, embossed, Brown & Bros. Winston, N.C.]
M145	MAGINTY TWIST	[round, embossed, Brown, Winston, N.C.]
M147	MAGINTY TWIST	[round, embossed, Brown & Williamson]
M149	**MAGNET**	**[oval, embossed, F.M. & Co.]**
M150	MAGNOLIA	[rectangular, embossed]
M151	**MAGNOLIA**	**[round, white on green]**
M152	MAGNOLIA	[round, black on red]
M153	MAGNOLIA	[oval, black on red]
M154	MAGNOLIA GRADE	[octagonal gold on green, Garr Bro]
M155	**MAGPIE**	**(a) [round, black on gray, picture of man & bird, letters right side]** **(b) [black on light gray, letters are on top left side]**
M157	MAHOGANY	(a) [round, red on white, W.T. Burton] (b) [red on yellow]
M159	MAHOGANY	[die-cut leaf, black on red, Geo O. Jones & Co.]
M161	MAHOGANY TWIST	[oval, embossed, Brown & Bros. Winston, N.C.]
M163	MAHOGANY TWIST	[oval, black on red]
M164	MAHOMET	[rectangular, black on red, picture of a ram]
M165	MAIDEN	[round, black on yellow]
M167	**MAID OF ATHENS**	**(a) [round, black on orange, picture of a girl]** (b) blue on white
M169	**MAID OF CLINTON**	**[round, embossed, picture of a girl]**
M170	MAID OF ORLEANS	[round, brown on yellow & green, picture of girl, Sam Blackburn & Co.]
M171	MAIL CARRIER	[rectangular, picture]
M172	MAKE NO DUST	[rectangular, black on yellow, two men on bikes, E.J. & A.J. Stafford]
M173	MALE GIRL TOBACCO	[oval, black, red on white, picture of girl, Stall & Finley]
M174	**MALLET**	**[die-cut mallet, gold on black, S.W. & Co.]**
M175	MALLET	[rectangular, embossed, picture of mallet]
M177	MALTESE	[die-cut cross, black on red]
M179	MAMIE'S	[diamond, black on yellow]
M181	**MAMIE'S CHOICE**	**[rectangular, embossed]**
M183	MAMIE'S TWIST	[round, embossed]
M184	**MAMMOTH CAVE**	**[round, black on white, picture of cave & men]**
M185	MAMMOTH CAVE TWIST	[round, black on red, picture of cave]
M186	**MAN IN THE MOON**	**[die-cut face in crescent moon**
M187	MANAKIN	[round, embossed]
M189	**MANAOLA**	**(a) [rectangular, scroll, embossed, Butler & Bosher]** (b) Butler & Bosher & The American Tob. Co.
M191	MANAOLA	[oval, embossed]
M193	MANAOLA	[rectangular, embossed]
M195	MANET	[round, black on red]
M197	MANHATTAN	[octagonal, black on gold & red]
M199	**MAN'S PRIDE**	**(a) [rectangular, black on yellow, picture of a girl, P.H. Hanes & Co.]** **(b) [different girl]** (c) [different girl, R.J.R. Tob.Co.]
M200	MAPLE LEAF	[die-cut a leaf embossed, L & S]
M201	**MAPLE LEAF**	**[die-cut leaf embossed]**
M203	MAPLE SUGAR	(a) [die-cut leaf, black on light green, O.H. & Co.] (b) black on dark green
M204	MAR & Co.	[rectangular, bow-tie shape, black on red]
M205	MARCH	[hexagon small, black on yellow, picture]
M206	MARECHAL NEIL	[round, red on yellow, Iredell Tobacco Co]
M207	**MARIE ROZE**	**[rectangular, red on yellow, picture of a girl's head]**
M208	**MARIE STUART**	**[round, embossed]**
M209	MARIGOLD	[oval, black on yellow]
M211	**MARINE**	**[round, embossed picture of ship]**
M213	**MARITANA**	**[die-cut shield embossed]**
M214	**MARITANA**	**[die-cut shield embossed, W.T. Hancock]**
M215	MARK TWAIN	[rectangular, black on yellow, picture of man]
M216	MARKER	[die-cut type of barrel embossed, P. & E. Cov. Ky.]
M217	MARLINE SPIKE	[rectangular, embossed]
M219	**MARQUIS**	**[rectangular, embossed]**
M220	MARROW BONE	[round, yellow on red & black, picture of dog head, Schoolfield]
M221	**MARROW BONE**	**[die-cut a bone black on yellow, Schoolfield]**
M223	**MARSALIS**	**[banner embossed]**
M225	MARSHALL'S CHOICE	[round, black yellow]
M227	MARSHALL'S BEST	[rectangular, black on yellow]
M229	**MARTHA WASHINGTON**	**[oval, black on red, picture of woman, Arnett**

		Snellings & Co.]
M231	MARTHA WASHINGTON	[oval, black on red, picture of woman, Snellings & Sparrow]
M332	MARTHA FORD	[round, black on red, picture of a woman]
M233	**MARTIN'S M NAVY**	**[diamond, black on red]**
M234	MARTIN'S OK TWIST	[rectangular, black on yellow]
M235	MARTINI	[oval, black on yellow]
M236	MARY ANDERSON	[oval, red on yellow]
M237	**MARY ANN**	**[hexagon black on yellow, picture of a girl washing clothes]**
M238	MARY ANN	[oval, embossed]
M239	MARY JANE	[round, embossed]
M240	MARY VIRGINIA	[round, black on red, picture of girl, Job Deshazo, Manfr.]
M241	**MARY LEE**	**(a) [small round, black on yellow Bynum Cotten & Co., a girl facing right]** (b) larger round
M242	**MARY LEE**	**[large round black on yellow, girl facing left Bynum Cotten & Co.]**
M243	**MARY NUNN 4 M.L.O.**	**[round, embossed]**
M244	MARY SWIFT	[rectangular, black on yellow, picture of a woman]
M245	MARYLAND CLUB	[rectangular, embossed]
M246	MARYLANDER	[rectangular, black on yellow]
M247	MASCADA RIZAL	[die-cut man's bust, black on red]
M248	MASCOT	(a) [round, black on yellow] (b) [round, blue on white]
M249	MASTER MASON	[oval, white on blue]
M250	**MASTERPIECE**	**[rectangular, gold on red Liggett & Myers]**
M251	**MASTER WORKMAN GENUINE**	**[round, The American Tobacco Co.]**
M253	**MASTER WORKMAN GENUINE**	**[round, Continental Tobacco Co.]**
M255	MASTER WORKMAN GENUINE	[round, Imperial Tobacco Co.]
M256	MASTIFF	[rectangular, embossed]
M257	**MATCH IT**	**[round, embossed with a star]**
M258	MATHEWS	[rectangular, embossed]
M259	MATINEE	[round, black on red]
M260	MAUD HARRIS	[rectangular, black on yellow, picture of a girl]
M261	MAUD HARRIS	[round, black on yellow, picture of girl, Robt Harris & Bro]
M262	**MAUD, S.**	**(a) [hexagon black on yellow-white]** (b) [black on green]
M263	MAUD, S.	[round, black on green]
M265	MAUD, S. TWIST	[rectangular, black & yellow on white]
M267	MAUD, S. PLUG	[round, black & brown on yellow]
M269	MAUL	[rectangular, red on black, picture of maul]
M271	MAXIMUM	[round, embossed]
M273	MAVERICK	[round, yellow on red]
M274	**MAY APPLE**	**[round, gold on black, T.C. Williams Co.]**
M275	MAY APPLE	[round, green on white & red, picture of apple, Carhart & Brother]
M276	MAY BELL	[round, embossed]
M277	**MAY BELL**	**[rectangular, embossed]**
M278	MAY BELLE	[round, black on yellow, Traylor, Spencer & Co.]
M279	MAY BLOSSOM	[rectangular, black on red]
M280	MAY CHERRY	[round, black on yellow & red, picture of cherries, Iredell Tob Co.]
M281	MAYFIELD	[rectangular, black on red]
M282	**MAY FLOWER**	**[rectangular, banner & leaves embossed]**
M283	**MAY FLOWER**	**[triangular, embossed, Vaughn]**
M284	**MAY FLOWERS NATURAL TWIST**	**[rectangular, black on white]**
M285	MAYINE	[round, embossed, picture of ship]
M286	**MAY POLE**	**[octagonal embossed]**
M287	MAYO'S HOLLY	[rectangular, yellow on black, sweet chewing tobacco, trademark 1878]
M289	**MAYO'S**	**(a) [rectangular, picture of a Rooster, red on black, yellow & brown frame]** **(b) [red on black, yellow & black frame]**
M295	**MAYO'S TOBACCO**	**[rectangular, red & white on black & brown, "Is Always Good"]**
M299	MAY QUEEN	[rectangular, embossed]
M301	**MAZY**	**[rectangular, black on white, picture of a girl with large hat on]**
M303	**McALPIN & Co**	**(a) [die-cut shield embossed, red]** **(b) gold** **(c) blue** (d) silver (e) green (f) black
M305	McAYER & Co	[curved banner embossed]
M307	THE McCAULEY TOB. CO	[round, embossed]
M309	**McCORD'S CHOICE**	**[rectangular, picture of a man, Arnold & McCord]**
M311	McCORD'S TWIST	[rectangular, black & tan on yellow]
M313	McLAUGHLIN & CO	[die-cut shield shape embossed]
M314	**McDOWELL**	**[round, black on red]**
M315	MEADOW BROOK	[rectangular, embossed]
M316	**MECHANICS DELIGHT**	**[round, embossed]**
M317	**MECHANICS DELIGHT**	**[oval, black on yellow, picture of two men, P. Lorillard Co.]**
M319	**MECHANICS DELIGHT**	**[round, black on yellow, P. Lorillard & Co.]**
M320	MEDLEY	[oval, red on yellow]
M321	MEIG'S BIG 5	[round, embossed]
M322	MEMPHIS	[round, black on yellow, Gilbert & Conner, Merray, Ky.]
M323	MELLOW POUNDS	[round, yellow & black on red]
M325	MELON	[round, embossed, picture of melon]
M327	MERCHANTS CHOICE	[round, black on yellow]
M329	MERCHANTS DELIGHT	[round, embossed]
M331	MERCHANTS TOBACCO CO. Pat. 1869	[rectangular, embossed]
M333	MERRIMAC	[round, black on yellow]
M335	**MERRY WAR**	**[rectangular, black on red, picture of men, Wilson & McCallay]**
M337	MERRY WIDOW	[oval, black on yellow, picture of lady]
M338	MERRY WIDOW	[oval, red, black & white]
M339	MESSENGER	[oval, black on yellow, picture of a man, Snellings & Sparrow]
M340	MESS-MATE	[Wilson-McCallay Tob. Co.]
M341	**METROPOLITAN**	**[rectangular, small, black**

		on yellow]
M342	MI CHU	[rectangular, black on red]
M343	**MICHIGAN**	**[rectangular, odd shield embossed]**
M347	**MICKY**	**(a) [oval, green on white, picture of three leaf clover]** (b) green on yellow
M349	MICKY	[oval, black on red]
M350	MICMAC	[rectangular, yellow on red, tag made in U.S.]
M351	MIDNIGHT	[round, white on black, picture of a owl]
M352	MIDNIGHT	[round, black on red, picture of owl's head]
M353	MIKADO	[die-cut shape, balloon, Arbuckles & Co.]
M354	MILD CHEW	[round, black on yellow]
M355	MILD CHEW	[round, black on yellow, Brown]
M356	MILD CHEW	[round, black on yellow, L. W. Davis]
M357	MILD & MELLOW	[round, red & yellow on blue]
M358	MILD & MELLOW	[rectangular, embossed, Ogburn & Hill & Co.]
M359	**MILD SMOKE**	**[round, black & red on yellow, Dill]**
M361	MILLER-BALTO	[round, embossed]
M363	MILLER'S TWIST	[oval, black on yellow]
M363	MILLER'S TWIST	[oval, white on red]
M365	MILLER'S PLUG	[round, white on red]
M366	MILLIONAIRE	[rectangular, picture of two $]
M367	MILLIONAIRE	[rectangular, picture, red on white]
M369	MINER	[oval, embossed]
M370	**MINER'S**	**[oval, embossed]**
M372	MINER'S TWIST	[rectangular, embossed]
M373	**MINNIE OGBURN**	**[rectangular, black on yellow, picture of girl, Ogburn Hill & Co.]**
M374	MINT JULEP	[round, red on white, multicolored picture of glass, "it's so good"]
M375	MINTMORE	[rectangular, bar, white on black]
M376	MIRA	[round, black on red & yellow, The Biggest of Them All, La Plus Grosse de Toutes]
M377	MISS MARION	[round, black on red, Kimbrough & Duncan]
M378	MISSING LINK	[rectangular, round embossed]
M379	MISSING LINK	(a) [oval, black on red, picture of chain link] (b) [oval, yellow on red]
M380	**MISSING LINK**	**[oval, embossed, picture of chain link]**
M381	MISSISSIPPI STANDARD	[rectangular, bar, red on black]
M382	MITCHELL 5c TWIST	[oval, black on cream]
M383	MITCHELLS NATURAL LEAF	[round, black on yellow]
M384	MOCK TURTLE	[round, embossed]
M385	MODEL	[die-cut leaf embossed]
M387	**MODEL MAN**	**[round, black on yellow-white, picture of a church-man]**
M389	**MODERN NAVY**	**[oval, black on yellow, ship Oregon, Spilman, Ellis Tob. Co.]**
M391	**MODOC**	**[oval, embossed, Poske, 123 W. Pratt St]**
M393	MODOCK	[oval, embossed, H.F. Poske, 221 Pratts]
M394	**MOGUL**	**[round, embossed]**
M395	**MOGUL**	**[die-cut crescent embossed]**
M396	MOHUN	[die-cut banner white on black]
M397	MO KAN PLUG	[round, black on red]
M398	MOLLY COTTON TAIL	[rectangular, black on cream, picture of rabbit]
M399	**MONARCH**	**(a) [round, black on yellow, red lion facing right, Ogburn Hill & Co.]** (b) smaller lion facing left
M401	**MONARCH**	**[die-cut lion facing left embossed]**
M403	**MONARCH & AURORA**	**[rectangular, embossed, F.M. & Co.]**
M405	**MONEY**	**[round, embossed]**
M406	MONEY HUNTER	[oval, black on yellow, J.G. Dill]
M407	MONEY KING	[oval, yellow on green]
M408	MONEY KING	[rectangular, blue on white]
M409	**MONITOR**	**[round, embossed, picture of ship, H. Wirgman. Lou.]**
M411	MONITOR	[rectangular, embossed, S & Co]
M413	MONKEY	(a) [oval, black on red] (b) [oval, black on green]
M415	MONKEY WRENCH PLUG	[rectangular, embossed]
M417	MONOCACY	[rectangular, black on yellow]
M418	MONOPOLE	[rectangular, black on yellow, J.W. Gravely]
M419	**MONOPOLE**	**[die-cut shield, black on white, picture of eagle crest]**
M420	MONTAGUE'S ANCHOR	[rectangular, black on red]
M421	**MONT BERNARD**	**[rectangular, embossed]**
M422	**MONTENEGRO**	**[oval, black on white & red, blue, white, flag]**
M423	MONTI CRISTO	[rectangular, banner embossed]
M424	**MONUMENTAL CITY TWIST**	**[oval, embossed, B.F. Hanes]**
M425	**MOONSHINE**	**[round, white on red]**
M426	MOONSHINE	[rectangular, embossed, Morotock, Henry County]
M427	MOONSHINE TWIST	[round, white on blue & black, picture of moon & trees]
M428	MOON SHINE TWIST	[round, black on red]
M429	MOORE & CALVI	[oval, black on yellow, picture of a tiger]
M430	**MOOR'S RED LEAF TWIST**	**[round, black on yellow, picture of man in center]**
M431	MOOSE HEAD	[round, red & black on yellow, picture of a moose, U.S. Tob. Co]
M432	**MOOSE HEAD**	**[round, black & red on yellow, picture of a moose, M. B. Tob. Co.]**
M433	**MOOSE HEAD**	**[die-cut embossed]**
M434	MOOSHLA	[round, black on yellow]
M435	MORGANTON	[rectangular, shield, black on red]
M437	MORGANTON TOB. Co.	[round, black yellow on red]
M438	MORNING GLORY	[triangular white on blue, picture of a flower]
M439	**MORNING GLORY**	**[triangular, yellow on black, picture of a flower]**
M441	MORNING TIME	[round, embossed]
M443	MOROCCO	[rectangular, embossed]
M444	MORRIS LEADER	[oval, embossed]
M445	**MORTON**	**[rectangular, gold on**

		black]
M449	A.H. MOTELY & CO	[rectangular, embossed]
M451	**MOTOR**	**[oval, black on yellow]**
M452	MOUNTAIN ROSE	[round, white on blue, R.A. Patterson Tob. Co. Richmond, Va.]
M453	MOUNTAIN DEW	[oval, blue on white]
M455	MOUSAH	[oval embossed]
M457	MOXIE	[round, black on red]
M459	**MUFFIN**	**[round, black on red]**
M460	MUFFIN REGISTERED	[round, black on red, Casey & McLevin Tobacco Co.]
M461	MULE EAR	[octagonal embossed]
M463	MULE'S HEAD	[die cut shape embossed, B.F. Hanes]
M465	MUMM'S EXTRA	[rectangular, embossed]
M466	**MURRAY'S OWN**	**[oval, embossed, Anngough]**
M467	MUSCADINE	[oval, black on yellow]
M468	MUSCADINE	[oval, blue on white]
M469	MUSCADINE SUNCURED	[oval, black & red on yel low]
M470	**MUSIC**	**[round, embossed]**
M471	MUSK MELON	[rectangular, red on yellow, picture of melon, Irvin & Poston]
M472	MUSK MELON	[rectangular, scalloped, Irvin & Poston, Statesville, N.C.]
M473	**MUSK MELON**	**[octagonal black on yellow & red, Turner Powell & Co.]**
M474	MUSSELMAN & CO	[square, black on red, Lou., Ky]
M475	MUSSELMAN'S HINDOO	[rectangular, black on red, picture of a horse]
M476	**MUTH'S EST 1829**	**[die-cut horse running, embossed]**
M477	MUTUAL BENEFIT	[round, white on red, Mathews]
M479	MY EAGLE	[die-cut Eagle embossed]
M481	**MY-LO**	**(a) [oval, small gold on red, Penn]** (b) [gold on black]
M482	MY-LO	[rectangular, black on yellow, Scotten]
M483	**MY MAMIE**	**[oval, embossed]**
M485	**MY MARYLAND**	**(a) [rectangular, black on yellow, B.F. Gravely & Sons]** (b) blue on white
M486	**MY NIGHT**	**[die-cut round black on red, picture of owl, Wilson & Mc.Call]**
M487	**MY OWN SWEET**	**[die-cut shield red on yellow, picture of a heart]**
M488	MYRTLE	[diamond, black on red]
M489	MYRTLE YOUNG	[rectangular, black on yellow]
M491	MYSTIC CHARM	[round, embossed]
M493	MYSTIC TIE	[octagonal, P.B. Gravely & Co.]

N

N101	**N A**	**[round, embossed]**
N102	**N B**	**[round, black on red]**
N104	N.B.	[round, black on yellow, picture of a man with papers]
N105	N C	[die-cut letters]
N106	N E PLUS ULTRA	[hexagon embossed]
N108	N N	[round, black on red]
N110	N.P.R.R.	[rectangular, embossed]
N112	**N.S.**	**[rectangular, embossed]**
N113	N.T.E.	[oval, green on yellow]
N115	**N T W**	**[round, embossed very small]**
N116	**N T W**	**[round, embossed small]**
N118	**N.T.W.**	**[round, embossed]**
N119	**N.T.W.**	**[round, embossed larger]**
N120	**N & W**	**[small round black on red]**
N121	N & W	[large round black on red]
N122	N. & W.	[rectangular, yellow on red]
N123	N. & W. KING	[square, embossed]
N124	**N. & W.T.**	**[round, black on red]**
N126	N. & W. T. Co	[round, black on red]
N128	**NABOB**	**[round, embossed]**
N129	**NABOB**	**[rectangular, black on yellow, picture of a man's head]**
N130	NABOB	[rectangular, yellow on black]
N131	NABOB	(a) [rectangular, bar, black on green] (b) [white on green]
N136	**NANCY GRAVELY**	**[round, black on yellow, P.B. Gravely & Co.]**
N137	NANCY GRAVELY	[rectangular, black on white, P.B. Gravely & Co.]
N139	NANCY HANKS	[oval, embossed]
N140	**NANCY HANKS**	**[rectangular, embossed]**
N141	NANCY HANKS	[round, picture of a horse, Penns, Time 204, Best of all]
N142	NANCY HANKS	[round, yellow on blue, J. Wright Co.]
N145	NAPOLEON	[die-cut shape, heart, W C MacDonald, Montreal]
N147	**NARROW GAUGE**	**[round, embossed also train tracks]**
N149	NASHVILLE GREENVILLE	[diamond, embossed]
N150	NASHVILLE GREENVILLE	[die-cut shape, a bell em bossed]
N152	**NATIONS CHOICE**	**[die-cut shield embossed]**
N154	NATIONAL CHEW	[round, multicolored, picture of eagle & flag, Liipfert, Scales & Co.]
N155	NATIVE SONS	[round, black on yellow picture of a bear]
N157	NATIVE TOBACCO	[rectangular, black on yellow, F.R. Penn & Co.]
N158	NATIVE TOBACCO	[rectangular, black on yellow, The F.R. Penn Tob. Co.]
N160	**NATURAL**	**(a) [die-cut leaf, red on yellow Taylor Bro.]** **(b) [black on yellow]**
N161	NATURAL BEAUTY	[round, black, blue & red on white, Sparger Bros Co.]
N162	**NATURAL BEAUTY**	**[die-cut leaf black on yellow]**
N163	NATURAL BEAUTY	[round, black on yellow, picture of girl]
N164	NATURAL LEAF	[rectangular, embossed, T.C. Williams, Co.]
N165	NATURAL LEAF	[die-cut leaf embossed, Liipfert, Scales & Co]
N166	NATURAL LEAF	[oval, black on yellow, picture of a leaf, T.C. Williams]
N167	NATURAL LEAF	[rectangular, black on yellow & red, J.W. Co. trademark]
N168	NATURAL LEAF	[die-cut leaf, black on yellow, Manufactured by Job Deshazo]
N169	NATURAL LEAF	[rectangular, red on gray, Finzer Bros.]
N170	NATURAL LEAF	[die-cut leaf red on yellow, Taylor Bro.]

N171	NATURAL LEAF	[die-cut leaf red on yellow, Everhart & Co.]
N172	**NATURAL LEAF**	**[die-cut leaf black on yellow, Doss Tob. Co.]**
N173	**NATURAL LEAF**	**(a) [rectangular, gold on green Scotten Dillon Company]** **(b) [gold on dark blue]** (c) [gold on red]
N174	**NATURAL LEAF**	**[rectangular, gold on red, Tinsley]**
N175	**NATURAL LEAF**	**(a) [round, black on yellow & yellow on red, F.M. Bohannon]** (b) [Natural Leaf, upside down]
N176	**NATURAL LEAF**	**[octagonal embossed, Peper]**
N177	NATURAL LEAF	[rectangular, embossed, Austin]
N178	NATURAL LEAF	[oval, black on red, L & M]
N179	NATURAL LEAF	[diamond, black on red, Axton]
N180	NATURAL LEAF	[round, black on brown & yellow, picture of leaf, Bailey Bros]
N181	**NATURAL LEAF**	**[die-cut leaf, black on brown, B-W. Tobacco Co.]**
N182	**NATURAL LEAF**	**[die-cut leaf black on yellow Taylor, Spencer, & Co.]**
N183	NATURAL LEAF	[rectangular, black on yellow, Cliffords]
N184	NATURAL LEAF	[rectangular, red on yellow, picture of leaf, Dalton Farrow & Co.]
N185	NATURAL LEAF	[rectangular, black 0n brown & yellow, I & P]
N186	NATURAL LEAF	[rectangular, hexagon red on yellow, J.W. Daniel]
N187	NATURAL LEAF	(a) [rectangular black on yellow, picture of leaf, S.A. Ogburn] (b) [black on orange]
N188	NATURAL LEAF	[die-cut banner, black on red, Strater]
N189	**NATURAL LEAF**	**[rectangular, embossed]**
N190	NATURAL LEAF	[round, black on yellow, picture of a man, Moore]
N191	NATURAL LEAF	[round, yellow on red picture of leaf, T.L. Vaughn & Co.]
N192	NATURAL LEAF	[round, black on red, Noel Tobacco Co.]
N193	NATURAL LEAF	[die-cut leaf, black on yellow, B.F. Hanes]
N194	**NATURAL LEAF**	**(a) [rectangular, black on brown & yellow, picture of leaf, P.H. Hanes & Co.]** (b) [P. H. Hanes & Co., R.J.R. Tob. Co. Suc'sr.]
N195	**NATURAL LEAF**	**[die-cut leaf black on yellow]**
N196	NATURAL LEAF	[die-cut leaf, black on yellow, Geo. O. Jones & Co.]
N197	NATURAL LEAF	[die-cut leaf, black on yellow, Robt. Harris & Bro.]
N198	NATURAL LEAF	[oval, black on yellow, Gravely]
N200	**NATURAL LEAF TWIST**	**[die-cut shield embossed, Strater.]**
N201	NATURAL LEAF TWIST	[oval, black on red, Newborgh]
N204	NATURALLY SWEET	[die-cut leaf, black on yellow]
N206	NAVAL CADET	[round, black on yellow, picture of Cadet]
N207	NAVY	[round, red on yellow, triangular shape in center]
N208	NAVY	[rectangular, banner embossed, The United States Tob Co.]
N210	NAVY JACK	[rectangular, black & red on green, Globe Tob Co.]
N212	NAVY TOBACCO	[oval, blue on cream, picture of navy boy]
N214	NE PLUS ULTRA	[hexagon embossed]
N216	**NECTAR**	**[round, black on red]**
N217	**NECTAR**	**[round, embossed Hudsons]**
N218	NECTAR	[oval, black on yellow, picture of bee]
N220	**NECTAR LEAF**	**[rectangular, embossed, Bohannons]**
N222	**NECTAR NATURAL LEAF**	**[oval, black on yellow & yellow on black, T.C. Williams Co]**
N223	NELLIE BLY	[round, black on yellow, picture of girl, Mt. Airy, N.C.]
N225	NEMO	[oval, black on brown]
N227	**NEOSHO**	**[oval, embossed]**
N229	NEPTUNE	[rectangular, embossed]
N230	NEPTUNE	[rectangular, red on yellow]
N232	NERO	[rectangular, embossed]
N234	**NERO NAVY**	**[round, embossed]**
N236	**NERVE**	**[rectangular, white on red flag, picture of a sailor in blue, Globe Tob Co.]**
N237	**NESTOR**	**[hexagon embossed]**
N239	NET WEIGHT 3 OZ WHEN PACKED	[diamond, black on red]
N240	NETTED GEM	[round, black on green & orange, picture of cantelope, Crews]
N242	L.H. NEUDECKER	[round, black on yellow, picture of Peacock]
N244	NEUDECKER'S OWN	[oval, embossed, hole in center]
N246	**NEVER DID**	**[round, embossed]**
N248	NEW BULLY	[round, black on tan, picture of a cat's head, H. Clarke & Sons]
N250	NEW CENTURY	[round, red on black]
N252	**NEW COON**	**[rectangular, black on red & yellow, W.A. Brown Tobacco Co.]**
N253	NEW COON	(a) [rectangular, yellow on black] (b) [white on blue]
N255	**NEW EDITION**	**[diamond, white on blue, J.A. Tolman Co.]**
N256	NEW EDITION	[rectangular, embossed]
N258	**NEW ENGLAND PLUG SMOKE**	**[round, yellow & black on red & yellow, H.N. Martin]**
N259	**NEW HAT**	**[die-cut hat gold on black]**
N261	NEW HOPE	[round, black on red]
N263	**NEW MOON**	**[die-cut embossed man face in the moon]**
N265	NEW MOVE	[oval, embossed]
N267	NEW SMOKE	[round, black on red, J. Wright Co.]
N269	NEW SOUTH	[oval, embossed]
N270	NEW SOUTH	(a) [die-cut N, black on red, N & W] (b) [gold on red]
N272	NEW STAMP	[rectangular, multicolored, picture, Farrow & Co.]
N274	NEW TAX	[round, red on yellow]
N275	NEW TAX	[round, black on yellow]
N276	**NEW WRINKLE**	**[rectangular, embossed]**
N278	NEWPORT	[diamond, black on green]

N280	**NEWSBOY**	**(a) [round, white on blue & red National Tobacco Works]**
		(b) [Continental Tob.]
		(c) [The American Tob.]
N282	NEWS-GIRL	[rectangular, white on red]
N284	NEW SOUTH	[die-cut letters, black on red, N. & W.]
N286	NEW STAMP	[round, embossed]
N287	NIAGARA	[round, black on red]
N289	**NICKEL CHEW 1884**	**[round, embossed]**
N290	NICKEL JACK	[round, embossed]
N292	NICKEL POCKET PIECE	[round, embossed]
N294	NICK-ON-TIN	[rectangular, banner black on cream]
N296	NICK, RUDY, FRED	[triangular, black on yellow]
N298	NICKEL ANTE	[round, embossed]
N300	NICKEL JACK	[rectangular, picture of black man and a donkey]
N301	NICKEL PLATE	[round, embossed]
N302	NICKEL PLUG	[round, embossed]
N304	**NICKEL TWIST**	**[rectangular, scalloped white on red, Scotten Tobacco Co.]**
N306	**NICKLEBY V**	**[round, embossed stars and wreath]**
N308	NICKLEBY 5¢	[round, embossed]
N310	NIGGER BABY	[round, picture of black baby, Pegram & Penn]
N311	NIGGER BABY	[rectangular, embossed, Pegram & Penn]
N314	NIGGER HAIR	[round, embossed]
N316	NIGHT CAP	[round, black on yellow, picture]
N318	**NIMROD**	**[half round embossed]**
N320	**9 INCH**	**[rectangular, blue on blue, picture of two men Gravely & Miller]**
N322	**1903**	**[rectangular, red on yellow]**
N324	97 RR	[round, embossed]
N325	99	[round, red on green & yellow]
N327	99 PLUG	[rectangular, red on green & yellow]
N329	900 CHAPMAN'S TOBACCO	[round, black & red on cream]
N331	NIP & TUCK	[rectangular, black on white, picture of two horse's, J.W. Daniel Tob Co.]
N332	NIP & TUCK	[rectangular, embossed]
N333	**No. 1 HARD**	**[round, black on red]**
N334	No. 1 TOBACCO	[octagonal black & red on gold, F.R. Penn & Co.]
N335	No. 1 TWIST	[round, embossed, Gravely & Miller]
N336	No. 1 PLUG	[round, embossed]
N337	No. 5	[round, red on black]
N338	No. 10	[round, embossed]
N340	**No. 11**	**[shield embossed]**
N341	**No. 44**	**[shield embossed]**
N342	NO JOKE	[round, black on yellow]
N343	NO REDEMPTION	[rectangular, red on yellow]
N344	NO TAG	[round, red on brown]
N345	NO TARIFF	[die-cut banner embossed]
N346	**NO TAX**	**[die-cut shoe, black on red]**
N347	**NO TAX**	**[die-cut shoe, black on red, Wilson & McCallay]**
N348	NO TAX	[round, embossed]
N349	**NOBBY**	**[small die-cut banner]**
N350	**NOBBY**	**[die-cut banner, embossed]**
N351	**NOBBY**	**[rectangular shape, embossed]**
N353	**NOBBY SPUN ROLL**	**[round, black on yellow, white on red]**
N355	NOBBY TWIST	[rectangular, black & yellow on brown]
N357	NOELL'S TWIN BROTHERS	[rectangular, black on white, picture of two men, Keith, Kenneth]
N359	NOON DAY	[round, black on red, Sun Cured]
N360	NORTH AMERICA	[round, black on yellow, picture of a Indian]
N362	**NONNIE V**	**[rectangular, black on orange, picture of a girl]**
N364	NORTH AND SOUTH	[oval, picture of flag, Winfree Sons & Maupin, Lynchburg, Va.]
N365	NORTH AND SOUTH	[rectangular, picture of two flags, multicolored]
N368	NORTH PIEDMONT	[rectangular, embossed]
N370	NORTH POLE	[round, blue on white, picture of a polar bear & seal, United States Tobacco Co.]
N372	NORTH STAR	[round, white on blue]
N374	NOSEGAY	[oval, white on dark blue]
N376	NOT TRUST MADE	[round, small black on red]
N378	NOVELTY	[round, embossed]
N379	NOVELTY	[round, black on red]
N381	NOW 5¢ OK TWIST	[oval, black on blue]
N382	NOX	[oval, black on yellow]
N384	NUMBER 1	[round, black on yellow, Taylor, Spencer & Co.]
N386	**NUT**	**[die-cut oval shape embossed nut]**
N388	NUTCRACKER	[round, black on yellow & brown, picture of Squirrel, J.N. Wyllie & Co.]
N389	NUTCRACKER	[round, white on red multicolored picture of squirrel, J.N. Wyllie & Co.]
N390	NUTCRACKER	[round, black on red, picture of squirrel]
N392	NUT-MEG	[rectangular, yellow on blue]
N395	**NUTMEG**	**[round, embossed picture of a nut and stars]**
N396	**NUTMEG**	**[round, black on red]**
N397	NUTMEG	[die-cut a nut, embossed]
N398	**NUTMEG**	**[rectangular, embossed]**
N401	NUTMEG TWIST	[oval, embossed]
N403	**NUTWOOD**	**[octagonal black on red]**

O

O101	**O**	**[die-cut embossed]**
O102	**O**	**[die-cut oval shape]**
O103	O.B.	[die-cut key, embossed]
O105	**O C**	**[rectangular, embossed]**
O107	**O C**	**[three round shapes rectangular, embossed, eye in center]**
O109	**O.C.T.W.**	**[rectangular, round, embossed]**
O110	O D	[round, embossed O over D]
O112	**O.J.**	**[die-cut bottle embossed]**
O114	O on K	[octagonal black on yellow, Bailey Bros]
O116	O K	[die-cut letters]
O117	**O.K.**	**[square, embossed]**
O118	**O.K.**	**[small rectangular, cut-out]**
O119	**O.K.**	**[larger rectangular, cut-out]**
O120	**O.K.**	**[square, black on red]**
O121	O K	[round, white on blue, Martins twist]
O124	O.K. TWIST	[round, white on blue, Hasting]
O125	O & M	[oval, embossed]
O126	**O N**	**[rectangular, cut-out O N,**

Code	Name	Description
		embossed Old Nick]
O128	**O.N.P.**	**[rectangular, black on yellow]**
O130	**O.N.T.**	**[octagonal red on yellow, P.H. Hanes & Co.]**
O131	**O.N.T.**	**[round, white on blue, Flunrer Tob Co.]**
O133	O.N.T. SMOKING	[round, embossed, Penn]
O135	**O P**	**[rectangular, embossed]**
O137	OBRECHTS KING DICK	[die-cut bow tie embossed]
O139	OCEAN WAVE	[round, embossed]
O141	OCTOROON	[large rectangular, black on tan, T.C. Williams Co.]
O143	**OCTOROON**	**[small rectangular, black on yellow, T.C. Williams Co.]**
O145	ODD FELLOWS	[rectangular, black on yellow, picture of cat & dog]
O147	OGBORN	[rectangular, black on red, picture of a man]
O148	OGBURN	[round, black on yellow, Manf'g by N.D. Sullivan Co.]
O150	**OGBURN HILL & Co**	**(a) [die-cut leaf shape, black on yellow]** (b) [black on tan]
O152	OH HOW NICE	[odd shape, red on yellow]
O154	OH MY	[oval, embossed]
O156	OLD BILL	[round, black on red, picture of goat]
O158	OLD BILL	[rectangular, black on yellow, picture of old goat]
O160	OLD BIRD	[oval, gold on black, picture of a bird]
O162	**OLD BOB**	**[octagonal black on yellow picture of old man]**
O163	OLD BOB	[round, black on yellow, picture of old man with pipe]
O165	OLD BOURBON	[round, red on yellow, picture of a chicken]
O166	OLD BOY	[rectangular, embossed]
O167	OLD BRIAR	[round, embossed]
O169	OLD BRICK	[rectangular, embossed]
O170	**OLD BUCK**	**[rectangular, embossed]**
O171	**OLD BUCK**	**[round, embossed picture of buck]**
O173	OLD CHIEF	[round, black on red, A.C. & Co. Danville Va.]
O175	OLD CHUM	[die-cut a hat, red on yellow, D. Rikhie Co. Trademark]
O176	OLD COMMANDER	[shield shape embossed]
O177	OLD CONGRESS	[round, black on red L.B. Tob. Co.]
O179	**OLD CONGRESS NAVY**	**[oval, black on red]**
O181	**OLD COON**	**[rectangular, white on red, Wetmore]**
O183	OLD CORN	[rectangular, yellow on red, picture of ear corn]
O184	OLD CORN	[rectangular, embossed]
O185	OLD COUNTRY TWIST	[rectangular, banner red on yellow & tan]
O187	**OLD CROW**	**[rectangular, yellow on black, D.H. Spencer & Sons]**
O189	OLD CUBA	[round, black on yellow, picture of man & horse, Davis]
O190	OLD DECK	[oval, embossed]
O191	**OLD DOMINION**	**[rectangular, embossed picture of a ship, J.N.W. & Co.]**
O192	**OLD DOMINION**	**[octagonal red on black]**
O195	**OLD 5¢ TIMES**	**[oval, embossed]**
O197	OLD FIELD	[rectangular, octagon, yellow on red, Van Cleve]
O199	OLD FORT	[round, embossed]
O201	OLD FOX	[round, picture of a fox's head]
O202	OLD FOX	[round, red on yellow, picture of fox, tag Made in U.S.A.]
O205	OLD GENTLEMAN TOBACCO	[round, black on white, picture of a man]
O207	OLD GLORY	[die-cut shield white & blue on red, white & blue]
O209	**OLD GUARD**	**[rectangular, red on cream, picture of a dog, Chatham Tobacco Co.]**
O211	**OLD HAM**	**[oval, embossed]**
O213	OLD HEDGE	[rectangular, embossed]
O214	OLD HICKORY	[round, embossed]
O215	OLD HOME SPUN	[square, black on red]
O216	**OLD HONESTY**	**(a) [small die-cut H black on red, Finzers]** **(b) [large die-cut H black on red, Finzers]**
O217	**OLD HUNDRED**	**[round, embossed]**
O218	OLD HUNK	[octagonal, black on yellow]
O219	OLD INDIAN CHIEF	[rectangular, picture of a Indian, W.E. Paherson]
O220	**OLD JOE**	**[round, black on yellow, picture of a black man's head]**
O221	OLD KAIN-TUK	[round, black on red]
O222	OLD KENTUCKY	[round, yellow on red & brown, picture of the colonel]
O223	**OLD KENTUCKY**	**[round, embossed]**
O224	**OLD KENTUCKY**	**[round, embossed, black on red in center H W T Co.]**
O225	OLD KENTUCKY HOME	[oval, black on yellow, picture of a cabin]
O226	OLD KENTUCKY HOME SPUN	[round, yellow on red]
O227	OLD KY. HOMESPUN	[round, black on red, Brodies]
O229	**OLD LOYALTY**	**[die-cut shield red, white & blue]**
O230	OLD LOOM	[rectangular, black on red, picture of a loom]
O232	OLD MEDFORD	[die-cut barrel, black on yellow]
O234	OLD MINT TWIST	[round, green on yellow, Rowlett's]
O236	OLD MEDFORD	[rectangular, yellow on black, picture of barrel]
O238	**OLD NAVY**	**[oval, embossed arrow in center]**
O239	OLD NAVY	[oval, embossed]
O242	**OLD NAVY ZAHM**	**[oval, embossed arrow in center]**
O244	OLD NICK	[round, black on red]
O246	OLD N. C. PEACH	[round, embossed]
O248	OLD NORTH STATE	[hexagon embossed]
O250	OLD NUT	[round, embossed]
O252	OLD OGBURN	[rectangular, black on yellow, picture of an old man]
O254	**OLD PEACH**	**[round, black on orange & yellow, picture of peach, sun cured]**
O256	OLD PEACH	[round, black on yellow, red & green picture of peach, Butler]
O257	OLD PEACH	[die-cut a peach, green on yellow & red]
O258	OLD PEACH	[die-cut peach, black on yellow & red, Butler]
O259	**OLD PEACH**	**[die-cut peach, green on yellow & red, Butler]**
O260	**OLD PEACH**	**[round, embossed, H & D]**

O264	OLD PEACH N.C.	[round, embossed]
O268	OLD PEACH SUN CURED	[round, black, orange & green on yellow, picture of peach]
O270	OLD PICKER	[round, black on red, picture of man & dog, Rucker Wihen & Morris]
O272	**OLD PIONEER**	**[round, black on red, picture of a man sitting]**
O274	OLD PLANTATION	[square, black on red]
O276	**OLD PORT**	**[die-cut bottle black on red, Hancock]**
O277	**OLD PORT**	**(a) [die-cut bottle, black on white & red]** (b) black on white & orange
O280	OLD QUINCY	[round, white on green, S.S.C. & Co.]
O282	OLD RABBIT GUM	[rectangular, black on red, N.S. & T.J. Wilson, Winston, N.C.]
O284	**OLD RELIABLE**	**[oval, embossed]**
O285	OLD RELIABLE	[round, embossed]
O286	OLD RELIABLE	[die-cut leaf embossed, Henry Co.]
O288	OLD RAT	[round, black on yellow, picture of rat, Bitting & Hay]
O289	OLD RIP	[round, yellow on black]
O290	OLD RIP	[round, gold on blue]
O291	OLD RIP VAN WINKLE	[oval, embossed]
O293	OLD ROMAN	[round, black on yellow, picture of man, Penn & Watson]
O295	**OLD ROVER**	**[rectangular, black on red picture of dog, Vaughn]**
O297	**OLD RYE**	**[round, black on red, picture of bundle rye]**
O298	OLD RYE	[round, embossed, picture of bundle rye]
O300	**OLD RYE PLUG**	**[die-cut embossed barrel, chew]**
O302	OLD SLEDGE	[round, yellow on blue, picture of hand & cards 4 Aces, United States Tob Co.]
O303	OLD SLEDGE	[round, picture of hand of cards, Edel Tob Co.]
O304	OLD SLEDGE	[round, picture of hand of cards, Edel, Pace & Co.]
O308	OLD SLUG	[oval, embossed]
O310	OLD SLICK	[octagonal black on red]
O312	OLD SMOKE HOUSE TWIST	[round, black on yellow, picture of smoke house]
O314	**OLD SOL SUN CURED**	**[round, yellow on green, picture of a sun face]**
O315	**OLD SOL SUN CURED**	**[round, yellow on green, picture of the sun]**
O316	**OLD SOLDIER**	**[rectangular, white & blue on red, white & blue]**
O318	OLD SPORT	[round, embossed]
O320	**THE OLD STATESMAN**	**[round, black on yellow, picture of man on horse, Nashville Tob.]**
O322	**OLD TAVERN**	**[oval, red on green, picture of old Tavern, Berdan & Co.]**
O324	OLD TAYLOR TWIST	(a) [round, black on yellow] (b) [yellow on red]
O326	OLD TENNESSEE	[long rectangular, band, gold on red, J.D. Bs.]
O328	OLD TIMES	(a) [rectangular, large, picture of horse & wagon, black on yellow] (b) [smaller, black on white]
O330	OLD TIMES 5¢	[oval, embossed]
O332	OLD TOM	[die-cut shield, embossed]
O334	**OLD TOM**	**[oval, black on white picture of a cat]**
O336	**OLD TOM**	**[round, black on tan, picture of a cat, Ebert, Payne & Co.]**
O340	OLD TONY	[rectangular, embossed]
O342	OLD TRAY	[die-cut dog embossed]
O344	OLD VA. PEACH	[oval, embossed, B.F. Gravely & Co. Ltd.]
O346	**OLD WHEAT**	**[die-cut wheat bundle embossed]**
O348	OLD WINE TWIST	[oval, yellow on red]
O349	OLD VIRGINIA	[die-cut clover leaf embossed]
O350	OLD VIRGINY	[round, embossed]
O351	**OLE VIRGINIA**	**[oval, black on red, Boone's Weed]**
O352	**OLE VIRGINY**	**[oval, gold on dark blue, Venable]**
O354	OLIVE	(a) [die-cut leaf, yellow on red] (b) [black on green]
O355	OLIVE BELLE	[round, embossed]
O356	OLIVE BRANCH	[octagonal yellow on black, picture of leaves]
O358	OLIVER TWIST	[rectangular, black on yellow, McAlpin]
O360	OMAHA DETROIT	[round, black on yellow, Michigan Tob. Co.]
O362	**OMEGA**	**[die-cut foot embossed]**
O364	ON CALL	[round, embossed]
O366	ON DECK	[oval, embossed]
O368	ON TAP	[banner embossed]
O370	ON THE SQUARE	[diamond, red & black on cream]
O372	**ON TOAST**	**[rectangular, embossed]**
O374	**ON TOP**	**[diamond, embossed]**
O376	**ON TOP**	**[die-cut top red on yellow, G & B]**
O378	**ON TRACK**	**[rectangular, green on yellow, McNamara's Trade Mark Pat. June 3, 84]**
O380	**ONE**	**[rectangular, 1 in black on orange]**
O382	ONE	[round, black on red 1, P & P chewing]
O384	**ONE CENT NAVY**	**(a) [round, black on red]** (b) [black on white]
O386	ONE CENT PLUG	[round, black on red]
O388	**ONE DIME**	**[round, black on gold]**
O390	**ONE DIME**	**[octagonal, gold on black]**
O392	**1 DOLLAR 1881**	**[round, embossed]**
O393	ONE IN A HILL	[rectangular, embossed]
O394	**ONE OF THE BOYS**	**[rectangular, embossed]**
O395	1/2 ACRE	[round, embossed]
O396	**1/2 BUSHEL TOBACCO**	**[rectangular, black on red]**
O398	ONLY GENUINE ALABAMA	[rectangular, embossed, The]
O400	ONTARIO	[rectangular, orange on white]
O402	ONWARD	[rectangular, embossed]
O404	OPAL	[oval, black on red]
O406	OPEN DOOR	[round, yellow on red]
O408	OPEN FIELD FAIR PLAY	[round, black on red, picture of horse]
O410	**OPEN HAND**	**[die-cut open hand embossed]**
O412	OPEN HAND	[die-cut open hand, black on brown]
O414	**OPERA GLASS**	**[rectangular, black on red picture of Opera Glass]**
O415	OPHIR	[oval, embossed]
O416	OPTION	[square, black on yellow, Walker Bros.]
O417	ORATOR	[round, embossed]
O418	ORANGE	[round, black on yellow, picture of orange on a twig]
O419	**ORANGE**	**[round, embossed]**
O420	ORANGE	[die-cut orange, embossed]

O421	**ORANGE**	**[round, black on orange, P. Lorillard & Co]**
O423	ORANGE	[round, black on yellow]
O425	ORANGE	[round, black on yellow, Watson, Winston, N.C.]
O426	**ORANGE FREE STATE**	**[oval, picture of a flag, black on white, red & blue on yellow]**
O428	ORANGE GIRL	[odd shape, embossed, King]
O430	ORANGE GROVE	[rectangular, yellow on black]
O432	ORIOLE	[oval, embossed]
O433	ORIOLE	[rectangular, black on yellow]
O434	ORIOLE	[round, white on red]
O436	**ORLANDO**	**[rectangular, embossed]**
O438	ORLEANDER	[rectangular, embossed]
O440	OROMMONO	[round, embossed]
O442	ORONOKE	[odd shape, rectangular]
O444	OSAKA	[odd shape, multicolored]
O446	OSBURN HILL & CO	[die-cut leaf, black on yellow]
O450	OUR ADVERTISER	[rectangular, embossed, Winston, N.C.]
O452	OUR BABY	[rectangular, bar, black on red]
O454	OUR BEST	[oval, embossed, M.P. & Co.]
O456	OUR BEST TWIST	[rectangular, embossed]
O458	OUR BIRDIE	[round, picture of girl, black on yellow]
O460	OUR BOY	[rectangular, embossed]
O462	OUR BOY	[rectangular, hexagon, embossed]
O464	**OUR BROWN BARS**	**[round, black on gold & red, Scotten Tobacco Co.]**
O468	OUR CHAMPION	[rectangular, multicolored, picture of dog]
O470	OUR CHARLIE	[round, embossed]
O472	OUR DAISY	[rectangular, hexagon white on dark blue]
O474	**OUR DARLING**	**[round, black on red]**
O476	OUR FANNIE	[round, black on white, picture of a little girl sitting]
O478	OUR FINEST	[rectangular, black on red, Henry Co.]
O480	OUR FLAG	[round, small, blue on white, red, white & blue, picture of American Flag]
O482	OUR FOUR	[rectangular, bow tie shape, black on yellow]
O484	OUR FRITZ	[rectangular, embossed]
O486	**OUR HOBBY**	**[round, embossed, J.F. & Co.]**
O487	OUR IDEA	[round, embossed]
O488	OUR JACK	[round, blue on yellow, Penn]
O490	OUR JIM	[rectangular, black on red]
O492	**OUR LEADER**	**[rectangular, banner, gold on red]**
O494	OUR LITTLE FANCY	[round, embossed]
O496	OUR LOG CABIN	[round, yellow on black]
O498	OUR MAMIE	[rectangular, odd, embossed]
O500	**OUR NATION**	**[round, white on red & blue, Globe Tob. Co.]**
O502	OUR NEW PLUG	[die-cut hat, black on red]
O504	**OUR OWN**	**[round, black on yellow, Manufactured by N.B. Sullivan]**
O505	OUR OWN TOBACCO CO.	[round, black on red]
O506	OUR PET	[rectangular, embossed]
O507	OUR PRIDE	[round, embossed, picture of bundle of tobacco]
O508	**OUR PRIZE**	**[half round embossed]**
O509	OUR Q	[octagonal embossed]
O510	OUR Q	[round, embossed]
O512	OUR ROD	[hexagon embossed]
O514	**OUR SENATOR**	**[round, embossed, by B.F. Hanes]**
O515	OUR STANDARD	[round, embossed]
O516	**OUR TOM**	**[rectangular, embossed]**
O518	OUR TRADEMARK	[die-cut shape, pickaxe, embossed]
O520	OUTCAST	[rectangular, black on yellow]
O522	**OVER ALL**	**[round, yellow on black & yellow, P. Lorillard & Co.]**
O524	**OVERLAND**	**[oval, embossed]**
O525	**OWL**	**[small die-cut shape embossed]**
O526	**OWL**	**[small round black on yellow, picture of a owl, Lorillard]**
O527	**OWL**	**[die-cut owl standing embossed]**
O528	OWL	[round, black on multicolored, picture of owl sitting on the moon, R.A. Patterson Tobacco Co.]
O530	OWL	[die-cut owl, black on white, B.F. Hanes]
O532	**OWL CLUB**	**[die-cut O embossed, T.C. & Co.]**
O534	**OX**	**(a) [round, black on yellow, picture of ox, Brown & Williamson Tob Corp.]** (b) [red on white] (c) [black on red]
O536	OX	[round, black on yellow, picture of ox, J.G. Flynt Tob Co.]
O537	OX	[round, white on red, picture of a ox head]
O539	**OYSTER PLUG**	**[die-cut embossed oyster shell]**
O542	OYSTER TWIST	[round, embossed]
O544	OZARK TWIST	[round, black on yellow, Rowlett's]

P

P101	**P**	**[square, embossed]**
P103	**P**	**[die-cut, embossed]**
P105	P	[die-cut, plane]
P107	**P**	**[die-cut, small P embossed, choice]**
P108	**P. C.**	**[hexagon, white on green]**
P110	P D	[rectangular, embossed]
P111	P D	[rectangular, cut-out]
P112	**P.D. & Co.**	**[rectangular, embossed]**
P114	P.D.Q.	[oval, embossed]
P116	P-D'S TWIST	[round, black on red, Petey's]
P120	P & E	[rectangular, embossed]
P122	P. & E.T.	[odd rectangular, yellow on green]
P124	P G	[round, black on red]
P126	P.G.C.	[oval, black on red]
P128	P & G	[round, red on yellow]
P130	**P & H**	**[round, black on white]**
P132	P I	[rectangular, cut-out letters]
P134	P.J.	[rectangular, embossed]
P136	P J E	[banner embossed]
P140	**P.K.**	**[die-cut shield embossed]**
P142	P.K.	[round, embossed]

P144	**P L**	**[die-cut P on L]**
P146	P.L. & Co. 1877	[die-cut arrow, black on yellow]
P147	P.L. & Co. 1877	[die-cut anchor, black on red]
P149	P 1	[round, embossed]
P150	P P P	[die-cut letters, black on red & yellow, Penns Private Plug]
P152	P & R	[oval, embossed, C.F. Young & Bro.]
P153	P & R	[rectangular, banner embossed, Penn & Rison]
P154	P R R	[rectangular, embossed]
P156	P & R R R	[round, embossed]
P158	P & W	[rectangular, white & red on blue]
P160	**PA-LO**	**[oval, gold on dark blue, Scotten]**
P162	**PACE, J.B.**	**[banner embossed]**
P164	**PACE, J.R.**	**[rectangular, round embossed]**
P166	PACE, TALBOTT & Co.	[rectangular, odd embossed]
P168	**PACE'S BEST**	**[crescent embossed]**
P170	**PACE'S DIANORA**	**[round, black on yellow, picture of a girl]**
P172	**PACE'S FANCY**	**[rectangular, banner embossed]**
P174	PACE'S PLUG	[rectangular, embossed]
P176	PACIFIC SLOPE SMOKE	[round, black on yellow, H.N. Martin & Co. Louisville, Ky.]
P178	PADLOCK	[die-cut embossed, Candler]
P180	PADLOCK	[round, picture, black on yellow]
P182	PA LO	[oval, small yellow on green, Scotten]
P184	PALETTE	[die-cut shape, palette, white on black & yellow, P. Lorillard & Co.]
P185	PALM LEAF	[die-cut fancy leaf, black on yellow]
P186	**PALM LEAF**	**(a) [die-cut embossed leaf, large]** **(b) [small leaf]**
P188	**PALM LEAF**	**[die-cut leaf, black on yellow]**
P190	PALM NUT	[rectangular, black on yellow & brown, picture of a nut, R.H. Heindon & Co.]
P192	**PALME TO**	**[round, black on red, Lotterson]**
P194	PALMER'S BEST	[round, embossed]
P195	PALMER'S CHALLENGE	[round, embossed]
P196	PALMER'S DARK PLUG	[rectangular, embossed]
P198	PALMER'S ROSE BUD	[rectangular, embossed]
P200	PALMETTO	[round, red on yellow, picture of palm leaf]
P202	**PAN CAKE**	**[round, black on red]**
P203	PAN CAKE	[round, red on yellow]
P204	**PAN ELECTRIC**	**[rectangular, embossed, Hanes]**
P206	**PANEL TAG**	**[odd shape, banner embossed]**
P208	**PANIC**	**[odd shape, cross embossed]**
P210	PANIC	[large shape, cross embossed]
P212	PANIC	[round, embossed]
P214	**PANSY**	**[die-cut letter P, black on yellow, G.S.S. & Co.]**
P216	**PANSY**	**[round, embossed]**
P218	PANSY	[round, multicolored, picture of a pansy]
P220	PANSY	[round, blue on white, picture of a pansy]
P222	PANTHER	[rectangular, picture of a panther]
P224	**PAPOOSE**	**[die-cut embossed]**
P226	PAPOOSE	[rectangular, black on green]
P228	**PARAGON**	**[round, embossed with two cut outs]**
P230	**PARAGON**	**[round, embossed with star]**
P231	PARKER GUN	[rectangular, black on orange]
P232	PAROLE	[square, black on red, picture of buggy racing]
P234	PAROLE	[rectangular, scalloped black on red, picture of buggy racing, Robt Harris & Bros]
P236	PARRAN'S BEST	[round, embossed]
P238	PARROT	[rectangular, picture of a parrot, green on yellow]
P239	**PARROT**	**[round, yellow on red, picture of parrot green on yellow, Talks for Itself]**
P240	**PARROT**	**[die-cut parrot, black on red & yellow]**
P242	PARROT	[die-cut of a fish, black on red]
P244	PARROT	[die-cut shield, black on yellow, picture of parrot, Lester & Gaines]
P246	**PARIS MEDAL**	**[round, green on yellow, T.C. Williams & Co.]**
P247	PASSPORT	[rectangular, black on red]
P248	PAT DUFFY	[rectangular, yellow on red]
P250	PAT HAND	[round, black on red, picture of cards]
P252	PAT JAN 4 1870	[rectangular, embossed]
P254	PAT'S TWIST	[odd shape, white on green]
P255	PATRIOT	[die-cut shield embossed]
P256	**PATROL**	**[rectangular, black on red]**
P258	**PATTERSON**	**[oval, black on red]**
P260	**PATTERSON**	**[rectangular, embossed]**
P262	PATTERSON'S BEST SMOKE	[round, red & orange on cream]
P264	PATTERSON'S DARK LIGHT	[round, black & tan on yel low]
P266	PATTERSON'S PLUG	[rectangular, black & red on yellow]
P268	**PATTI SING SING**	**[round, scalloped black on pink, picture of Oriental girl]**
P270	PATTYS CAKE	[round, black on red, picture of a girl eating]
P272	**PAUL-BROWN**	**[rectangular, embossed]**
P274	PAY CART	[round, embossed]
P276	PAY DAY	[octagonal, red on yellow, Harris Sons]
P278	PAYROLL	[die-cut snowshoe, black on red]
P280	PAYROLL	[die-cut snowshoe, black on red, printed in U.S.A.]
P282	PEACE AT HOME	[diamond, embossed]
P284	**PEACE & GOOD WILL**	**[diamond, black on red]**
P286	**PEACE AND GOOD WILL**	**[diamond, black on red, Weissinger]**
P288	**PEACE MAKER**	**[round, black on red]**
P290	PEACH	[rectangular, black on red, Greenville]
P292	**PEACH**	**[rectangular, embossed]**
P294	PEACH	[oval, black on red, Dominion Tob Co. Montreal]
P296	PEACH & HONEY	[rectangular, embossed]
P297	PEACH & HONEY	[die-cut leaf embossed]
P298	**PEACH & HONEY**	**[round, red on blue, picture of peach & bee hive, Taylor Bros.]**
P300	**PEACH & HONEY**	**[die-cut P embossed]**
P301	PEACH & HONEY	[oval, embossed]

P302	**PEACH & HONEY**	**[octagonal black on red]**
P304	PEACH & HONEY TOBACCO	[round, picture of peach & bee hive, Mfg. by Whitaker Harvey Co.]
P306	PEACH AND HONEY	[octagonal black on red]
P308	**PEACH AND HONEY**	**[oval, black on red]**
P309	PEACH PIE	[J.G.B. & Co.]
P310	PEACHEY	[round, red on yellow, red & green picture of peach, J.G. Geller Sons & Co.]
P312	**PEACHEY PLUG**	**(a) [large die-cut odd shape, red on green & yellow, picture of a peach] (b) [small on reverse of tags, Scotten Dillon Co.]**
P314	PEACHEY TWIST	[round, red on yellow & green, Scotten Co.]
P316	PEACOCK	[octagonal black on yellow, picture of Peacock]
P318	**PEACOCK**	**[round, black on yellow, picture of Peacock, L.H. Neudecker]**
P320	PEACOCK	[die-cut shape, peacock embossed]
P324	PEANUT	[rectangular, black on red, picture of a peanut]
P326	PEANUT	[die-cut shape, peanut, embossed]
P328	PEAR	[round, black on yellow, picture of pear, yellow on black, Patterson, Richmond]
P330	PEAR	[die-cut pear, embossed]
P332	PEAR	[die-cut pear, brown on yellow]
P334	PEAR TRADE MARK	[die-cut pear, embossed]
P336	**PEARL**	**[odd five-sided shape black on yellow, P. Lorillard & Co.]**
P338	PEARL	[round, yellow on red, P.E.J.]
P340	PEDRO	[round, embossed]
P342	**PEELER**	**[round, embossed, L.O.Mc.N P & Co.]**
P344	PEELER	[round, embossed, McN.S. & M.]
P346	PEELER	[round, embossed, W & G McN.]
P348	**PEEP**	**[banner embossed]**
P350	**PEEP O'DAY**	**[octagonal, black on green, picture of a clock face]**
P352	**PEERLESS**	**[round, embossed]**
P353	PEERLESS	[die-cut P, black on green]
P354	**PEERLESS**	**[oval, embossed]**
P356	PEERLESS	[oval, embossed, Danville, Va.]
P358	PEERLESS	[rectangular, embossed, Danville, Va.]
P360	PELICAN	[die-cut bird embossed]
P362	PEMBERTON T.W.	[round, embossed]
P364	PENINSULA	[oval, white on red]
P366	PENINSULA	[round, black on green, picture of Chesapeake Bay]
P368	**PEN MAR**	**[rectangular, black on red]**
P370	PENN & CO., F.R.	[rectangular, embossed, Reidsville, N.C.]
P372	PENN & Co., F.R. NATURAL LEAF	[die-cut leaf red on yellow]
P373	PENN RISON	[rectangular, black on green, picture of man & long gun]
P374	PENN TOB. Co., F.R. THE NATURAL LEAF	[die-cut leaf, red on yellow]
P376	**PEN POINT**	**[die-cut pen point]**
P378	**PENN'S**	**[rectangular, black on yellow, picture of a pen point]**
P380	**PENN'S**	**[die-cut pen point, black on yellow]**
P382	PENN'S JACK	[square, red on yellow & brown, picture of a mule]
P384	**PENN'S NO. 1**	**[die-cut pen point, gold on red]**
P386	**PENN'S NATURAL LEAF**	**[die-cut leaf red on yellow]**
P387	PENN'S PRIVATE PLUG	[die-cut letters, P P P, black on red & yellow]
P388	**PENN'S THICK**	**[die-cut leaf, red on yellow]**
P390	**PENN'S THICK**	**[die-cut leaf, red on yellow, Natural Leaf]**
P392	PENN'S STONEWALL	[rectangular, picture of man on a horse, The only Genuine]
P394	PENNY POST	[round, embossed]
P400	**PENNY TAG**	**[round, green on yellow, Nicholas Kunner, Davenport, Iowa]**
P402	**PEOPLES CHOICE**	**[round, red on yellow, P.D. & Co.]**
P404	**PEPER NATURAL LEAF**	**[rectangular, embossed]**
P406	**PEPER'S BEST**	**[round, embossed]**
P407	PEPER'S G C	[round, embossed]
P408	PEPER'S TWIST	[rectangular, embossed]
P410	**PERFECTION**	**[round, black on red, with a star]**
P412	**PERFECTION**	**[round, gold on red]**
P413	**PERFECTION**	**[diamond, black on red]**
P414	PERFECTION	[triangular, black on red]
P415	PERFECTION	[rectangular, embossed]
P416	PERFECTION	[odd shape, embossed, P T W]
P418	**PERKINS & ERNST**	**[round, with cut-out center, embossed]**
P419	PET	[oval, embossed, Turner]
P420	PET	[round, black on yellow, E.S.J.]
P421	PET	[square, embossed, J.F.W.]
P422	PETER JACKSON	[round, black on yellow, picture of boxers]
P424	PETERS	[oval, black on red, Jersey Lily, Lynchburg, Va.]
P426	PEYTON GRAVELY	[round, black on yellow]
P428	PFAFF'S OWN	[die-cut letter P embossed]
P430	**PIC-NIC**	**[oval, embossed]**
P432	PICAYUNE	[rectangular, embossed]
P434	**PICK**	**[die-cut small pick embossed]**
P436	**PICK**	**[die-cut small pick, engraved O U B trademark]**
P438	PICK	[die-cut pick, black on red & silver]
P440	**PICK, NATURAL LEAF**	**[die-cut pick black on red & silver]**
P442	**PICK & SHOVEL**	**[die-cut shapes embossed]**
P444	THE PICKLE	[die-cut a pickle embossed]
P446	**PICKWICK CLUB**	**[round, gold on red]**
P447	PICKWICK CLUB	[round, black on red]
P449	PICNIC	[rectangular, embossed]
P452	**PIC-NIC**	**[rectangular, large embossed]**
P454	**PIC-NIC TWIST**	**[rectangular, small, black on red]**
P456	**PICNIC TWIST**	**(a) large [round, white on blue & red, Liggett & Myers Tob Co] (b) small**
P458	**PICNIC TWIST**	**(a) small [round, white on blue & red, The American Tob Co.] (b) large**
P460	**PIE**	**[round, red & black on yellow, Taylor Bros]**

P462	PIE	[round, embossed, S.C. Co.]
P463	PIEDMONT	[oval, embossed]
P464	PIEDMONT	[oval, black on red]
P468	**PIEDMONT No. 11.**	**[rectangular, red on tan, J.R. Bryant Mfgr.]**
P470	PIEDMONT NATURAL LEAF	[diamond, brown on yel low]
P472	PIG IRON	[rectangular, black on yellow, multicolored picture of men working, Penn & Rison]
P474	PIG IRON	[rectangular, black on yellow, picture of red iron, Penn & Rison]
P476	PIGTAIL	[rectangular, odd shape, black & tan on yellow]
P478	**PIKE**	**[die-cut fish, embossed]**
P480	**PIKE**	**[die-cut fish embossed, Schoolfield]**
P482	PIKE'S PEAK	[round, embossed]
P484	**PILGRIM**	**[die-cut, black on blue & white, picture of a man]**
P486	**PILGRIM**	**[large die-cut, shape of a man, embossed]**
P488	PILOT	[die-cut heart, engraved, W C McDonald, Montreal]
P490	PILOT BUCK	[rectangular, embossed]
P491	**PILOT CROWN**	**[round, embossed]**
P492	PIN MONEY	[rectangular, yellow on red, Penn & Rison]
P494	PIN MONEY	[rectangular, yellow on red, Penn & Sons]
P496	PIN WHEEL	[round, embossed]
P500	**PINCHER**	**[diamond, black on red, Wilson & McCallay Tobacco Co.]**
P502	PINCHER	[rectangular, black on yellow]
P504	**PINE APPLE**	**[rectangular, black on white, F.A. Davis & Sons]**
P506	PINE APPLE	[rectangular, small black on yellow, F.A. Davis & Co.]
P507	PINE APPLE	[rectangular, large black on orange, F. A. Davis & Co.]
P508	PINE APPLE	[round, embossed]
P510	PINE APPLE	[rectangular, black on yellow]
P512	PINE APPLE	[hexagon embossed, picture of pine apple]
P514	PINEAPPLE TWIST	[red on yellow, picture of pineapple, G. Penn Sons Tob Co.]
P516	PINE KNOT	[round, black on red, picture of wood with knot]
P517	PINE KNOT	[oval, red on yellow, Bitting & Hay, Winston, N.C.]
P519	PINE TREE	[round, black on red, picture of tree, The Key. Co.]
P521	PINE TREE	[round, green on white, picture of tree, Key & Co.]***?check name w/ above**
P522	PINEY WOODS	[oval, embossed, Lowery]
P524	PINK PLUM	[rectangular, bar embossed]
P526	**PIONEER**	**[rectangular, embossed]**
P528	**PIPER HEIDSIECK**	**[oval, black on red, Champagne Flavor]**
P530	**PIPPIN**	**[round, yellow on black, picture of apple, J.G. Flynt Tob Co.]**
P532	PISTOL	[die-cut pistol, embossed]
P534	PIXIE	[diamond, black on red]
P536	PLANTATION	[rectangular, round ends, black on yellow]
P538	**PLANET**	**[round, cut-out B L, white on blue, B & L]**
P540	PLANK ROAD	[rectangular, embossed]
P542	PLOW BOY	[round, black on yellow, picture of man & horse plowing, J.N. Wyllie & Co.]
P543	PLOW BOY	[round, black on cream, picture of man & horse plowing, J.N. Wyllie & Co.]
P546	PLOW MAN TOBACCO	[rectangular, embossed]
P548	PLUCK	[oval, embossed]
P550	PLUG HAT	[die-cut hat, black on white, The P.J. Sorg Co.]
P552	**PLUG HAT**	**[rectangular, black on gray, picture of top hat, Gravely & Millers]**
P556	**PLUM**	**[round, black on purple & white, picture of plum, T.C. Williams Co. The color of the plum also comes in, blue, orange, blue green, red & brown]**
P557	**PLUM**	**[round, black on purple & white, T.C. Williams Co.& Continental Tob Co]**
P559	PLUM	[round, black on purple & white, T.C. Williams Co. & The American Tob Co.]
P560	**PLUM**	**[round, gold on green]**
P564	**PLUMS BOB**	**[die-cut Plum Bob, yellow on red & black]**
P566	PLUMB GOOD	[round, red on white, Taylor & Spencer & Co.]
P568	**P. OF R.**	**[oval, embossed, G.F. Young & Bro.]**
P570	POCKET KNIFE	[rectangular, black on yellow, picture of pocket knife]
P571	**POCKET KNIFE**	**[die-cut shape, open blades, embossed]**
P574	**POINTER**	**[die-cut embossed shape of a dog, facing left]**
P575	POINTER	[round, black on green, picture of a dog facing left]
P576	POINTER	[diamond, embossed, Flora & Co.]
P580	**POKER 10¢ CHIPS**	**[round, embossed]**
P582	POLE AXE	[round, black on red]
P584	**POLICE**	**[rectangular, blue on red, white & blue]**
P586	POLICE CAP	[die-cut, embossed]
P588	**POLICE HAT No. 10.**	**[die-cut hat, embossed]**
P590	**POLO**	**[die-cut P black on green, G.S.S. & Co.]**
P592	**POND LILY**	**[diamond, gold on black]**
P594	POND LILY 9 IN 5	[oval, embossed, Bittings & Hay]
P596	**PONY**	**[die-cut, embossed pony, running left]**
P598	PONY	[rectangular, black on yellow, Timberlake, Jackson & Co. Lynchburg, Va.]
P600	**PONY NAVY**	**[rectangular, embossed picture of a pony]**
P602	POOR MAN'S COMFORT	[rectangular, black on yel low, picture of man sitting]
P604	POPPY	[round, yellow on green, picture of yellow poppy]
P605	**POPULAR BLOCK**	**[round, black on red]**
P606	PORCUPINE	[round, picture of porcupine, P. Lorillard & Co.]
P608	**PORK**	**[die-cut, embossed pig]**
P610	**PORK & BEANS**	**[die-cut can, white on red & black, Wetmore]**

P612 PORK & BEANS [round, embossed]
P614 POSSUM [rectangular, yellow on red]
P616 POSSUM [rectangular, embossed]
P618 POSSUM HOLLOW [round, embossed]
P620 POST BOY [oval, yellow on dark brown]
P622 POT LUCK [die-cut pot, embossed]
P624 POUND*CAKE [rectangular, embossed]
P626 POUND CAKE [round, white on blue, The G. Penn Sons Tob Co., Danville, Va.]
P628 1/4 POUND B [square, embossed]
P630 1/4 POUND D [rectangular, embossed]
P632 1/4 POUND V [odd square, embossed]
P633 1/4 SECTION 10 [square, embossed]
P634 PRAIRIE BELLE [round, black on yellow, picture of girl]
P636 PREMIUM [round, blue on silver, Tinsley]
P638 PREMIUM [round, embossed]
P640 PREMIUM NATURAL LEAF [die-cut leaf, yellow on red, A.D. Reynolds]
P642 PRESIDENT TWIST [round, blue on white, picture of Washington]
P644 PRETZEL [oval, black on yellow, picture of pretzel, Schoolfield & Watson]
P646 PRIDE [oval, embossed, Meyer]
P648 PRIDE OF ATHENS [round, black on red, picture of a leg, Geo. O. Jones, Ridgeway, Va.]
P650 PRIDE OF DIXIE (a) [round, black on yellow]
(b) green on white
P652 PRIDE OF DIXIE [octagonal black on white]
P653 PRIDE OF LEATHERWOOD [rectangular, red on black, Montaque]
P654 PRIDE OF THE NAVY [round, picture of a ship, United States Tob. Co.]
P656 PRIDE OF VIRGINIA [round, black on yellow]
P658 PRIDE OF WINSTON [oval, black on yellow, picture, B & W]
P659 PRIME PLUG [round, embossed]
P661 PRIME TWIST [rectangular, embossed]
P664 PRINCE [die-cut shield, silver on black]
P666 PRINCE OF WALES [die-cut heart engraved, W C McDonald, Montreal]
P668 PRIVATE CHEW [rectangular, black on red, Winfree]
P670 PRIVATE STOCK [round, black on yellow, picture of man, J.B. Pace]
P672 PROTECTIVE UNION [round, green on white]
P673 PROVIDENCE [die-cut shield embossed, McAlpins]
P674 PRUNE JUICE [oval, small gold on red, Scotten]
P675 PRUNE JUICE [D.S. & Co]
P676 PRUNE JUICE [oval, blue on dark blue]
P680 PRUNE NUGGET [oval, embossed]
P681 PUCK [round, embossed]
P682 PUFF [rectangular, scalloped, embossed]
P684 PUG [diamond, embossed]
P686 PUG [oval, black on white]
P688 PULASKI [rectangular, yellow on black]
P690 PULLMAN [rectangular, embossed]
P692 PUNCH [round, small embossed cut-out P]
P693 PUNCH [round, large embossed cut-out P]
P694 PUNCH [round, black on red, picture of a man sitting]
P696 PURE CREAM [rectangular, scalloped, yellow on red, picture of horse]
P698 PURE HONEY [oval, red on yellow, B. Ash]
P700 PURE GOLD [round, black on gold]
P702 PURE GOLD [round, embossed]
P703 PURE GOODS [oval, black on yellow, Rankin Bros]
P704 PURE GRAPE [rectangular, yellow on red]
P706 PURE GRAPE-WINE FLAVOR [rectangular, gold on red]
P708 PURE LEAF [rectangular, black on yellow, Walker's Chew Smoke]
P710 PURE SUNCURED [round, red & black on yellow, Bailey Bros]
P712 PURITY [die-cut, shape a bird embossed]
P713 PURITY [rectangular, embossed]
P714 PURITY [oval, embossed]
P715 PURITY [round, white on blue, picture of bird]
P716 PUSH [rectangular, scalloped embossed]
P718 PUSSY [round, embossed picture of a cat]
P720 PUSSY [round, black on red picture of a cat]
P722 PUZZLE [round, black on red]
P724 PYLON [die-cut, embossed, a pylon]

Q

Q101 Q [die-cut letter Q]
Q103 QT [rectangular, embossed]
Q105 Q STICK [rectangular, white on blue, picture of Q Stick, T. C.W. Co.]
Q106 QUAIL [rectangular, red on black]
Q108 QUALITY & QUANTITY [round, red on yellow & black]
Q110 QUAKER CITY TWIST [oval, embossed, B.F. Hanes, Winston, N.C.]
Q112 QUARTER POUND [rectangular, round end, embossed]
Q114 QUARTER POUND BUSY BEE [round, embossed]
Q116 QUARTETTE [rectangular, embossed, S.R. & Co.]
Q118 1/4 SECTION 10 [diamond, embossed]
Q120 QUEEN ANNE [oval, black on white, picture of girl]
Q121 QUEEN OF BEAUTY [rectangular, black on yellow, picture of a girl]
Q122 QUEEN OF GEORGIA [round, black on yellow, picture of girl]
Q123 QUEEN OF HEARTS [die-cut heart shape, black on red]
Q126 QUEEN OF THE RANCH (a) [rectangular, black on yellow, picture of a girl, Irvin & Poston]
(b) black on red
Q128 QUEEN RUBY [rectangular, black on red, specialty manufactured for Lamb & Co.]
Q130 QUEEN'S NAVY [round, embossed]
Q132 QUICK SALE [round, embossed]
Q134 QUICK STEP [round, embossed with a star]
Q136 QUICK TWIST [rectangular, embossed]
Q138 QUICK STOP [round, embossed]
Q140 QUICK STEP [rectangular, black on yellow]
Q142 QUINCY [rectangular, embossed]

R

R101	**R**	**[die-cut plane]**
R103	R	[die-cut embossed]
R106	R	[round, large R silver on black & in red La Grange]
R108	R.B.	[die cut half round, black on red]
R110	R BRO'S	[diamond, embossed]
R112	R & B	[die-cut barrel shape, yellow on red]
R113	R & B	[die-cut barrel shape, white on red]
R114	R & B	[rectangular, yellow on red]
R116	**R E A**	**[round, black on gold]**
R118	**R.F.**	**(a) [die-cut heart shape embossed, red]** (b) [gold]
R120	**R H**	**[rectangular, cut-out]**
R122	R.I.	[round, black on red, Geo. F. Young & Bros, R.I.]
R124	**R.I.P.**	**[rectangular, black on yellow]**
R126	R.J. CHRISTION	[round, embossed, picture of a star in center]
R128	R J R	[die-cut letters red]
R130	**R J R**	**[die-cut letters black on red, R.J. Reynolds Tob Co. Trade Mark]**
R132	**R J R**	**[die-cut letters black on red, white & green, M'F'G'D. by R.J. Reynolds Trade Mark]**
R134	R.K.	[oval, embossed]
R136	**R L C**	**[die-cut litters black on yellow, Made by R.L. Candler & Co. Winston, N.C.]**
R138	**R. P.**	**[round, embossed]**
R140	**R R R**	**[rectangular, cut-out]**
R142	R & S	[round, black on yellow]
R143	R & S	[rectangular, embossed]
R144	R.T.	[round, red on white, Rough Country Twist]
R146	R & W	[round, black on yellow, Ogburn, Hill & Co. Winston]
R148	**R & W**	**[hexagon black on yellow, Ogburn Hill & Co. Winston]**
R150	R W T	[round, black on red]
R152	**RABBIT**	**[die-cut a rabbit, embossed, facing left]**
R153	RABBIT	[rectangular, black on red, picture of a rabbit]
R154	RABBIT	[oval, black on red, picture of a rabbit running]
R156	RABBIT	[oval, black on yellow, picture of a rabbit]
R158	RABBIT FOOT	[round, black & brown on yellow, picture of a rabbit]
R160	RACE HORSE	[round, black on red, picture of jockey on horse]
R162	RACER	[die-cut shape, striped cap, Penn]
R164	**RACKET**	**[round, embossed, Lowrey]**
R166	R.A.D's NATURAL LEAF	[die-cut leaf, black on yellow, Mfd by R.A. Deshazo]
R168	RAGING BULL	[die-cut bull embossed, W.L. & Co.]
R170	RAILROAD	[rectangular, gold on blue, Mfg. by Robt. Harris & Bro]
R172	RAILROAD	[rectangular, embossed, red on black]
R174	RAILROAD PLUG	[die-cut train engine, embossed]
R176	**RAILROAD TWIST**	**(a) [round, black on red, Ryan-Hampton Tob. Co.]** (b) [round, white on red]
R178	**RAINBOW**	**[round, black on multicolors, Nall & Wiliams Tob Co.]**
R180	RAINBOW	[round, black on multicolors, Nall & Williams Tobacco Co.]
R182	RAINBOW	[rectangular, odd, Foree's Best Chew Made]
R184	RAINBOW	[round, embossed]
R186	RALPH'S	[oval, black on yellow, picture of man]
R188	**RALPH'S SUCCESS**	**[oval, embossed]**
R190	**RAM HEAD**	**[round, small red on black, picture of a ram's head]**
R192	**RAM'S HORN**	**(a) [round, orange on yellow & black, picture of Rams Head, Taylor Bros.]** (b) [red on yellow & black] **(c) [red on yellow & black, large center prong]**
R194	**RAM'S HORN SUNCURED**	**[round, black on yellow & red, picture of a ram's head]**
R196	RANKINS STANDARD	[rectangular, black on red]
R198	RASPBERRY	[round, embossed]
R200	**RATTLER**	**[rectangular, shield embossed]**
R202	**RATTLER**	**[round, embossed]**
R204	RATTLER	[rectangular, black on red, Traylor Spencer & Co.]
R205	RATTLER	[rectangular, embossed, Traylor Spencer & Co.]
R206	**RAVEN**	**[die-cut bird embossed, facing right]**
R208	RAVEN	[round, black on gray, picture of bird]
R210	RAZOR	[round, embossed]
R212	**RAZOR**	**[round, black on white, picture of a straight razor]**
R214	R BRO'S	[diamond, large embossed]
R216	R BRO'S	[diamond, small embossed]
R218	READY CASH	[rectangular, black on red]
R220	READY SALE	[round, embossed]
R221	REAL STUFF	[die-cut letters J H, black on red Hilton]
R222	REAL-STUFF TWIST	[round, yellow on red, Highest Type Leaf, J. H. Hilton]
R224	**REAL THING**	**(a) [large, round white on green & red]** (b) Small
R226	REAPER	[die-cut shape, reaper embossed]
R228	REBATE PLY	[round, embossed]
R230	**REBECCA**	**[hexagon black on red, P. Lorillard & Co.]**
R232	REBEL	[round, embossed]
R234	REBEL BOY	[round, black on yellow, picture of boy, F.R. Penn & Co.]
R235	REBEL GIRL	[octagonal black on orange, picture of girl, The F.R. Penn Tob. Co.]
R236	**REBEL GIRL**	**[round, black on red picture of a girl, M'F'G'D.**

by F.R. Penn & Co.]

R238	RECEPTION	[rectangular, shield, black on red]
R240	**RECORD**	**[oval, black on red, A.B.C. & Co.]**
R242	RED	[round, red on yellow, picture of a lion]
R244	RED	[diamond, red on black, picture of a horse]
R246	**RED ANT**	**[round, red on yellow, picture of an ant]**
R248	RED B TWIST	[round, red on yellow]
R250	**RED APPLE**	**[round, black on red & white, picture of a apple, T.L. Vaughn & Co.]**
R252	RED BANDANNA	[round, red on white, picture of a bandanna]
R254	RED BELL	[die-cut a bell, brown & red on cream]
R256	**RED BIRD**	**(a) [octagonal small, gold on dark green Garr Bros.]** (b) [gold on dark blue]
R258	**RED BIRD**	**(a) [small oval, black on white & red, picture of a bird] (b) [larger]**
R260	RED BIRD	[die-cut bird, black on red]
R262	RED BIRD	[rectangular, black on red, picture of a bird facing left]
R264	RED BOW	[rectangular, banner embossed]
R266	**RED BOX**	**[square, black on red]**
R268	RED BOX	[die-cut box black on red]
R270	**RED BUD**	**[oval, embossed]**
R272	RED BUD	[round, red on white, picture of flower bud]
R274	RED BUD	[round, black on red]
R276	RED BUD TWIST	[round, red on yellow, picture of flower]
R278	RED BULL	[die-cut standing bull, red]
R280	RED CEDAR	[round, black on red, U.S. Tob Co.]
R282	RED CHIEF	[round, black on red]
R284	RED CLOUD	[die-cut a ship embossed]
R286	**RED CLOVER**	**[round, red on green & white, picture of three-leaf clover]**
R288	**RED COON**	**[die-cut coon, yellow on red, Taylor]**
R290	**RED COON**	**[rectangular, yellow on red, picture of a coon, Taylor]**
R292	**RED COON**	**[rectangular, yellow on red picture of a coon, Whitaker-Harvey Co]**
R294	**RED COON**	**[rectangular, picture of a red coon on white]**
R296	RED COON	[rectangular, embossed]
R298	RED COON	[round, embossed]
R300	RED COW	[rectangular, black & red on yellow, picture of cow]
R302	**RED CROSS**	**[small round black on red & white, P. Lorillard & Co.]**
R304	RED CROSS	[round, black on red picture of a cross on white]
R306	RED CROSS	[round, black on yellow picture of a cross on white]
R308	**RED CROSS**	**[round, Red Cross reading left to right & bottom to top]**
R310	**RED CROW**	**[die-cut a Crow, black on red**
R312	**RED CROW**	**[round, red & black on white, picture of a bird]**
R314	RED CROWN	[round, red on white, picture of crown]
R316	RED CROWN	[round, black on yellow, picture of red crown, Jno. J. Bagley & Co., Detroit, Mich.]
R318	RED DEER	[round, black & brown on yellow, picture of deer head, J.H. Jenkins & Co.]
R319	RED DEER	[round, black & red on yellow, Southern Tob Co.]
R320	**RED DIAMOND**	**[diamond, shape black on red]**
R322	**RED DIAMOND**	**[die-cut diamond, red]**
R324	**RED DOT**	**[die-cut round, small, red]**
R326	RED DOT	[die-cut round large, red]
R328	RED EAGLE	[round, black on yellow & red, picture of a eagle]
R330	**RED ELEPHANT**	**[die-cut elephant, embossed]**
R332	**RED ELEPHANT**	**[rectangular, black on red, picture of an elephant]**
R334	RED ELEPHANT	[die-cut picture of elephant, black on orange, red & yellow]
R336	**RED ELEPHANT TOBACCO**	**[round, red on yellow, pic ture of an elephant, Brown Bros. Co.]**
R338	**RED ELEPHANT TOBACCO**	**[round, red on yellow, el ephant picture, H.H. Reynolds]**
R340	**RED ELK**	**(a) [round, black on yellow & brown, elk head facing left, Penn & Watson] (b) [elk head facing straight on]**
R342	**RED EYE**	**[round, white on black, picture of Red Eye, Taylor Bros.]**
R344	**RED EYE**	**[round, black on yellow, picture of Red Eye, Taylor Bros.]**
R346	RED FISH	[die-cut fish, black on red]
R347	RED FOX	[round, red on black]
R348	**RED FOX**	**[oval, red on black, picture of a fox, Spencer Bros.]**
R350	**RED FOX**	**[rectangular, black on white & brown picture of a fox, Arnold & McCord]**
R352	RED FOX	[round, red on white & white on blue, picture of fox head, F.M. Bohannon]
R354	RED FOX	[round, red on black, in center picture of a fox, F.S.J.]
R356	**RED FOX TWIST**	**[oval, black & red on white, picture of a fox, F.M. Bohannon]**
R358	RED GOOSE	[die-cut bird, yellow on red]
R360	RED HAM	[oval, black on red & yellow, picture of ham, W.P. Pickett & Co., High Point, N.C.]
R362	RED HEAD	[round, black & red on white, L.W. Davis, Norfolk Va.]
R364	RED HEAD 5¢ TWIST	[round, white on red]
R366	RED HEART	[round, blue on cream, picture of red heart, Hancock Bros & Co.]
R368	RED HEAT	[rectangular, green & black on white, picture of building, Penn Watson & Co.]
R369	RED HORSE	[round, black & red, on white, picture of horse,

		The G. Penn Sons Tob Co.]
R370	RED HORSE	[rectangular, red on yellow, picture of horse]
R372	**RED-HOT**	**[rectangular, embossed]**
R374	RED HOT	[diamond, black on red, A.G. Fuller & Co. Tobacco]
R376	RED HOT	[diamond, embossed]
R378	**RED J.**	**[rectangular, red on yellow, Penn]**
R380	RED J.	[rectangular, red on yellow, The F.R. Penn Tob Co.]
R382	RED JACKET	[round, black on yellow, Musselman]
R383	RED JACKET	[rectangular, red on black, picture of horse, Pegram & Penns]
R384	RED JUICE	[rectangular, with round center, red on yellow, picture of glass of juice]
R386	**RED JUICE**	**[round, yellow & red on green, picture of a glass]**
R388	RED JUICE	[round, red on yellow, picture of glass of juice]
R390	**RED LEAF TWIST**	**[round, black on yellow, picture of a man, Moore]**
R392	RED LEAF TWIST	[square, red on yellow, Carthage Tobacco Works]
R394	RED LETTER	[die-cut letter W, black on red, Walker Bro.]
R395	RED LETTER	[octagonal yellow on green, Liipfert Scales & Co.]
R396	RED LIGHT	[diamond, black on red]
R397	RED LIGHT	[die-cut cross, black on red]
R398	RED LIGHT	[diamond, red on white]
R400	RED LINE	[rectangular, picture of a train]
R402	**RED LION**	**(a) [large oval black on yellow & red, embossed lion]** **(b) [small oval]**
R404	**RED LION**	**[hexagon red on white, picture of a Lion]**
R406	RED MAN	[oval, black & red on yellow]
R408	RED MEAT	[round, multicolored, picture of a slice of watermelon, Liipfert Scales & Co.]
R410	**RED MEAT**	**[die-cut half a watermelon, black on red & green]**
R412	RED MEAT	[round, picture of a slice of watermelon]
R414	RED MEAT PLUG	[round, black on red, picture of a watermelon slice, Lockett, Vaughn & Co.]
R416	**RED MOON**	**[round, red on white, Union Made in center]**
R417	RED MOON	[round, red on yellow, picture of crescent moon]
R418	RED MOUSE	[rectangular, red on yellow, picture of mouse, Liipfert & Jones]
R420	RED NICKEL	[round, embossed]
R422	RED OAK	[round, small, white on red]
R424	RED PEACH	[round, black, red & green on white, picture of peach, D. Marion]
R426	RED PEACH PLUG	[round, multicolored, picture of peach]
R428	RED PLUM	[round, multicolored, picture of plum]
R430	RED POODLE	[rectangular, embossed]
R432	RED RABBIT	[round, small, black & red on yellow, picture of rabbit, B.F. Hanes]
R434	**RED RABBIT**	**[round, large, black & red on yellow, picture of rabbit, B.F. Hanes]**
R436	RED RABBIT	[round, large, yellow & red on green, picture of rabbit, B.F. Hanes]
R438	**RED RAMBLER**	**[round, red on yellow, picture of a fox, Fluhrer Bros.]**
R440	RED RAVEN	[round, red on white, picture of red bird]
R442	RED RAVEN	[oval, red on yellow, picture of bird]
R444	**RED RIVER**	**[round, black on red, picture of a river]**
R446	RED ROCK	[round, white & blue on red, The U.S. Tob. Co.]
R448	RED ROOSTER	[die-cut, embossed shape of a rooster, red]
R450	RED ROSE SUN-CURED	[black on yellow, picture of red rose, J.G. Flynt Tob Co.]
R452	RED ROSE TWIST	[rectangular, black & tan on yellow]
R454	**RED SEAL**	**[round, scalloped, black on red]**
R456	RED SEAL	[round, yellow on red, R.L. Candler Co.]
R458	RED SEAL NAVY	[round, large black on red]
R460	RED SHIELD	[round, embossed]
R462	RED SHOE	[rectangular, multicolored picture, Tobacco, G. Penn Sons Tob. Co.]
R464	RED SHOES	[die-cut shoe, yellow on red, Brodnax & Co.]
R466	REDSKIN	[diamond, red on yellow]
R467	RED SNAPPER	[rectangular, black on red]
R468	**RED SNAPPER**	**[round, embossed picture of a fish]**
R469	RED SQUIRREL	[round, black, red & yellow, picture of squirrel, Adams, Powell & Co.]
R470	**RED STAR**	**[round, black on red & black]**
R472	RED STAR	[round, embossed picture of star]
R474	**RED STAR BRAND**	**[die-cut shape, black on red]**
R476	RED WHEEL	[round, red on yellow, picture of spoke wheel, Dalton, Farrow Tob Co.]
R477	RED WHEEL	[rectangular, red on yellow, Dalton, Farrow & Co.]
R478	**RED, WHITE & BLUE**	**[round, red on blue, blue on white, white on red, F.M. Bohannon]**
R480	**RED WILKES**	**[rectangular, white on red]**
R482	RED WINE	[rectangular, embossed with two stars]
R484	RED WING	[round, embossed]
R486	RED WRAPPER	[rectangular, black on yellow, picture of girl, Liipfert & Jones]
R488	REED BIRD	[oval, embossed]
R500	REED'S BEST	[round, embossed]
R502	REED'S TWIST	[rectangular, odd embossed]
R504	REEL FOOT	[rectangular, Flournoy Tob Co., Paducah, Ky.]
R506	**REGATTA**	**[die-cut sail boat embossed]**
R508	REGULAR	[odd rectangular shape, yellow on red, picture of a soldier]
R510	REGULAR	[rectangular, embossed]
R512	REID'S EXTRA	[octagonal red & black on yellow]

R514	RELIABLE	[rectangular, embossed]
R516	RELIABLE	[round, red on black, picture of horse]
R518	REMEMBER THE MAINE	[rectangular, picture of the American flag]
R520	REPORTER	[oval, black on red]
R522	RESCUE	[diamond, black on red, picture of dog]
R524	RETURN 20 TAGS FOR KNIFE	[rectangular, Wissinger Tob Co. Louisville, KY.]
R526	REX	[rectangular, yellow on black, A.J. Long, Macon, Ga.]
R527	REX	[oval, embossed]
R528	REX	[round, silver on red]
R530	**REYNOLDS NATURAL LEAF**	**[oval, white on blue]**
R532	REYNOLDS TWIST	[round, yellow on red]
R534	R.J. REYNOLDS 8 oz	[round, embossed]
R536	**RIC RAC**	**[round, embossed]**
R538	**E. RICE GREENVILLE**	**[rectangular, embossed]**
R539	E. RICE GREENVILLE	[rectangular, black on yellow]
R540	RICE'S BEST	[diamond, black on red]
R541	RICE & VAUGHAN	[oval, red on black]
R542	RICHLAND	[oval, yellow on black]
R544	**RICH & RIPE**	**(a) [die-cut leaf, yellow on dark brown]** **(b) [yellow on orange]**
R546	**RICH & RIPE**	**[round, yellow on red, Taylor Brothers, Inc.]**
R548	RICHMOND	[round, black on red, J. Wright Co. Richmond, Va.]
R550	RICHMOND BEST NAVY	[round, black & red on yellow]
R552	RICHMOND-CAVENDISH-COX	[oval, black on yellow]
R554	RIFLE	[die-cut, embossed, a long rifle]
R556	RIGHT OF WAY	[round, black on multicolored, picture of American flag, Taylor Spencer & Co.]
R558	RING	[die-cut shape of a ring, embossed]
R560	RING	[round, embossed]
R562	RING COIL HOT CAKE	[rectangular, red & green on yellow]
R564	RING LEADER	[round, black on red]
R566	RINGER	[round, embossed hole in center, Greenville]
R568	THE RINK	[oval, red on tan]
R570	RIO	[diamond, embossed]
R572	RIPE AND JUICY	[die-cut leaf, yellow on brown]
R574	**RIPE ORANGE**	**[round, black & orange on yellow, picture of an orange]**
R576	RIPE ORANGE	[round, black & orange on yellow, picture of an orange, Turner Powell & Co.]
R578	RIPE ORANGE	[round, black & orange on yellow, picture of orange, Adams Powell & Co.]
R580	**RIPE PEACH**	**[round, black on yellow & orange picture of peach, Rob't. Harris & Bro.]**
R581	RIPE PEACHES	[rectangular, yellow on red, picture of peaches, Taylor Bros.]
R582	**RIPE PEACHES**	**[rectangular, yellow on red & green, picture of peaches, Taylor Bros.]**
R584	RIPE PEACHES	[rectangular, yellow on red, picture of peaches, Ogburn]
R586	RIPE PEACHES	[round, yellow on red, picture of peaches]
R588	RIPE & READY	[round, black on yellow, picture of fruit, J.B. Taylor Tob Co. Leaksville, N.C.]
R590	**RIPPER**	**[large oval embossed]**
R592	**RIPPER**	**[small oval embossed]**
R594	RIPPLE	[round, embossed]
R596	RIPPLING BROOK	[round, black on yellow & brown, picture of a horse, G.R.T. Co.]
R598	**RISE & SHINE**	**[rectangular, square, white on red]**
R600	**RISE & SHINE**	**[oval, white on red]**
R604	**RISE & SHINE**	**[oval, red, yellow & green, picture of sunrise, Wilson & McCallay Tob]**
R606	RISE & SHINE	[oval, red, yellow & green, picture of sunrise]
R608	RISING SUN	[octagonal embossed]
R610	RIVET	(a) [round, silver on blue, picture of rivet, J.B. Pace Tob. Co.] (b) [gold on blue]
R612	ROANOKE	[rectangular, embossed]
R614	ROANOKE	[rectangular, white on red, Butler & Bosher]
R616	**ROB ROY**	**[round, embossed]**
R618	ROB ROY	[oval, black on red, Booker Tob. Co. Lynchburg, Va. U.S.A.]
R620	ROB ROY	[round, black on red]
R622	ROBT. HARRIS	[round, picture of canary]
R624	**ROBIN RUFF**	**[rectangular, embossed]**
R626	ROCK BOTTOM	[diamond, embossed]
R628	**ROCK AND RYE**	**[round, embossed]**
R629	**ROCK AND RYE**	**[rectangular, black on yellow, Bailey Bros.]**
R630	**ROCK AND RYE**	**[rectangular, embossed, Bailey Bros.]**
R632	**ROCK AND RYE**	**[round, red on green & white, picture of a clover leaf]**
R634	ROCK AND RYE	[die-cut bottle shape black on red]
R636	**ROCKET**	**(a) [die-cut rocket, black on yellow, Trade Mark D. Scotten & Co.]** **(b) [black on red]** **(c) [black on green]** (d) [black on white]
R637	ROCKET TWIST	[rectangular, embossed, picture of rocket]
R638	ROCKY FORD TWIST	[round, black on yellow]
R640	ROCKY HILL TWIST	[round, red on black]
R642	ROD	[rectangular, octagon, black on yellow, Rosenfeld]
R644	ROD	[oval, red on tan]
R646	RODA	[rectangular, red on yellow, N.C.]
R648	ROGER WILLIAMS	[round, red & black on cream]
R650	**ROLL CALL**	**[round, red on yellow]**
R652	**ROLL CALL**	**[round, red on yellow, Strater Bros.]**
R654	ROLL CALL	[round, black on yellow]
R656	ROLL CALL SMOKE	[oval, black on red, Strater]
R658	**ROLL ON**	**[round, embossed wheel]**
R660	ROLLY BOLLY	[round, black on yellow, picture of a pony]
R662	**ROOSTER**	**[die-cut, embossed rooster]**
R664	ROOSTER	[round, multicolored picture of a rooster]
R666	**ROOTER**	**[die-cut R black on red, S.F. Hess & Co.]**
R668	**ROPE**	**[die-cut shape, brown & black]**
R670	ROPE TWIST	[round, black on blue]

R672	ROSE	[round, embossed, Leidy]
R673	ROSE BUD	[octagonal, embossed]
R674	ROSE BUD	[round, black on red, 9 inch 5s, by Geo. O. Jones, Ridgeway, Henry County, Va.]
R675	ROSEBUD	[round, red on yellow, picture of green leaves & red rose]
R676	**ROSEBUD**	**(a) [round, large black on red, picture of Rose, Tatlor Bros. Inc.]** **(b) [round, small]**
R677	ROSEBUD	[oval, embossed, picture of rose]
R678	**ROSEBUD**	**[die-cut of a rosebud, red & green]**
R680	ROSEBUD	[rectangular, embossed, Palmer]
R681	ROSEBUD	[round, picture of rosebud, S.A. Ogburn]
R682	**ROSEBUD**	**[rectangular, gold on red]**
R683	ROSEBUD	[round, red on white, picture of picture of green leaves & red rose]
R684	ROSE LEAF	[large rectangular, black on yellow, P. Lorillard & Co., Pat. 1888]
R686	ROSE LEAF	[die-cut leaf, black on green, P. Lorillard Co.]
R688	**ROSE LEAF BRIGHT**	**[large rectangular, black on yellow, P. Lorillard & Co., Pat. 1888]**
R690	ROSE RED LEAF	[large rectangular, black on yellow, P. Lorillard & Co. Pat. 1888]
R692	ROSE RED LEAF	[round, black on yellow]
R694	**ROSE RED LEAF**	**[square, black on yellow]**
R696	**ROSE TREE**	**[rectangular, small black on red, C. & E.]**
R697	ROSE TWIST	[round, embossed]
R700	ROSEMARY	[round, red on cream]
R702	ROSENFELD'S VALUE	[rectangular, black on yellow]
R704	ROUGH COUNTRY FIRED	[rectangular, black on yel low]
R706	ROUGH & READY	[rectangular, red on black, Deshazo, Spencer, Va.]
R707	ROUGH & READY	[rectangular, hexagon, black on yellow, Bailey Bros.]
R708	ROUGH & READY	[McAlpin]
R710	ROUGH RIDER	[round, picture of man on horse, with flag, Knaffl]
R712	**ROUGH & TOUGH**	**[rectangular, embossed]**
R714	**ROUMANIA FLAG**	**[oval, blue, yellow & red flag on white]**
R716	ROUND TOP	[rectangular, red on black, picture of soldier]
R718	ROUSER	[rectangular, embossed, B & L]
R720	ROVER	[diamond, black on green]
R722	ROVER	[round, gold on red]
R724	ROWLETT'S BOURBON FIRED	[rectangular, octagon, black on red]
R726	**ROWLETT'S PEACH**	**[round, black on yellow, red peach in center]**
R728	ROWLETT'S TWIST	[Taylor Co]
R730	ROX	[rectangular, black on yellow]
R732	**ROYAL**	**[round, embossed, L.S. & Co.]**
R733	ROYAL	[square, black on yellow, Bendall, picture of tiger]
R734	ROYAL	[diamond, embossed, Robert Harris & Bros., Reidsville, N.C.]
R736	ROYAL	(a) [rectangular, black on red] (b) [red on yellow]
R738	ROYAL AAAA	[rectangular, embossed]
R740	ROYAL ARCH	[rectangular, shield, black on red]
R742	ROYAL ARMS	[die-cut heart, W.D. McDonald & Co. Montreal]
R744	ROYAL BLUE	[rectangular, picture of train]
R746	**ROYAL BUMPER**	**[rectangular, scalloped, black on red, picture of a goat]**
R748	**ROYAL COLORS**	**[octagonal black on yellow, T.C. Williams Co.]**
R750	**ROYAL CROWN**	**[round, black on red]**
R751	ROYAL CROWN	[round, embossed]
R752	**ROYAL EAGLE**	**(a) [die-cut oval heart, black on red]** **(b) [red on red black out line]**
R754	**ROYAL GEM**	**[round, embossed]**
R756	ROYAL HENRY Co.	[rectangular, bar, black on red]
R758	ROYAL MINT	[round, white on green, gold center, picture of man]
R760	**ROYAL OAK**	**[round, black on dark green]**
R762	**ROYAL OAK**	**[round, embossed]**
R764	ROYAL OAK	[oval, embossed]
R766	**ROYSTERS BEST**	**[round, red on yellow, picture of a man, G.R.T. Co.]**
R768	**R.R. PLUG**	**[die-cut, embossed, railroad locomotive]**
R770	**RUBBER**	**[round, embossed]**
R771	RUBY	[oval, embossed]
R772	RUBY	[oval, black on red, Henry Co., Va.]
R773	RUBY	[round, embossed]
R774	**RUDDER**	**(a) [round, blue on white, picture of a rudder, Scotten Dillon Company]** (b) [black on yellow]
R776	**RUDY**	**[round, black on yellow, picture of a ship's wheel, Trade Mark]**
R777	**RUDY**	**[round, gold on red, picture of a ship's wheel, Trade Mark]**
R778	**RUDY**	**[large round, gold on black]**
R780	**RUDY**	**[smaller round, gold on black]**
R782	RUDY-NICK-FRED	[triangular, black on yellow, circle in center]
R784	RUFUS SANDERS	[rectangular, black on red]
R786	RULER	[oval, gold on red]
R788	RULER LIGHT	[round, yellow on red]
R790	**RUM**	**[die-cut barrel, black on red]**
R792	RUM PUNCH	[round, yellow on red, picture in center, United States Tob Co.]
R794	**RYE**	**[small rectangular, embossed]**

S

S101	**S**	**[die-cut embossed]**
S102	**S**	**[die-cut S black on red, Finzer]**
S104	**S**	**[small round embossed]**
S105	**S**	**[rectangular, black on**

		green]
S106	S	[round, red on green, S, est. 1873]
S107	S	[round, large embossed S in center, dots around rim]
S108	S.A.L.	[oval, red on cream]
S110	S.E.	[round, embossed]
S112	**S H H**	**[die-cut shield embossed]**
S114	S.M. & Co.	[rectangular, banner, yellow & black on green]
S116	**S O**	**[small rectangular, cut-out letters]**
S117	S O	[large rectangular, cut-out letters]
S118	**S. R.**	**[round, embossed]**
S120	**S & S**	**[banner cut-out]**
S122	S S S	[rectangular, embossed]
S124	S.T.	[round, embossed, black on yellow, Sweet Chewing]
S125	S & W	[die-cut letters, red, white & blue]
S126	S & W	[die-cut letters, red, yellow & blue]
S128	S & W.	[die-cut letters, black on red & yellow, Schoolfield & Watson]
S130	S & W	[round, Sanford & Williams]
S131	SABRE	[oval, red on white, picture of sabre sword]
S132	SABRE	[oval, black on yellow, picture of sabre sword]
S134	**SABRE SWORD**	**[die-cut shape of sword, embossed]**
S136	SABER SWORD	[rectangular, embossed, picture of sword]
S138	SAILOR BOY	[die-cut shield, embossed]
S140	SAILOR JACK	[round, black on yellow, picture of sailor]
S141	SAILOR JACK	[round, green on white, picture of sailor]
S142	SAILOR KNOT	[round, tan on yellow]
S144	**SAILOR'S DELIGHT**	**[round, black on green, P. Lorillard & Co.]**
S146	SAILOR'S DELIGHT	[round, red on yellow]
S148	SAILOR'S HOPE	[rectangular, white on yellow]
S150	SAILOR'S JOY	[round, blue on cream]
S152	**SAILOR'S PRIDE**	**[round, white on black banner]**
S154	SAILOR'S SOLACE	[rectangular, embossed]
S156	SAILOR'S TWIST	[rectangular, brown on yellow]
S158	**SALESMAN**	**[rectangular, black on red, picture of a man]**
S160	SALLIE JAY	[rectangular, black on yellow & red]
S162	SALLIE JAY	[rectangular, embossed]
S164	SALLIE STULTZ	[rectangular, silver on black]
S166	**SALLY JAY**	**(a) [small octagonal black on gold & red, Pegram & Penn.]** **(b) [larger]**
S168	**SALLY JAY**	**[octagonal black on gold & red, Pegram & Penn Tobacco]**
S170	SALLY JAY	[rectangular, odd, black on yellow]
S172	SALM'S NATURAL LEAF	[round, black on green, Fulton Hill Tobacco Co.]
S174	SALMON HANCOCK & CO	[die-cut shield, embossed]
S176	**SALVADOR**	**[oval, red & blue flag on white]**
S178	SAM BASS	[rectangular, black on yellow]
S180	SAM JONES	[round, black on red, picture of man]
S182	SAM JONES BEST	[rectangular, black on red]
S184	SAM JONES TWIST	[rectangular, red on yellow]
S186	**SAM REID**	**[round, embossed, 5 in circle of stars, Tob M'F'G. Co.]**
S188	**SAMBO**	**[die-cut banner, embossed]**
S190	SAMBO	[rectangular, embossed]
S192	SAMSON	[diamond, black on red]
S194	SAMSON	[diamond, black on red, Dick Middleton & Co.]
S196	**SAMSON'S BIG 4 TWIST**	**[round, red on white]**
S198	**SANTA CLAUS**	**[rectangular, black on red, McAlpin]**
S200	SANTA CLAUS	[rectangular, embossed]
S202	**SATIN SLIPPER**	**(a) [oval, blue on white picture of a slipper, United States Tob. Co.]** (b) [green on white]
S203	SATISFACTION	[short rectangular, embossed]
S204	**SATISFACTION**	**[rectangular, embossed, D.H. Spencer & Son]**
S205	SATISFACTION	[long rectangular, embossed]
S206	SATURDAY NIGHT	[round, black on yellow & red]
S208	SAUNDERS	[round, scalloped, black on red]
S210	SAUNDERS	[hexagon, black on yellow]
S212	**SAW**	**[die-cut, buck-saw, embossed]**
S213	**SAW**	**[die-cut, hand-saw, small handle, embossed]**
S214	SAW	[die-cut, hand-saw, large handle, embossed]
S216	**SAW BUCK**	**[die-cut two X X & rectangular, banner embossed]**
S218	SAW LOG	[round, embossed, picture of saw in a log]
S219	SAW MILL	[round, embossed]
S220	S. BAER & SONS	[octagonal, embossed]
S222	SCALPING KNIFE	[rectangular, embossed]
S223	SCAT	[hexagon embossed]
S224	SCAT	[die-cut, cat face, black on red]
S226	**SCHNAPPS**	**[rectangular, embossed, red on black]**
S228	**SCHNAPPS**	**[rectangular, red on black]**
S230	SCHOOLFIELD'S JOE BOWERS	[die-cut banjo, black on red, Schoolfield & Watson, Bowers]
S232	SCHOOLFIELD & WATSON	[rectangular, black & red on green]
S234	SCHOONER	[die-cut, embossed, ship]
S236	SCHOSSER'S	[rectangular, yellow on red]
S238	SCRAPPER	[oval, black on yellow, picture]
S240	SCRAPPER	[round, embossed]
S242	SCOTCH LASSIE	[round, multicolored picture of girl, J.N. Wyllie & Co.]
S244	SCOTT & Co.	[oval, embossed]
S245	SCOTT'S HEEL TAP	[round, yellow on red]
S246	**SCOTTEN**	**[rectangular, embossed]**
S248	**SCOTTEN**	**(a) [rectangular, blue on white]** (b) [red on white] (c) [red on yellow] (d) [yellow on black]
S250	**SCOTTEN'S**	**(a) [rectangular, small, blue on white]** **(b) [black on red]** **(c) [black on yellow]** **(d) [rectangular, very small, black on red]**
S252	SCOTTEN TOBACCO	[round, red & green on yel

No.	Name	Description
	Co.	low]
S254	SCOTTSVILLE TWIST	[round, black on red]
S256	SCOUT 12	[die-cut shield, black on silver]
S258	**SCOUT W & W**	**[round, embossed]**
S260	**SCREW**	**[die-cut large screw embossed]**
S262	SCREW	[round, embossed]
S263	SCREW	[round, black on red, picture of screw]
S264	**S-E FIG**	**[round, black on yellow, Spilman-Ellis Tob Co.]**
S266	SEA ISLAND	[odd round yellow on brown]
S268	SEAL	[round, black on yellow, Perfection Tobacco]
S270	SEAL BRAND	[round, embossed]
S272	SEAL OF DURHAM	[die-cut barrel, black on yellow, Whitted]
S274	SEAL OF N.C.	[round, red on blue & white]
S276	SEAL OF TENNESSEE	[hexagon, black on blue & white]
S278	SEAL ROCK SLUG	[round, embossed]
S280	**SEAL SKIN**	**[round, black on red]**
S282	SEAL SKIN	[rectangular, odd, black on red, picture of seal]
S284	SEAL SKIN DARK	[round, black on red]
S286	SECOND BATTLE 1900	[round, picture, W.J. Bryan]
S288	**2ND BASE**	**[die-cut shield blue & red on white, picture of two bats & a mask]**
S289	SELECT	[oval, embossed, A.M.W.]
S290	**SELECT**	**[rectangular, round embossed]**
S292	**SELECT**	**[rectangular, black on yellow, Lash]**
S294	SELECT CHEW	[oval, black on red]
S296	**SELECT 3 PLY**	**[oval, embossed, J. Herget]**
S298	SEMPER IDEM	[die-cut leaf, gold on green]
S300	SENATE	[round, black on red, L.B. Tob Co.]
S302	**SENATE**	**[round, white on blue, picture of a building]**
S304	SENATE	[round, black on red]
S306	**SENATE PLUG**	**[round, white on blue]**
S308	**SENATOR TWIST**	**[round, embossed]**
S310	SENSATION W T H	[round, embossed]
S311	SENSATION	[rectangular, odd embossed]
S312	SENSIBLE	[round, embossed]
S314	**SENTINEL**	**[octagonal black on red, picture of a owl, J.H. Harrissons]**
S316	SENTINEL	[round, black on red picture of a owl, J.H. Harrissons]
S318	SETTER	[oval]
S320	7	[die-cut 7]
S321	7¢ BOSTON CREAM	[diamond, black on red, H.H. Kohlsaat]
S322	7¢ BOSTON CREAM	[oval, embossed, H. Piper]
S324	7 oz.	[odd shape, embossed, Obrechts]
S326	7-11	[round, red on yellow]
S327	76 PLUG	[die-cut a shield embossed, C.W. Allen]
S328	SEVEN UP	[round, black on yellow, picture of playing cards, Pegram & Penn]
S330	SHAME	[hexagon black on yellow]
S332	SHAME LIGHT	[rectangular, brown on yellow]
S334	SHANGHAI	[round, embossed, picture of chicken]
S336	SHAWNEE SUN CURED	[oval, black & orange on yellow, Gravely Tob Co.]
S338	SHEEP	[oval, black on red, picture of a sheep]
S340	SHEEP'S HEAD	[round, white on blue & black, picture of a sheep's head]
S342	**SHEILA**	**[rectangular, banner embossed]**
S344	SHELL ROAD	(a) [rectangular, black on green, picture of buggy & horse, R.A. Patterson Tob Co.] (b) [same smaller tag]
S346	SHELL ROAD	[rectangular, multicolored, picture of buggy & two horses, R.A. Patterson Tob Co.]
S348	SHENANDOAH	[diamond, black on yellow, W. & Co. 12 in 4s]
S350	**SHENANDOAH**	**[diamond, black on yellow]**
S352	SHEPHERD PLUG	[rectangular, black on red]
S354	**SHEPHERD TWIST**	**[round, black on yellow, picture of a dog]**
S356	SHERRIL TOBACCO COMPANY	[round, embossed, 16 to 1, picture of eagle]
S358	**SHIELD**	**[die-cut, large shield]**
S360	SHIELD	[round, embossed]
S363	**SHINER**	**[round, embossed, scalloped edge, B.S. & Co.]**
S364	SHINERS	[rectangular, octagon, embossed]
S366	SHINERS TWIST	[round, red on black]
S368	SHIP AHOY	[rectangular, shield embossed]
S370	**SHIP MATE**	**[die-cut, shield, embossed]**
S372	SHIP'S WHEEL	[die-cut, embossed, a ship's wheel]
S374	SHOO FLY	[rectangular, embossed]
S376	SHORT BIT	[oval, red on cream]
S378	SHORT BIT	[round, embossed]
S380	SHOT	[octagonal embossed]
S382	**SHOT**	**[octagonal, red on yellow]**
S383	SHOT TOWER	[die-cut tower embossed]
S384	SHOVEL	[die-cut a shovel embossed]
S386	**SHOW DOWN**	**[rectangular, gold on red, Hancock Bros & Co.]**
S388	SHRIMPS	[rectangular, picture of a shrimp, white on green]
S390	SHRIMPS	[rectangular, black on red]
S392	**SHURE MIKE**	**[rectangular, yellow on black]**
S394	**SICKLE**	**[large die-cut shape, sickle, black on red]**
S396	SICKLE	[round, black on green, picture of sickle, P. Lorillard & Co.]
S397	SIFE	[oval, red on cream]
S398	SIGHT DRAFT	[die-cut, hand & banner]
S400	**SIGNAL**	**(a) [round, black on white, red dot in center]** (b) [black on yellow]
S402	SIGNET TOBACCO	[round, black on yellow, British Aust Tob Co. Sydney]
S404	SIKE'S BRAG	[rectangular, embossed]
S408	**SILK**	**[rectangular, embossed]**
S410	SILK TIE	[die-cut bow tie, black on yellow]
S412	SILK TIE	[die-cut bow tie, black on white, Hickok]
S414	**SILK VELVET**	**[rectangular, black on red, Robard's Natural Leaf]**
S415	**SILVER AGE**	**[round, embossed, picture of man's head & two stars]**
S416	SILVER BILL	[rectangular, embossed]
S417	SILVER COIN	[round, embossed]
S418	**SILVER COIN**	**[round, embossed, Liberty**

No.	Name	Description
		& Stars]
S419	SILVER COIN	[rectangular, shield embossed]
S420	**SILVER COIN**	**[rectangular, blue & white on red, white & blue, Lottier's Navy]**
S422	SILVER COIN	[rectangular, blue & white on red, white & blue, Lottier's Clubs]
S424	SILVER COIN NAVY	[round, gold on red, Lottier]
S425	SILVER CORD	[rectangular, embossed]
S426	**SILVER CORD**	**[rectangular, embossed, W.C. Hamilton & Co.]**
S427	**SILVER CORD**	**[round, black on red]**
S428	SILVER CORD	[round, black on cream]
S430	**SILVER CRESCENT**	**[rectangular, black on red, picture of silver crescent moon]**
S432	SILVER DICK	[oval, black on red]
S434	**SILVER DIME**	**[round, embossed]**
S435	**SILVER DIMES**	**[round, black on yellow, picture of three coins]**
S436	SILVER DIMES	[round, silver on blue, picture of three coins]
S437	SILVER FORK	[oval, embossed]
S438	SILVER KEY	[die-cut key, embossed]
S440	**SILVER LAKE**	**[octagonal, gold on black]**
S442	**SILVER MOON**	**[die-cut half moon, silver on black]**
S444	SILVER NAVY	[diamond shape, black on yellow]
S448	SILVER QUARTER	[round, embossed]
S450	SILVER QUARTER	[round, white on blue, picture of a coin]
S452	**SILVER SEAL**	**[round, embossed scalloped]**
S454	**SILVER SEAL**	**[small round cut-out, scalloped]**
S455	**SILVER SEAL**	**[large round cut-out, scalloped]**
S456	**SILVER SPOON**	**[oval, embossed, picture of a spoon]**
S458	SILVER SPOON	[rectangular, black on red, picture of silver spoon, Made by Hancock Bros & Co.]
S460	**SILVER TAG**	**[triangular, embossed]**
S462	**SILVER TIP**	**[octagonal, embossed]**
S464	**SILVER WING**	**[diamond, embossed]**
S466	SIR JOSEPH K. C. B.	[rectangular, shield, embossed]
S458	SIR TOBY	[round, black on red, Butler & Bosher, Richmond, Va.]
S469	6¢ Lbs. BOSTON CREAM	[round, embossed]
S470	16 TO 1	[odd shape, embossed]
S472	**66**	**[rectangular, cut-out]**
S474	**SKATING RINK**	**[rectangular, black on red, picture of a man skating]**
S476	SKI	[rectangular, blue on silver]
S478	**SKINNER**	**(a) [small rectangular, black on yellow, picture of a man standing] (b) [large]**
S480	**SKYLARK**	**[oval, embossed, picture of bird]**
S482	SKYLARK	[oval, black on red]
S484	SKYLIGHT	[die-cut quarter round, red on yellow, red trim]
S486	**SKYLIGHT**	**[die-cut quarter round, red on yellow]**
S488	SKY LIGHT	[round, yellow on blue]
S490	SLABS	[oval, black on red]
S492	SLAP JACK	[odd shape, black on red]
S494	**SLAP JACK**	**[die-cut B black on red]**
S496	**SLEDGE**	**[die-cut sledge hammer, black on silver & yellow]**
S497	SLEDGE	[die-cut shape, sledge hammer, yellow on brown]
S499	SLEDGE HAMMER	[round, black on red, picture of arm hammer in hand, J.W. Daniel Tob. Co.]
S500	SLEEPY	[rectangular, banner, embossed]
S502	**SLEEPY TOM**	**[rectangular, odd shape, embossed]**
S504	SLEEPY VERN	[rectangular, black on yellow]
S506	SLIDE KELLY SLIDE	[round, red on black, Jno H. Flood, Lynchburg, Va.]
S508	SLIGHT DRAFT	[round, embossed]
S510	SLIM JIM	[round, Penn & Rison]
S512	SLIM MAN	[round, embossed]
S514	**SLY COON**	**[rectangular, embossed]**
S516	SMACK	[oval, embossed]
S518	SMILAX	[round, yellow on green, P. Lorillard & Co.]
S519	**SMILAX**	**[round, large black on red]**
S520	SMOOT'S CHOICE	[rectangular, black on red]
S521	SNAG	[round, black & red on yellow]
S522	**SNAKE ROOT**	**[oval, brown on yellow, picture of snake]**
S523	SNAKE'S	[round, picture of snake's face]
S524	**SNAP**	**[round, embossed]**
S525	SNAP	[round, embossed rope edge]
S526	SNAPPER	[oval, white on red]
S528	SNAP SHOT	[rectangular, odd embossed, W.C. Hort, Tob. Co., Danville, Va.]
S530	SNIPE	[round, embossed, Gravely & Miller]
S532	**SNOW FLAKE**	**[round, embossed]**
S534	**SNOWDEN**	**[round, black on green]**
S536	SNOWDEN	[round, scalloped black on red]
S538	SNOW DOWN	[rectangular, gold on red, Hancock & Bros]
S540	SNOWSHOE	[die-cut dog's head, brown & yellow]
S542	SNUFFY	[small round scalloped, black on yellow, cartoon picture of man]
S544	**SOLDIER BOY**	**[round, black on red]**
S546	SOLDIER'S & SAILOR'S PLUG	[rectangular, picture of monument, R.H. Boykin & CO.]
S548	**SOLID**	**[oval, embossed, Herget's 69 O'Donnell]**
S550	SOLID	[oval, black on yellow]
S552	**SOLID COMFORT**	**[round, black on green, picture of a man smoking a pipe]**
S554	SOLID MEAT	[round, white on red, Crews, Walkertown, N.C.]
S556	SOLID MEAT	[round, red on yellow]
S558	SOLID PLUG	[round, embossed]
S560	SOLID SHOT	[rectangular, odd, embossed]
S562	SOLID SOUTH	[rectangular, black & green on white]
S564	SOLID TWIST	[rectangular, embossed]
S566	SOMETHING BETTER	[hexagon, black on red]
S568	**SOMETHING EXTRA**	**[octagonal, gold on dark blue]**
S570	SOMETHING GOOD	[round, embossed]
S572	**SORG & Co. P.J.**	**[die-cut banner, embossed]**
S574	THE SORG & Co. P.J.	[die-cut shape, anchor, black on red]
S576	**THE SOUTH**	**[oval, embossed]**
S577	SOUR MASH	[rectangular, banner em-

		bossed]
S578	SOUTH BOUND	[round, black on red]
S579	SOUTH LAND	[round, embossed]
S580	SOUTH LAND	[octagonal embossed]
S582	SOUTH SEAS	[round, blue on cream & black]
S584	**SOUTH BOUND**	**[round, black on red]**
S586	SOUTH CAROLINA	[rectangular, embossed]
S588	SOUTHDOWN	[rectangular, banner embossed]
S590	SOUTHERN BEAUTY	[rectangular, black on yellow, picture of girl]
S592	SOUTHERN COMFORT	[round, embossed]
S594	SOUTHERN FAVORITE	[oval, black on yellow, J.L. King & Co.]
S596	SOUTHERN GEM	[round, black on red, D.F. Kings]
S598	SOUTHERN LEAGUE	[diamond, red on yellow, L.V. & Co.]
S600	SOUTHERN PRIDE	[round, embossed]
S601	SOUTHERN PRIDE	[round, red on green & white]
S602	SOUTHERN QUEEN	[round, picture of woman & children, Kerner Bros.]
S604	**SOUTHERN ROSE**	**(a) [oval, black on green]** (b) [black on yellow]
S606	SOUTHERN SEAL	[die-cut leaf, black on yellow]
S608	SOUTHERN SEAL	[round, picture of man on horse, Jenkins]
S610	**SOUVENIR**	**[die-cut banner, embossed]**
S612	SPADE	[die-cut black spade]
S614	SPANKER	[round, black on yellow & red, picture of mom spanking baby, J.G. Flynt Tob Co.]
S616	SPANKER	[rectangular, black on red, L.H.R. & Co.]
S618	SPANKER TOBACCO	[rectangular, black on yellow, picture of animals, J. Wilson]
S620	SPARGER BROS. NATURAL LEAF	[die-cut leaf, black on yel low]
S622	SPARGER BROS. PLUG	[round, black on tan]
S624	**SPARK PLUG**	**[rectangular, white on red]**
S626	**SPARKS 5 CENTS**	**[oval, black on red]**
S628	SPARROW BROS.	[rectangular, round in center, picture of bird]
S630	SPARROW & GRAVELY	[rectangular, round in center, picture of bird, Tob Co.]
S631	SPARROW'S MESSENGER	[oval, black on yellow, pic ture of man running]
S632	SPARROW'S POCKETPIECE	[rectangular, black on yel low, picture of a bird]
S634	**SPEAR HEAD**	**(a) [die-cut spear head, black on red, The P.J. Sorg Co.] (b) [The P.J. Sorg Co., Contl. Tob. Co. Suc.] (c) [The P.J. Sorg Co., The American Tob Co. Suc.]**
S636	SPEAR HEAD	[die-cut shape, spear head embossed]
S638	SPECIAL HENRY COUNTY	[rectangular, odd, black on red, Geo O. Jones & Co.]
S639	SPECIAL DELIVERY	[rectangular, picture of man, C.W. Coan & Co.]
S640	SPECIAL	[rectangular, black on red, Weissinger]
S642	**SPECIAL**	**(a) [rectangular, red on yellow, B.F. Gravely & Sons] (b) [red on white]**
S643	SPECIAL	[round, embossed]
S644	SPECIAL DRIVE	[round, embossed]
S645	SPECIAL DRIVE	[octagonal, embossed, B.F. Hanes]
S646	**SPECIAL DRIVE**	**[round, white on red, blue on white, F.M. Bohannon]**
S648	**SPECKLED BEAUTY**	**(a) [oval, black on yellow & blue, picture of a fish, P.H. Hanes & Co.] (b) [black on yellow & orange]**
S650	**SPENCE S/L**	**[square, embossed]**
S652	**SPENCER NATURAL LEAF**	**[rectangular, white on dark blue]**
S654	**SPENCER BROS.**	**[die-cut S black on red]**
S656	SPENCER BROS. NATURAL LEAF	[rectangular, black on yel low & brown, picture of a leaf]
S658	SPENCER'S BEST	[round, embossed]
S660	SPENCER'S TWIST	[round, embossed]
S662	SPIKE	[die-cut shape, large spike, black on red]
S664	SPILMAN-ELLIS	[rectangular, black on yellow]
S665	SPINNER	[rectangular shield embossed]
S666	**SPLENDID**	**[oval, black, blue, red, picture of two flags, one American, P. Lorillard & Co.]**
S667	SPLENDID	[oval, picture of one flag, P. Lorillard & Co.]
S668	**SPLENDID**	**[oval, black, red, white on green, picture of two flags, P. Lorillard & Co.]**
S670	SPOON	[die-cut spoon, black on red]
S672	SPORT	[rectangular, red on yellow, A. Davis & Co.]
S674	STOT CASH	[rectangular, embossed]
S676	**SPOT CASH**	**[round, orange on yellow, picture of a hand with a coin, J.N. Wyllie & Co.]**
S678	SPOTTED DOG	[rectangular, black & brown on white]
S680	**SPOTTED FAWN**	**[round, embossed, picture of a deer]**
S682	SPOTTED FAWN	[die-cut fawn, embossed]
S684	SPOTTED HOUND	[round, embossed]
S686	SPOTTED SETTER	[rectangular, black on yellow, picture of dog, R. Everhart & Co.]
S688	**SPOTTED TAIL**	**[round embossed, picture of an arrow & tomahawk]**
S690	SPOTTED TIGER	[die-cut tiger, embossed]
S692	**SPRING**	**[rectangular, small, black on red]**
S694	**SPRING CHICKEN**	**[rectangular, black on red, picture of chick coming out of egg]**
S696	**SPRING CHICKEN**	**[round, black on yellow, multicolored picture of a chicken]**
S697	**SPUR**	**[large, die-cut spur]**
S698	**SPUR**	**[die-cut a spur]**
S699	**SPUR**	**[die-cut spur & S embossed]**
S700	SPURR'S BEST	[round, embossed]
S702	SPURR'S LEADER	[rectangular, embossed]
S704	**SPY**	**[die-cut cross, embossed]**
S706	**SQUARE**	**[die-cut shape, embossed framing square]**
S708	**SQUARE & COMPASS**	**[die-cut square & compass]**
S710	**SQUARE DEAL**	**[square, embossed, red dot in center]**
S712	SQUARE DEAL	[square, black on red]
S713	SQUARE DEAL	[odd square, embossed, picture of square]
S714	**SQUARE AND**	**[square, yellow on red,**

No.	Name	Description
	HONEST	**Penn]**
S716	**SQUARE KNOT**	**[die-cut, embossed rope knot]**
S718	SQUARE MEAL	[round, embossed]
S718	STAFF OF LIFE	[round, black on red]
S720	STAFFORD E.J. & A.G.	[oval, picture, Greensboro]
S722	STAFFORD'S LEADER	[oval, embossed]
S724	**STAG**	**[diamond, black on red, picture of a deer]**
S725	**STAG**	**[square, embossed]**
S726	**STAG**	**[rectangular, odd, black on red, picture of snow shoes, E.T.Co. LTD.]**
S728	STAG	[rectangular, odd, black on red, picture of a deer, tag Made in U.S.A.]
S730	STAGE HORSE	(a) [round, black on red, picture of horse & boy, Blackburn Harvey & Co.] (b) [horse & girl]
S732	STAGE HORSE	(a) [round, black on red, Sam Blackburn & Co. picture of horse & boy] (b) [round, black on yellow]
S734	STALLION	[rectangular, black on green, picture of a jockey & horse]
S736	**STAND PAT**	**[oval, white on red]**
S738	**STANDARD A**	**[diamond, embossed]**
S739	STANDARD	[diamond, embossed, Marvin]
S740	**STANDARD**	**[rectangular, shape with frame embossed]**
S741	**STANDARD**	**[die-cut, embossed knife]**
S742	**STANDARD**	**[rectangular, embossed]**
S743	**STANDARD**	**[round, black on red, L. Waterbury & Co., New York]**
S744	STANDARD	[round, embossed, Gravely]
S746	STANDARD	[rectangular, black on red, Geo. O. Jones]
S748	STANDARD PLUG	[round, embossed]
S750	STANDARD UNION	[round, red, white & blue]
S752	**STAPLE**	**[die-cut, embossed large staple]**
S754	**STAPLE**	**[rectangular, odd embossed, W. & D. Tob Co.]**
S756	**STAR**	**(a) [large die-cut star] (b) [medium]** (c) [small]
S758	**STAR**	**[die-cut, embossed]**
S760	STAR	[round, silver on blue, picture of star]
S762	STAR BANNER	[rectangular, banner embossed]
S764	STAR BRAND	[rectangular, multicolored, P. Lorillard & Co.]
S766	**STAR OF HENRY**	**[rectangular, red on yellow, B.F. Gravely & Sons]**
S768	**STARS & BARS**	**(a) [round, blue on white & red, picture of Confederate flag, Taylor Bros] (b) [blue on yellow & red]**
S770	STARS & BARS	[oval, small black on yellow]
S772	**STARRY CROWN**	**[rectangular, gold on red, picture of a crown]**
S774	STARTER	[rectangular, black on red]
S776	**STRATER BROS**	**[diamond, black on red]**
S778	**STRATER BROS. TOBACCO CO. INC.**	**[diamond, black on red]**
S780	**STRATER BROTHERS**	**[square, black on red]**
S782	STRATERS NATURAL LEAF	[die-cut rectangular, banner black on red]
S784	**STRATERS NATURAL LEAF TWIST**	**[die-cut shield embossed]**
S785	STATE SEAL	[round, embossed]
S786	STAUNTON BELLE	[oval, white on blue]
S788	ST. ELMO	[round, embossed]
S790	**ST. GEORGE**	**[round, gold on red, Venable]**
S792	ST. GEORGE	[round, embossed, Venable]
S794	ST. LAWRENCE	[rectangular, embossed]
S796	ST. NICHOLAS	[small oval, black on yellow]
S798	**STEAM BOAT**	**[round, embossed]**
S800	**STEAM-SHIP**	**[round, black on yellow, picture of a ship facing right]**
S801	STEEL RAIL	[round, embossed]
S802	**STEEL TRAP**	**[die-cut of a animal trap, round embossed]**
S804	STEER HEAD	[die-cut shape, steer head embossed, K. & R.]
S806	STEP QUICK	[oval, green on yellow]
S808	STERLING	[round, black on red]
S810	**STERLING NAVY**	**[oval, black on red, picture of an eagle]**
S812	STERLING TWIST	[rectangular, embossed]
S814	**STEWARTS**	**[die-cut shaft of tobacco, embossed]**
S816	STIRRUP	[round, embossed]
S818	STOCKHOLDER	[rectangular, black on green & red]
S820	**STONE FENCE**	**[rectangular, black on yellow]**
S824	STONEWALL	[round, black on yellow, picture of Jackson, J.N. Wyllie & Co.]
S825	STONEWALL	[octagonal, black on red, J.N. Willie & Co.]
S826	**STOUT BOY**	**[round, white on green]**
S828	STRAIGHT TIP	[round, embossed]
S830	**STRATER**	**[small round, black on red]**
S831	**STRATER**	**[rectangular, black on red]**
S832	**STRATER**	**[medium round, black on red]**
S834	**STRAWBERRY**	**[round, red on white, picture of a strawberry]**
S835	**STRAWBERRY**	**[round, green on yellow, picture of a strawberry, R.J. Reynolds Tob Co.]**
S836	STRAWBERRY	[die-cut strawberry, black on red]
S837	STRAWBERRY	[die-cut half round, embossed, R.J.R. Tob Co.]
S838	STRAWBERRY	[round, black on red & yellow, picture of strawberry, R. J. Reynolds Tob.Co.]
S840	STRAWBERRY	[round, embossed picture of strawberry]
S841	STRAWBERRY	[die-cut strawberry, embossed]
S842	STRIKES MY EYE ALL RIGHT	[round, picture of a large eye]
S844	STRIPPED	[round, embossed]
S846	STRING	[rectangular, black on yellow, picture of horse]
S848	STRONG	[round, red on white]
S850	**STRONG BOY**	**[round, black on white, It's Pure Tobacco]**
S852	**STRONGHOLD**	**(a) [large die-cut shape hinge, yellow on red] (b) [small]**
S854	STRONGHOLD	[rectangular, embossed]
S858	STUART	[oval, green on white]
S860	STUD HORSE	[die-cut shape, horse, embossed]
S862	STUD TURNER	[rectangular, embossed]
S864	**STUNNER**	**[oval, embossed]**
S865	SUCCESS	[oval, embossed, Ralphs]
S866	**SUCCESS**	**[rectangular, odd embossed]**

S868	SUGAR BEET	[oval, black on red, picture of sugar beet]
S870	SUGAR CANE	[round, black on yellow, picture of plant, Martinsville, Henry Co., Va.]
S872	SUGAR CANE	[round, black on yellow, picture of sugar cane]
S874	**SUGAR CURED**	**[die-cut odd ham, black on yellow]**
S876	SUGAR CURED	[die-cut odd ham, yellow on black round on yellow]
S878	SUGAR LUMP	(a) [round, black on yellow] (b) [blue on white]
S880	SUGAR PEACH	[round, multicolored, picture of peach]
S882	SUGAR PLUM	[octagonal, black on yellow, picture of plum]
S883	SUGAR PLUM	[octagonal, blue & red on white, picture of plum]
S884	**SULTAN**	**[rectangular shield, embossed]**
S886	**SUMMIT**	**[die-cut Indian standing with flag, embossed]**
S888	**SUMMIT**	**[round, embossed]**
S890	**SUN**	**[round, embossed picture of the sun]**
S891	**SUN**	**[round, embossed]**
S892	SUN	[die-cut S black on red]
S893	SUNBEAM	[round, yellow on red]
S894	SUN CURED	[oval, black on red, Davis]
S895	SUN CURED	[round, black on red]
S896	**SUN CURED**	**[oval, red on yellow, in black Taylor]**
S897	SUN CURED	[diamond, black on red, Flynts]
S898	**SUN CURED**	**[round, yellow on black & red, F.M. Bohannon]**
S899	**SUN CURED**	**[oval, gold on black, Reynolds]**
S900	**SUN CURED**	**[round, black on yellow, Hancock Bros & Co.]**
S901	**SUN CURED**	**(a) [large round black on orange & blue, Brown & Williamson] (b) [small round]**
S902	SUN CURED	[round, yellow & black on black & red, Brown & Williamson]
S903	SUN CURED	[oval, black on gold & red, Allen Bros. Tobacco Co.]
S904	**SUN CURED**	**[oval, orange on black, Liipfert Scales & Co.]**
S906	**SUN CURED**	**[rectangular, black on green, H.C.T. Co.]**
S908	SUN CURED	[rectangular, black on red, S. Watkins]
S910	SUNFLOWER	[round, multicolored, picture of a sunflower]
S912	**SUN-LIGHT**	**[round, red on dark green, picture of a ships wheel, Trade Mark]**
S914	**SUN LIGHT**	**[oval, black on red]**
S915	SUN LIGHT	[round, black on red]
S916	**SUNLIGHT**	**[die-cut S embossed]**
S917	**SUNLIGHT**	**[die-cut S black on red]**
S918	SUNMAID	[die-cut crescent embossed]
S920	**SUNNYSIDE**	**[rectangular, odd, black on red, picture of setting sun]**
S922	SUNOL	[rectangular, gold on red]
S924	SUN PERCH	[rectangular, hexagon, embossed]
S926	SUNRISE	[oval, black on red, Davis]
S928	**SUN-SET**	**[die-cut S black on red]**
S930	**SUNSET**	**[die-cut shield, gold on red, Sullivan & Early Rd]**
S932	**SUNSHINE**	**[half round embossed, picture of setting sun]**
S934	SUNSHINE LIGHT	[round, embossed]
S936	**SUPERB**	**[round, embossed]**
S938	SUPERB	[round, embossed, P. Lorillard & Co.]
S940	SUPERFINE NATURAL LEAF	[die-cut leaf, black on yel low, P. Lorillard & Co]
S942	SUPERIOR 36'S	[round, black on yellow, Gazell Twist]
S943	SUPERIOR SUN CURED	[oval, black on yellow, Piedmont]
S944	SURE GOOD	[round, black on yellow]
S946	SURE HIT	[rectangular, embossed]
S948	SURE POP	[oval, black on green, picture of man with bow & arrow, W.H.I. Hayes]
S949	SURE SHOT	[rectangular, red on yellow, picture of a Indian]
S950	SURE THING	[oval, red & tan on yellow]
S951	SURPRISE	[round, black on yellow]
S952	SURPRISE	[rectangular, embossed]
S954	SURRY	[rectangular, embossed, J.F.L.A. & Co.]
S956	**SUSQUEHANNA**	**[rectangular, embossed]**
S958	SUSIN'S EXCELSIOR	[rectangular, embossed]
S959	**SWAN**	**[die-cut swan embossed, H. Hudson]**
S960	**SWAN**	**[round, embossed picture of a swan]**
S962	SWAN	[round, black on cream, picture of swan]
S964	SWANSON	[round, red & green on yellow, J.N. Wyllie & Co.]
S966	SWANSON'S PRIDE	[rectangular, red & yellow on green]
S968	SWEDEN	[oval, black on white, flag of Sweden, red, yellow & blue]
S970	SWEEP STAKE	[rectangular, picture of horse racing, Stafford]
S972	SWEEP STAKE	[rectangular, black on red, Hanes]
S974	**SWEEP-STAKES**	**[die-cut S, black on white]**
S976	SWEEPSTAKES	[rectangular, black on red]
S977	SWEET ALABAMA	[round, embossed, W.H.W.]
S978	SWEET APPLE	[hexagon black on red]
S980	**SWEET APPLE**	**(a) [large oval black on red]** (b) [small oval]
S982	**SWEET BITE**	**[rectangular, embossed]**
S984	**SWEET BRIER**	**[round, white on dark green]**
S985	SWEET BRIER	[round, embossed]
S986	SWEET BURLEY	[round, embossed]
S988	**SWEET BUY & BUY**	**[round, embossed]**
S990	**SWEET CHUNK**	**[oval, black on red, Scotten Tobacco, Co.]**
S992	SWEET CLOVER	[die-cut leaf, yellow on green, smoke & chew]
S993	**SWEET CLOVER**	**[die-cut three-leaf clover embossed]**
S994	SWEET CLOVER	[die-cut, four-leaf clover, J. Wright Co.]
S995	SWEET CLUSTER	[round, embossed, picture of grapes]
S996	**SWEET CORE**	**[rectangular, black on red & gold, Scotten-Dillon Company]**
S998	SWEET CREAM	[die-cut shield picture of cow, brown on white, Dodson Bros]
S1000	SWEET CUBA	[oval, green on black]
S1002	SWEET FERN	[round, small, red on yellow]
S1004	SWEET FERN	[rectangular, black on yellow]
S1006	SWEET FODDER	[round, black on yellow, pic-

		ture of cornstalks]
S1010	SWEET GIRL	[octagonal, black on red, picture of a girl, C.R.T. Co.]
S1012	**SWEET GUM**	**[octagonal, black on yellow, picture of a bee hive, W.P. Pickett & Co]**
S1014	**SWEET GUM**	**[die-cut leaf, black on green]**
S1015	SWEET GUM	[rectangular, embossed]
S1016	SWEET GUM	[oval, white on red]
S1018	SWEETHEART	[round, black on yellow, picture of a girl's head, G. Penn Sons Tob. Co.]
S1020	SWEET HUT	[rectangular, black on red, picture of log cabin]
S1021	SWEET ISAAC	[round, black on yellow]
S1022	**SWEET & JUICY**	**[round, blue on black, picture of a orange in center]**
S1024	**SWEET & JUICY**	**[round, embossed letters, blue on black, picture of a orange]**
S1026	**SWEET & JUICY**	**[die-cut fruit & leaf, black on orange, red & green sun cured]**
S1028	**SWEET LEAF**	**[oval, orange on yellow & blue, picture of a leaf, Carhart & Bro.]**
S1030	SWEET MARIE	[round, black on yellow]
S1032	SWEET MARIE	[rectangular, black on yellow, picture of girl, Mt. Airy, N.C.]
S1034	**SWEET MASH**	**[rectangular, black on red]**
S1036	**SWEET MASH**	**[rectangular, embossed]**
S1038	**SWEET MASH**	**[rectangular, embossed, T.W.O.C.]**
S1040	SWEET MASH	[rectangular, black on yellow, picture of a girl, Mfg. by Irvin & Poston]
S1041	SWEET MASH	[rectangular, black on yellow, picture of a girl, Irvin & Poston]
S1042	SWEET MASH	[octagonal, black on yellow, picture of a girl, Adams Powell Co.]
S1043	SWEET MASH	[octagonal, black on yellow, picture of a girl, Turner Powell Co.]
S1044	SWEET MASH	[round, yellow on black, picture of a barrel, Stenberg & Sons]
S1046	SWEET ME	[oval, embossed]
S1047	SWEET ME	[round, embossed]
S1048	SWEET MEAT	[round, black on yellow, picture of a nut, G. Penn Sons & Co.]
S1050	SWEET MIST	[banner embossed]
S1052	SWEET NICK	[round, white on red]
S1054	SWEET ORANGE	[round, black on yellow, multicolored picture of orange, Hardgrove & Co]
S1056	SWEET ORANGE	[round, black on yellow, multicolored picture of orange]
S1058	SWEET OWEN	[oval, embossed]
S1060	**SWEET PEACH**	**[round, black on red]**
S1062	SWEET PEA	[round, black on yellow, red picture of flower, T.C. Williams Co.The American Tob, Co]
S1064	**SWEET PEA**	**[round, black on white & red, picture of a flower, T.C. Williams Co.]**
S1066	SWEET PENNY ROYAL	[round, black on red]
S1068	SWEET PLUM	[oval, embossed]
S1070	SWEET ROLL	(a) [round, white on red]
		(b) [white on blue]
S1071	SWEET ROSE	[Clark & Co., Elvira. O.]
S1072	SWEET ROSE	[round, embossed]
S1073	SWEET RUSSET	[round, brown on yellow, picture of an apple]
S1074	SWEET SIXTEEN	[round, black on red]
S1074	SWEET SIXTEEN	[round, embossed]
S1076	SWEET TOOTH	[rectangular, black on red, Carhart & Bro.]
S1078	SWEET TOOTH	[oval, embossed]
S1080	SWEET TIT TOP	[round, black on red, Andersons]
S1084	SWING	[die-cut banner, yellow on white]
S1086	**SWITZERLAND FLAG**	**[oval, red on white, picture of red flag with a white cross]**
S1090	SWORD	[die-cut sword, embossed]

T

T101	**T**	**[die-cut]**
T102	**T**	**[square, black on red]**
T104	**T C**	**[rectangular, cut-out, blue]**
T106	**T C J**	**[oval, cut-out]**
T108	T D	[round, embossed]
T110	T E & H	[round, black on red]
T112	T.F.	[round, embossed]
T114	T H	[rectangular, cut-out letters]
T116	T.I.C.	[oval, red on green]
T118	T.I.T.	[rectangular, black on orange]
T119	T-L	[oval, embossed]
T120	**T L**	**[rectangular, cut-out]**
T122	**T L**	**[diamond, embossed]**
T124	**T.P.**	**[round, black on white, clock face]**
T126	**T P A**	**[round, dark blue on white in triangles]**
T127	T R	[die-cut shield embossed]
T128	**T R**	**[diamond, embossed]**
T130	T. & S.	[rectangular, embossed, Henry Co., Va.]
T132	T. & S.	[round, black on yellow, Dalton Farrow & Co.]
T134	T T T	[round, black on yellow, P.H. Hanes & Co. Winston, N.C.]
T136	**T & W**	**[rectangular, embossed, red]**
T138	T.Z.	[oval, orange on yellow]
T140	**G.P. TALBOTT**	**[rectangular, embossed, red]**
T142	T.J. TALBOTT	[rectangular, embossed]
T144	TAG MADE IN U.S.A.	[round, green on white, picture of a three-leaf clover]
T146	**TAKE NO DUST**	**[rectangular, black on yellow, picture of two bicycles, Stafford]**
T148	**TAFFY**	**[die-cut cross embossed]**
T150	TALKER NATURAL LEAF	[die-cut leaf brown on yel low]
T152	**TALKER PLUG TOBACCO**	**[die-cut parrot, black on green, red & yellow]**
T154	TAMARACK	[rectangular, black on tan]
T156	**TAMMANY**	**[round, embossed]**
T158	TAMMANY	[octagonal black on red]
T160	**TANNER**	**[die-cut star, embossed]**
T162	**TAPE LINE**	**[round, yellow on red, picture of a tape measure]**
T164	TAR BUCKET	[rectangular, black on yellow, Moody & Olive, Mt. Airy, N.C.]

T166	TAR HEEL	[round, black on red]
T168	TAR HEEL	[round, embossed]
T170	**TARGET**	**[round, black on white]**
T172	**TARGET**	**[round, embossed]**
T174	**TARGET**	**[round, embossed, Cov., KY.]**
T176	TARGET TWIST	[round, black on yellow, picture of a target in center]
T178	TARRED CHEWING	[round, embossed]
T180	TATE'S XX	[rectangular, black on yellow]
T182	TATTOO TWIST	[round, black on yellow]
T184	**TAYLOR'S TOBACCO**	**[rectangular, yellow on black, picture of a rooster]**
T186	**TAYLOR'S RED COON**	**[die-cut coon, yellow on red]**
T188	TAYLOR BROS.	[die-cut leaf black on yellow]
T190	**TAYLOR BROS & Co.**	**[rectangular, embossed, Winston, N.C.]**
T192	**TAYLOR MADE**	**[round, white on blue]**
T194	**TAYLOR MADE**	**[oval, white on red]**
TI96	TAYLOR, SPENCER & Co.	[rectangular, embossed]
T198	TEA CART	[round, red on yellow]
T200	TEA TRAY	[rectangular, embossed]
T202	**TEAL**	**[oval, black on yellow, green & brown, picture of ducks]**
T204	**TEASER**	**[rectangular, embossed]**
T206	TEDDY B.	[oval, black & brown on yellow, picture of a bear, Hermitage Tobacco Works]
T208	**TEDDY'S TWIST**	**[round, black on red]**
T210	TELEGRAPH	[rectangular, oval, embossed]
T212	**TELEGRAPH**	**[rectangular, black on red]**
T213	TELEGRAPH	[round, black on red]
T214	TELEGRAPH	[rectangular, black on red, D.M. & Co.]
T216	TELEPHONE TOBACCO	[rectangular, embossed, B.F. Hanes]
T218	TEMPLOS GOLDEN	[diamond, embossed]
T220	**TEN ACRE LOT**	**[round, black on yellow, center big bass, P. Lorillard & Co.]**
T222	TEN O'CLOCK	[round, black on white, clock face]
T224	TEN PENNY	[oval, black on yellow]
T226	TENENT BIG 4	[rectangular, embossed]
T228	**TENDERLOIN**	**[round, black on yellow, picture of a knife & fork, Tenderloin top & bottom]**
T229	TENDERLOIN	[round, black on yellow, picture of a knife & fork]
T230	**TENDERLOIN**	**[round, black on yellow, picture of knife & fork, P. Lorillard & Co.]**
T232	TENN RIVER	[round, black on red, T.R.G. Greenville]
T234	**TENNESSEE CROSSTIE**	**[rectangular, embossed]**
T236	TENNESSEE LEAF	[round, embossed, picture of a leaf]
T238	**TENNESSEE LEAF**	**[rectangular, white on blue]**
T240	TENNESSEE LEAF	[rectangular, yellow on black]
T242	**TENNESSEE NATURAL LEAF**	**[rectangular, yellow on black, picture of leaf, East Tenn. Tob Co.]**
T244	TENNESSEE SUN CURED	[round, red on black]
T245	TENNESSEE STAPLE	[rectangular, black & brown on orange, picture of a leaf]
T246	**TERRAPIN**	**[oval, green on light green, picture of a turtle, P. Lorillard & Co.]**
T247	TEXAS	[oval, black on red, picture of a battleship]
T248	TEXAS COW-BOY	[round, red on yellow, picture of men on horses]
T250	TEXAS PACIFIC	[rectangular, embossed]
T252	TEXAS PONY	[rectangular, odd banner, embossed]
T254	TEXAS RANGER	[round, black on yellow, picture of man]
T256	TEXAS STEER	[rectangular, brown & black on cream, picture of bull]
T258	**THAT**	**[rectangular, embossed]**
T260	THAT	[round, embossed]
T262	THAT'S WHAT	[rectangular, embossed]
T264	THE B	[oval, black on red]
T268	**THE BARON**	**[oval, black on red]**
T270	THE BEST YET	[round, embossed]
T272	THE CELEBRATED ZEBRA	[rectangular, picture of a zebra running]
T274	THE C.E.M. CO'S	[rectangular, black on red, Winston Salem, Winston, N.C.]
T276	THE COMRADE PLUG	[round, blue on white & red, blue]
T278	THE DARK HORSE	(a) [round, black on red, picture of a man on a horse, J.H. Hargraves & Son] (b) [black on white]
T280	THE DERBY	[oval, black on cream, picture of hat]
T282	THE DRIVING PLUG	[round, scalloped, black on red]
T284	THE DURHAM	[oval, black on red]
T286	**THE EARTH FOR 5¢**	**[round, black on red]**
T288	THE EARTH TWIST	[round, white on red, L. & B. Tob. Co.]
T290	THE GOVERNOR	[round, embossed]
T292	THE HUNTER	[round, black on red]
T294	**THE KING**	**[round, black on red, picture of a king]**
T296	**THE OLD HOME STEAD**	**(a) [rectangular, black on green, J.H. Hargrave & Son]** (b) [black on green, J.H. Hargrave & Son]
T298	THE OLD HOME STEAD	
T300	**THE OLD OAKEN BUCKET**	**[cone shape, black on red, picture of a bucket]**
T302	**THE OLD STATESMAN**	**(a) [round, black on yellow, man on horse, Nashville Tob Co.]** (b) [Nashville Tobacco Works]
T304	THE OLD STATESMAN	[round, black on yellow, man on a horse]
T306	**THE ONLY GENUINE ALABAMA**	**[rectangular, embossed]**
T308	**THE PICKET**	**[round, & rectangular, picture of a horse, Winner of American Derby]**
T310	THE PILGRIM	[die-cut P black on red, G.S.S. & Co.]
T312	**THE PRIDE OF WINSTON**	**[round, black on yellow, picture of a plug, G. Flynt Tob. Co.]**
T314	THE PRIDE OF WINSTON	[round, black on yellow, picture of plug, Brown & Williamson]
T316	THE PRIME	[round, embossed]
T318	THE RING	[round, yellow on blue & yellow]
T319	THE RINK	[round, red on yellow]
T320	THE SOUTH	[oval, embossed]

T322	THE VESTIBULE	[rectangular, black on red, picture of train]
T324	THE VIRGINIANS	[octagonal, black on yellow, Camerons Virginia Factory]
T326	THE WIDOW	[oval, embossed]
T328	THE WORLD FOR A NICKEL	[round, embossed]
T330	**THIMBLE**	**[die-cut, embossed]**
T332	THIMBLE	[round, yellow on red, picture of a thimble]
T334	**THIS IS YOURS**	**[oval, embossed]**
T336	**THOROUGH MADE**	**[diamond, embossed]**
T337	THOROUGH BRED	[octagonal black on yellow, picture of a horse & girl]
T338	THOROUGH BRED	[octagonal, picture]
T339	THOROUGH BRED	[round, embossed]
T340	**THRASHER**	**[die-cut cross, black on red, W & McC. Tob Co.]**
T342	THRASHER	[round, multicolored, picture of bird, Nashville Tob. Co.]
T344	**3 BLACK CROWS**	**(a) [octagonal black on yellow, P. Lorillard Co. in script]**
		(b) [in block letters]
		(c) [in dark heavy block letters]
T346	3 PLY	[round, embossed, P.B. Gravely & Co.]
T348	3 PLY	[round, black on yellow, Rosenbrocks Union Factory]
T350	3 PLY FLUE CURED	[oval, embossed, Gravely]
T352	3 PLY PLUG	[banner embossed]
T354	3 PLY TWIST	[round, black on yellow, W.T. Burton]
T356	**3**	**[die-cut, large 3]**
T358	3	[round, embossed]
T360	**33**	**[rectangular, cut-out]**
T362	**THREE CRONIES**	**[round, green on yellow, picture of three black crows, Kerner Bros]**
T364	THREE PLY	[rectangular, embossed, Hancock]
T366	THREE SISTERS	[round, embossed]
T370	THROUGH MADE	[diamond, embossed]
T372	THUMPER	(a) [round, black on yellow, picture of an arm & hammer]
		(b) [brown on blue]
T374	THUNDERBOLT	[octagonal, black on yellow, picture of a thunderbolt, Blackburn & Co.]
T376	TIBBIT'S TWIST	[rectangular, banner embossed, J.B. Pace]
T378	TID BIT	[rectangular bar, black on red]
T380	TIGER	[oval, black on yellow, picture of a tiger]
T382	TIGER	[die-cut tiger, black on yellow]
T384	TIGER PLUG	[round, black on red, picture of a tiger head]
T386	**TIGHT ROPE**	**[rectangular, round ends, black on yellow, picture of a rope]**
T387	TILTON'S TOBACCO	[round, black & red on yellow]
T388	TIN HAT	[oval, embossed]
T390	TIN TAG	[rectangular, embossed, Lorillard]
T392	**TINSLEY, W. N.**	**[rectangular, black on light green, picture of a man]**
T394	**TINSLEY 5**	**[round, embossed, Tobacco Co.]**
T396	**TINSLEY'S**	**[rectangular, embossed]**
T398	TINSLEY'S MONOGRAM	[round, embossed]
T400	**TINSLEY'S NATURAL LEAF**	**[rectangular, gold on red]**
T402	TINSLEY'S PLUG	[round, embossed]
T404	**TINSLEY'S PREMIUM**	**(a) [round, blue on silver]**
		(b) [black on green]
T406	TINSLEY'S TOBACCO Co.	[round, green & black on silver]
T408	TIP ABBOTT	[round, embossed]
T410	**TIP TOP**	**[diamond, embossed]**
T411	**TIP TOP**	**[octagonal embossed]**
T412	TIP TOP	[die-cut shape, anchor.black on yellow, G. & S.]
T414	**TIPPECANOE**	**[die-cut canoe black on yellow, P. Lorillard & Co.]**
T416	TIPPECANOE	[round, embossed]
T418	**TIPPECANOE**	**[round, black on yellow]**
T420	TIPSY CHEW	[round, embossed]
T422	**TIT-BIT**	**[die-cut banner, flag embossed]**
T424	TIT-BIT	[oval, red on yellow]
T426	TOBACCO	[round, black on yellow, picture of bird, Robt Harris & Bro, Reidsville, N.C.]
T427	TOBACCO TWIST	[die-cut embossed, twist of tobacco]
T428	TOBACCO WORM	[rectangular, black on red]
T430	TOBACCO WORM	[rectangular, embossed]
T432	**TODDY PLUG**	**[round, white on blue]**
T434	TO DAY	[round, embossed]
T436	TO SUCCESS	[die-cut key]
T438	**TOKAY**	**[rectangular, black on red, picture of a bunch of grapes]**
T440	TOKAY	[round, black on red, picture of a bunch of grapes]
T442	TOKAY	[rectangular, red on cream, picture of a bunch of grapes]
T444	**TOLEDO**	**[oval, black on white & green, picture of a frog]**
T446	TOLEDO LIGHT	[round, green & white on orange]
T448	TOLL GATE	[rectangular, black on red, picture of a man on horse at gate]
T450	**TOM**	**[small round, black on white, picture of a cat]**
T552	TOM	[oval, brown on yellow, picture of a donkey]
T454	TOM BOY	[round, black on red]
T456	TOM COLE	[round, black on yellow]
T458	TOM JEFFERSON	[rectangular, black on yellow, picture of a man]
T460	**TOM & JERRY**	**[hexagon, embossed, S.R. & Co.]**
T462	TOMS APPLE	[oval, black on red]
T464	TOM'S	[die-cut shape, log cabin, embossed]
T466	**TOMAHAWK**	**[die-cut tomahawk, black on red, P. Lorillard & Co.]**
T468	TOMAHAWK	[die-cut tomahawk, embossed]
T470	TOM WATSON	[rectangular, bar, embossed]
T472	**TONIC**	**[rectangular, black on red, M.E.C. & Co.]**
T474	**TONIC**	**[round, gold on red]**
T476	TONIC	[octagonal, black on red]
T478	**TONY**	**[rectangular, embossed fancy]**
T480	TOOTHPICK	[round, embossed]
T482	**TOP**	**[die-cut top, blue on white, red, white & blue]**
T484	**TOP HAT**	**[die-cut, embossed, shape of a top hat]**
T486	TORCH LIGHT	[diamond, black on yellow, picture of a torch]

T488	**TORCH LIGHT**	**[oval, embossed]**
T490	**TORNADO**	**(a) [round, red on yellow, Hickok]** (b) [red on white]
T492	TOUGH & GOOD	(a) [round, blue on white] (b) [black on yellow]
T494	TOW BOAT TWIST	[round, white on red, Ryan Hampton Tob Co.]
T495	TOW LINE	[rectangular, embossed]
T496	TOWN CLOCK	[round, black on yellow, picture of clock, Liipfert & Jones]
T498	**TOWN TALK**	**[oval, red on yellow & red]**
T499	TOWN TALK	[round, embossed]
T500	TOWN TALK	[oval, embossed]
T502	**TRADE**	**[rectangular, embossed]**
T504	TRADE	[oval, embossed]
T506	TRADE MARK	[die-cut snowshoe, black on yellow]
T507	TRADE MARK	[die-cut snowshoe, black on yellow, tag made in U.S.A.]
T508	**TRADE MARK**	**[round, embossed, picture of a man's head]**
T510	**TRADE MARK**	**[rectangular, black on yellow, L.B. & T.]**
T512	TRADEMARK U & S	[round, black on red]
T514	TRAMWAY	[rectangular, embossed]
T516	TRAPPER	[round, red on cream]
T518	**TRAVELER**	**[rectangular, red on black]**
T520	TRAYLOR SPENCER & CO.	[octagonal, embossed]
T522	**TRIANGLE**	**[die-cut, red]**
T524	TRIANGLE	[die-cut, embossed triangle]
T525	TRICK	[round, embossed, Turner]
T526	TRILBY	[rectangular, black on yellow]
T528	**TRINKET**	**[oval, small, black on red]**
T530	**TRIO**	**[round, black on red, Musselman Tobacco Co.]**
T532	TRIO	(a) [rectangular, black on light green, picture of three men, Dalton, Farroe & Co] (b) [black on blue green]
T534	TROPHY	[round, yellow on red, picture of a bear]
T536	**TROUT**	**[rectangular, black on red, picture of a fish]**
T538	TRIP	[oval, embossed]
T540	TRIPLETS	[round, black on yellow, picture of three girls]
T542	TRIPLETTE	[octagonal black on red]
T543	TRIUMPH	[round, embossed]
T544	**TRUCK**	**[round, black on red, picture of a hand truck, Lottier]**
T546	TRUCK	[rectangular, yellow on black, picture of a hand truck, Lottier]
T548	TRUE CHEW	[round, black on blue]
T550	TRUMPET	[die-cut shape, trumpet, embossed]
T552	TRUMP CARD	[round, embossed]
T554	**TRUMPS**	**[round, gold on red]**
T556	TRUNK	[die-cut, embossed, shape of a trunk]
T558	**TRUNK LINE**	**[die-cut, embossed, shape of a trunk, Z.T. Co.]**
T560	**TRUST**	**[die-cut, embossed, shape of a safe]**
T562	TRUST ME	[round, embossed]
T564	TRUTHFUL JAMES	[round, black on orange, picture of a man sitting on the pot]
T566	TUBE ROSE	[rectangular, banner, black on yellow]
T568	TUCKAHOE	[octagonal, black on green, R.A. Patterson Tobacco Co.]

T570	TUCKAHOE PLUG	[rectangular, black on yellow & yellow on black]
T572	TUCKAHOE TWIST	[round, yellow on black]
T574	TUG RIVER	[oval, multicolored, picture of a boat]
T576	TUG RIVER	[oval, multicolored, picture of a boat, Dalton Farrow & Co.]
T578	TUG RIVER BOAT	[oval, black on yellow, picture of a boat, Dalton Farrow & Co.]
T580	TUG OF WAR	[rectangular, embossed, picture of two dogs, Tough & Waxy]
T582	**TULIP**	**[round, embossed]**
T584	**TULIP**	**[rectangular, blue on gold]**
T586	TULIP TIME	[round, yellow on red]
T588	TUNNEL CITY	[round, embossed]
T590	**TURF**	**[oval, black on red]**
T592	**TURF**	**[oval, black on red, Strater Brothers]**
T594	**TURKEY**	**[die-cut T, black on red, picture of a turkey, Musselman]**
T595	TURKEY	[rectangular, black on red, picture of a turkey]
T596	**TURKEY**	**[rectangular, black on red picture of a turkey, Musselman]**
T598	TURKEY PLUG	[round, black on yellow]
T600	TURKEY TWIST	[round, black on red, picture of a turkey]
T602	TURKEY TWIST	[round, black on red, picture of a turkey, Smith & Scott Tob Co.]
T604	TURNIP	[round, black on yellow, picture of a turnip]
T606	TURTLE	[die-cut large, embossed, shape of a turtle]
T607	**TURTLE**	**[die-cut small, embossed shape turtle]**
T608	TURTLE	[die-cut black on green]
T610	TURTLE	[round, black on green, picture of a turtle]
T612	**TWILIGHT**	**[round, black on red]**
T614	**TWIN CITY**	**[round, embossed**
T616	TWIN CITY	[rectangular, picture of the two cities, Winston & Salem]
T618	TWINS	[die-cut bow tie shape, black on red]
T620	TWISTNO	[rectangular, bar, embossed]
T622	2 BY 4	[rectangular, bar, white on black]
T624	TWO IN ONE	[round, embossed]
T626	TWO WHALES	[die-cut block, black on blue, red & yellow, Scotten]
T628	TWO WHATS	[die-cut T & W, red on green]
T630	**TYCOON**	**[oval, embossed]**
T632	TYCOON PLUG	[oval, brown on yellow]

U

U101	U	[die-cut embossed]
U102	**U B**	**[rectangular, embossed]**
U104	U AND I	[rectangular, embossed]
U106	**U & I**	**[rectangular, cut-out]**
U108	**U & I**	**[oval, embossed]**
U110	U & I	[rectangular, embossed]
U112	U-L	[round, white on blue,

		Chew, Smoke]
U114	U N	[round, embossed]
U116	**U.S.**	**[large die-cut, embossed]**
U118	U.S.	[round, black on red, trademark]
U119	U.S. STANDARD	[rectangular, red on yellow]
U120	**U.S.M.**	**[round, black on red]**
U121	U.S.M.	[rectangular, odd, embossed]
U122	U T C	[die-cut letters, red, yellow & green]
U124	**U. BET**	**[round, black on red]**
U126	**UCHUME**	**[rectangular, black on red & yellow, picture of a wigwam]**
U128	UNCLE ANDY	[round, black on yellow]
U130	UNCLE DAN	[round, black on white, picture of a man with a hat on]
U132	UNCLE EASY	[round, black on yellow, picture of a black man]
U134	UNCLE ESSEX	[rectangular, black on green, picture of a black man, Robt H. Herndon]
U136	UNCLE HENRY	[round, red on yellow]
U138	UNCLE JAMES	[round, embossed]
U139	UNCLE JOHN	[octagonal embossed]
U140	**UNCLE JOHN**	**[rectangular, curve, black on red]**
U142	UNCLE JOSH	[round, black on cream]
U144	UNCLE LI JE	[round, black on light green, picture of a black man]
U146	**UNCLE REMUS**	**[round, black on yellow, picture of a old black man, J.N. Willie & Co.]**
U148	UNCLE SAM	[rectangular, bar, red on black]
U150	**UNCLE SAM**	**(a) [triangular, gold on red $ in center]** **(b) [black on red]**
U151	**UNCLE SAM**	**[round, red, white & blue, picture of Uncle Sam]**
U152	**UNCLE SAM**	**[die-cut man, red, white & blue with stars]**
U154	UNCLE SAM	[die-cut man standing, red, white & blue on yellow base]
U156	UNCLE SAM	[triangular, gold on blue, S. Foree & Co.]
U158	**UNCLE WILLIAM**	**(a) [round, gold on black]** (b) [black on red]
U160	UNICORN	[rectangular, black on yellow, picture of a Unicorn]
U162	**UNION**	**[round, cut-out embossed, N.H. & Co. Cov., Ky]**
U164	UNION	[round, black on yellow]
U166	**UNION CARD**	**[round, black on brown, S. Coleman, Buffalo, N.Y.]**
U168	UNION CHEW	[round, yellow on red, Edson]
U170	**UNION CLUB**	**[round, black on silver]**
U172	UNION COMMANDER	[round, black on red]
U174	UNION FACTORY	[rectangular, embossed]
U176	UNION JACK	[rectangular, multicolored, picture of a sailor & flag, The Erie Tob. Co. Ltd.]
U178	**UNION KNOT**	**[diamond, blue on yellow, picture of a knot]**
U180	**UNION LABOR**	**[rectangular, yellow on red & yellow]**
U182	UNION LEADER	[oval, red on black]
U184	UNION MADE	[round, blue on cream]
U186	**UNION MADE RUSTIC**	**[round, black on red]**
U188	UNION NAVY	[rectangular, embossed]
U190	UNION NAVY	[rectangular, white on blue]
U192	UNION NAVY	[round, picture of the American Eagle]
U194	UNION PLUG	[round, embossed]
U196	UNION RUSTIC	[round, black on red]
U198	**UNION STANDARD**	**[die-cut U black on red]**
U199	**UNION STANDARD**	**[round, embossed, U in center]**
U200	**UNION STANDARD**	**[round, red on silver, U in center on blue dot]**
U202	UNION STRIKE	[die-cut U, yellow on green]
U204	UNION TWIST	[round, embossed]
U206	UNITED FARMERS	[rectangular, embossed]
U208	**UNITED STATES OF AMERICA 5¢**	**[round, embossed coin]**
U210	**UNITED STATES TOBACCO CO. U.S.**	**[die-cut shield, red white & blue]**
U212	UNIVERSAL	[banner embossed]
U214	**UNMATCHED**	**[die-cut cross, black on red]**
U216	**UNMATCHED**	**[rectangular, embossed]**
U220	**UMBRELLA**	**[die-cut shape of a girl holding an umbrella, white on brown, pink & blue]**
U222	UMPIRE	[oval, embossed]
U224	UMPIRE TWIST	[round, red on green]
U226	**UNOME**	**[rectangular, black on brown, Spilman, Ellis Tob. Co.]**
U228	UP TO DATE	[rectangular, scalloped shape, embossed]
U230	UPPER SEVEN	[round, embossed]
U232	UPPER TEN	[round, embossed]
U234	U.S. BONDS GOVERNMENT	[rectangular, & round cen ter, picture of eagle, Irving & Poston]
U236	U.S. GOVERNMENT BONDS	[rectangular, with round center, Turner Powell & Co.]
U237	U.S. NATURAL LEAF	[rectangular, multicolored, picture of a leaf]
U238	**U.S. STANDARD**	**[rectangular, blue on red & white flag]**
U240	**U.S. TRADEMARK**	**[round, black on red]**
U242	U.S. WORKER PLUG	[round, red on orange & yellow]
U244	UWANTA CHEW	[rectangular, bar, orange on black]

V

V101	VALLEY ROSE	[die-cut bow tie, embossed, Chew & Smoke]
V102	VALUE	[rectangular, black on yellow, Rosenfeld]
V104	VAN BIBBER	[round, embossed]
V106	VANGUARD	[rectangular, yellow on red]
V108	VANILLA CREAM	[oval, black on yellow]
V110	VANITY FAIR	[round, red on cream]
V112	**VARIETY**	**[rectangular, odd shape, embossed]**
V114	VARIETY	[banner embossed]
V116	VAUGHN'S TWIST	[odd rectangular, black on tan & cream]
V118	VELVET	[oval, black on brown]
V120	VENABLE'S	[rectangular, dark blue on yellow]
V122	VENABLE'S	[round, embossed]
V124	**VENABLE'S**	**[rectangular, black on yellow]**
V126	**VENABLE'S St. GEORGE**	**[round, embossed]**
V127	VENABLE'S St. GEORGE	[round, white on red]
V128	VENEZUELA	[oval, black on white, flag of Venezuela, yellow, blue & red]

V130	**VENUS**	**[oval, embossed]**
V132	VENUS	[rectangular, odd embossed]
V134	VENUS	[die-cut shield, multicolored, picture of a woman, B.A.T. Co. Pr Ltd Melbourne]
V136	VERMONT CREAM	[round, green on white]
V138	VETERAN	[oval, embossed]
V140	**VICTOR**	**[rectangular, embossed]**
V142	**VICTOR**	**[hexagon black on yellow, P. Lorillard & Co.]**
V144	VICTOR	[die cut shield, embossed]
V146	VICTOR	[round, black on red]
V148	VICTORIA	[die-cut heart, W.C. MacDonald, Montreal]
V150	VICTORIA DE BRITANNICA	[rectangular, multicolored, woman standing]
V152	VICTORY	[round, E. & M.]
V154	**VINCO**	**(a) [small round, black on orange]** (b) [larger round]
V156	**VINCO**	**[round, black on green, J. Livingston & Co.]**
V158	VINCO	[round, black on orange, J.C. Herman]
V160	**VINDEX**	**[rectangular, embossed]**
V162	VINO	[round, black on green, picture of a glass & large leaf]
V164	**VIOLET**	**[rectangular, round top, green on light green, picture of a flower]**
V166	**VIOLET GRADE**	**[octagonal gold on black]**
V168	VIOLET PLUG	[rectangular, yellow on red]
V170	VIRGIN	[round, black on yellow, T.C. Williams Co.]
V172	VIRGIN LEAF	[round, brown on yellow]
V174	VIRGINIA	[oval, black on gold, large T C, small R C]
V176	VIRGINIA BEAUTY	[odd rectangular, shape, black on red]
V178	VIRGINIA BOSS	[oval, embossed]
V180	VIRGINIA DARE	[round, black on yellow]
V182	**VIRGINIA GARNET**	**[round, embossed]**
V181	VIRGINIA LEAF	[die-cut, embossed leaf]
V184	VIRGINIA LEAF	[round, embossed, picture of leaf]
V186	**VIRGINIA SEAL**	**[round, black on yellow, picture of two men fighting]**
V188	VIRGINIA SUN CURED	[die-cut heart, yellow on green]
V190	VIRGINIA TWIST	[oval, black on red, Bonne]
V192	**V.T.C. TORONTO**	**[diamond, die-cut, cut-out engraved]**
V194	**VOGLER'S CHOICE**	**[oval, embossed, 809 Aliceanna]**

W

W101	**W**	**[small die-cut, embossed]**
W102	W	[round, black on red W in a diamond]
W104	W	[round, cut-out letter]
W106	W	[die-cut letter, black on yellow, Walker Bros. Winston, N.C.]
W108	W.A. BROWNS NATURAL	[die-cut leaf, red on brown & yellow]
W110	W.B.	[oval, cut-out letters]
W112	**W & B**	**[round, cut-out]**
W114	**W & B**	**[rectangular, cut-out]**
W116	W.D. Co	[die-cut heart shape, black on red]
W118	**W & D**	**[die-cut]**
W119	W F	[rectangular, cut-out letters]
W120	W L & Co	[die-cut buffalo, embossed]
W121	W L & Co	[round, embossed]
W122	W L & Co	[oval, embossed]
W124	**W.N.T.**	**[rectangular, embossed]**
W126	**W T C**	**[round, embossed]**
W127	W.W.W.W.	[rectangular, embossed]
W128	WAFFLE	[round, black on red]
W130	**WAKE UP**	**[rectangular, red on yellow & red]**
W132	WALKER'S BEST	[round, black on yellow]
W134	**WALKER'S PURE LEAF**	**(a) [rectangular, black on off-white, Chew-Smoke in center]** **(b) [black letters, red dot in center]** (c) [rectangular, with tabs]
W136	WALKER'S TWIST	[round, black on tan]
W138	WALLABY	[rectangular, banner, black on yellow]
W140	WALTER RALEIGH	[oval, embossed]
W142	WALTON	[round, black on red]
W144	WALTON	[round, scalloped black on red]
W146	WAR	[oval, black on red, Henry Co., Va.]
W148	WAR CLUB	[round, embossed]
W150	**WAR HAWK**	**[round, white on red]**
W151	WAR KING	[round, black on yellow, picture of man, Cumberland Tobacco]
W152	**WAR PATH**	**[half round, embossed]**
W153	WARPATH BANGOR	[round, embossed, T.R.S. & Co.]
W154	WAR PLUG	[round, embossed]
W156	**WAR TIME PLUG**	**[oval, black on red, Zahm's Old Navy, picture of a ship]**
W158	WAR SAW	[round, black on yellow, picture of a hand saw]
W160	WARD	[round, black on yellow]
W162	**WARD BRAND**	**[round, black on red, Hardgrove]**
W164	WARM BABY	[rectangular, red on black]
W166	WARNICK & BROWN	[round, embossed]
W167	WARNICK & BROWN	[rectangular, embossed]
W170	**WARREN COUNTRY TWIST**	**[round, yellow on red]**
W172	**WARWICK**	**[oval, black on red]**
W174	WASHBOARD	[die-cut, embossed washboard]
W176	WASHINGTON'S CHEW	[rectangular, embossed]
W178	WASHINGTON'S TWIST	[oval, embossed]
W180	WASP	[round, red on yellow, multicolored picture of a wasp]
W182	WASP	[round, red on yellow, multicolored picture of a wasp with six legs]
W184	WATCH	[oval, embossed]
W186	WATCH KEY	[rectangular, black on yellow, picture of watch key]
W188	WATER LILY	[round, green on blue & white]
W189	WATER LUO	[round, red on yellow]
W190	WATER MELON	[round, black, red & green on white, picture of a watermelon]
W191	**WALTER RALEIGH**	**[oval, embossed]**
W192	WATT PENN & Co.	[rectangular, embossed, Reidsville, N.C.]
W194	WATSON & McGILL	[rectangular, yellow on red]
W196	**WAVERLY**	**[rectangular, odd shield embossed]**
W198	**WAX**	**[round, black on yellow, picture of a bee]**

W200 W. BROS. [oval, black on yellow]

W202 WEAKLY WORKMAN [round, embossed]

W204 WEDDING CAKE [round, embossed]

W206 WEDDING CAKE [round, embossed, Weir's B.C.]

W208 WEDDING CAKE [round, black on white]

W210 WEDGE [round, red on yellow & green, picture of wedge, Trade Mark]

W211 WEIRS B.C. [octagonal, embossed]

W212 HARRY WEISSINGER TOBACCO Co. [diamond, black on red]

W214 WEISSINGER'S SPECIAL (a) [small rectangular, black on red]

(b) [larger]

W216 WELCOME [die-cut shield, blue on red white & blue]

W217 WELCOME 5 CENT PLUS [rectangular, odd embossed, T.N.W. & Co.]

W218 WELCOME NUGGET [odd round, embossed]

W220 WELCOME VISITOR [round, embossed]

W222 A.B. WELLS [die-cut shield, embossed]

W224 WEST TENNESSEE [rectangular, embossed]

W226 WEST VIRGINIA [round, embossed]

W228 WESTERN CAROLINA [rectangular, multicolored, picture of a landscape]

W230 WESTORE'S NATURAL LEAF [rectangular, black on yel low, picture of a leaf]

W231 WESTOVER [rectangular, black on yellow, picture of a horse, Patterson]

W232 WESTOVER (a) [rectangular, black on yellow, R.A. Patterson. Tob. Co., picture of a horse]

(b) [black & green on white]

W234 WETMORE'S BEST [rectangular, odd, white on red, A C Wetmore]

W236 WETMORE'S KING JACK [octagonal white on red]

W238 WETMORE'S LITTLE MAJOR [rectangular, odd, white on red, A C Wetmore]

W240 WETMORE'S NATURAL LEAF [rectangular, white on red, A C Wetmore]

W244 WETMORE'S OLD COON [rectangular, white on red, A C Wetmore]

W246 WETMORE'S TWIST [oval, white on red]

W248 WHAT KNOX [round, black on yellow, picture of a bucking horse]

W250 WHEAT BREAD [square, black on yellow]

W252 WHEAT BREAD SUN CURED [square, black on yellow]

W254 WHEEL [die-cut spoke wheel, embossed]

W255 WHEEL [die-cut curved spoke wheel, embossed]

W256 WHEEL OF FORTUNE [round, red on yellow, picture of a gambling wheel]

W258 WHEEL HORSE [round, embossed, J.F.L.A. & Co.]

W260 WHEEL HORSE [round, black on yellow, picture, Stafford]

W262 WHEEL HOUSE [round, black on red, picture of a house]

W263 WHEELING [rectangular, embossed, J. Speigel & Co.]

W264 WHEELMAN [rectangular, black on red]

W268 WHIST SMOKE [round, white on red]

W269 WHITAKER'S NATURAL LEAF [die-cut leaf, black on red]

W270 WHIT-LEATHER [rectangular, embossed]

W271 WHITE [die-cut leaf embossed]

W272 WHITE CHIEF [round, black on yellow]

W274 WHITE COON [diamond, green on yellow, picture of a coon, P. Henkel]

W275 WHITE COON [round, black on tan, picture of a coon]

W276 WHITE COON [round, black on yellow, picture of a coon]

W278 WHITE CROSS [round, cut-out cross in center, yellow on red, J.B. Pace Tob. Co.]

W280 WHITE ELEPHANT [die-cut elephant]

W282 WHITE FAWN [diamond, black on red, Actt]

W284 WHITE HORSE [round, white on blue, picture of a horse, Carhart & Brother]

W286 WHITE MONEY [round, embossed]

W288 WHITE MULE [die-cut mule, black on white]

W290 WHITE PACER [oval, black on light yellow, picture of a horse, Payne & Ramsay]

W292 WHITE PACER (a) [oval, yellow on black, picture of a horse, Ebert Payne & Co.]

(b) [white on green]

W294 WHITE ROSE [round, embossed]

W296 WHITE ROSE [die-cut flower & banner embossed]

W297 WHITE ROSE [rectangular, embossed, Worthington]

W298 WHITE ROSE [rectangular, white on red, Worthington]

W300 WHITE SAILS [rectangular, black on yellow, picture of a ship]

W301 WHITE SAILS [rectangular, red & white on blue, picture of a ship]

W302 WHITE SWAN [round, red on yellow, picture of swan, Greenville]

W304 WHITE TIPS [round, embossed]

W306 WHITE WAX [rectangular, embossed]

W308 WHITTED TOB CO. [rectangular, embossed]

W310 WHOLE CHUNK [round, black on yellow, brown picture of a plug, Pegram & Penns]

W312 WHOLE CHUNK [round, black on yellow, picture of a plug]

W314 WHOLE HOG [die-cut shape, pig, embossed]

W316 WHOLE PLUG [rectangular, embossed]

W318 WIDE GAUGE [rectangular, black on yellow, picture of track, Schoolfield & Watson]

W320 WIDE TIRE [round, black on yellow, picture of wheel]

W322 WILD BILL [round, black on red]

W323 WILD CAT [square, black on yellow, picture of cat]

W324 WILD DUCK [round, red on yellow, picture of a flying duck]

W326 WILD DUCK [rectangular, black on red]

W328 WILD DUCK [rectangular, embossed]

W330 WILD FIRE [oval, embossed]

W332 WILD FLOWER [round, embossed]

W336 WILD FLOWER (a) [rectangular, gold on black]

(b) [gold on blue]

W338 WILD FLOWER [rectangular, embossed]

W340 WILD FRUIT [round, black on red, picture of berries]

W341 WILD GOOSE [oval, picture of goose, L.W. Davis]

W342 WILD ROSE [rectangular, gold on red, Carhart & Bros]

W344 WILD ROSE [round, white on red, Carhart & Bro., New York]

W346 WILD TURKEY TOBACCO [round, brown on white & green, picture of turkey, B-

Code	Name	Description
		W. Tob.Co.]
W348	T.C. WILLIAMS	[round, gold on black]
W350	WILL OF THE PEOPLE	[rectangular, embossed]
W352	WILLIE G.	[oval, embossed]
W354	WILLIE HARRIS	[round, black on yellow, picture of a man on a horse, Robt. Harris & Bros.]
W356	**WILLS'S TWIST**	**[rectangular, embossed]**
W358	**WILSON McCALLAY TOBACCO CO.**	**[small embossed, rectangular & hexagon center]**
W360	**WILSON & McCALLAY TOB CO.**	**[die-cut man walking, black on red]**
W362	WILSON & McCALLAY TOB CO.	[rectangular, black on red, picture of a man walking]
W364	**W & McC.T. Co.**	**[rectangular, black on red]**
W366	**WINCHESTER**	**[rectangular, black on yellow, picture of a sword]**
W368	WINCHESTER	[rectangular, picture of a rifle]
W370	WINCHESTER	[rectangular, embossed]
W372	**WINCHESTER D-1**	**[oval, white on red, Pure Selected Leaf]**
W374	**WIND MILL BRAND**	**[round, embossed]**
W376	WIND MILL TWIST	[round, embossed]
W378	**WINE GOLD**	**[oval, black on gold banner on red, McHies. Tob. Co. plug]**
W380	WINESAP	[round, black on red, O & M]
W381	WINE SAP	[rectangular, embossed]
W382	WINE SAP	[octagonal, black on yellow, Ebert Payne & Co.]
W383	**WINE SAP**	**[octagonal, black on red, picture of a grape, The F.R. Penn Tob Co.]**
W384	WINE SAP	[octagonal, black on red, large W on S in center, F.R. Penn Tob Co.]
W385	WINE SAP	[octagonal, black on red, Penn]
W386	WINE SAP	[round, embossed, J.W. Daniel, Henry Co.]
W387	**WINE SAP**	**[square, black on yellow]**
W388	WINE SAP	[octagonal black on yellow, picture of red apple, W.T. Hancocks]
W389	WINE SAP	[octagonal black and red on orange, Ebert Payne & Co.]
W390	**WINE SWEET**	**[rectangular, black on yellow]**
W392	**WINEBERRY**	**[diamond, black on yellow, Monarch Tobacco Works]**
W393	WINNER	[rectangular, embossed]
W394	WINNER	[oval, embossed]
W395	WINNER	[rectangular, red on yellow]
W396	**WINNER**	**[oval, black on red]**
W398	WINNING NIKE	[round, black on red]
W400	WINSTON BOOM	[round, black on red]
W402	WINSTON HUMMER	[round, red on yellow, picture of bird, The C.E.M. Co.]
W404	WINSTON LEADER	[rectangular, embossed]
W406	WINSTON SALEM	[rectangular, The C.E.M. Co. Winston, N.C.]
W408	WINSTON SPORT	[rectangular, black on yellow, picture of a man in top hat, Dalton Farrow]
W410	WINTER NELIS	[die-cut pear, black on yellow]
W412	WISDOM TOOTH	[rectangular, black on yellow, Casey & Wright]
W414	**WISHBONE**	**[die-cut, black on red]**
W416	**WISHBONE**	**[die-cut, white on black & red]**
W417	WISHBONE	[round, white on black & red, picture of a wishbone]
W418	**WISHBONE**	**[round, black on red, white letters, picture of a wishbone]**
W420	WITNESS	[round, black on yellow, Crews]
W422	W.J. NORTHEN	[oval, black on cream, picture of a man with a beard]
W424	W.K. PREMIUM	[rectangular, odd embossed]
W426	WOLF	[die-cut wolf's head, embossed]
W428	WOLVERINE	[round, black on red, picture of wolverine]
W430	**WOMAN'S HEART CHEW**	**[die-cut heart, embossed]**
W432	WONDER	[oval, embossed]
W434	**WOODCOCK**	**[rectangular, odd red on yellow, picture of a bird, McAlpins]**
W436	WOODCOCK	[round, black on yellow, picture of a bird]
W438	WOODPECKER	[round, black on red, picture of a bird]
W440	WORKER	[rectangular, The United States Tobacco Co.]
W442	WORKER'S BEST	[round, embossed]
W444	**WORKER, LIGHT, DARK**	**[rectangular, white on blue, gold & red, Not Made By a Trust]**
W446	WORLD BEATER	[rectangular, black on yellow, multicolored picture of a man & horse, racing]
W448	**WORLDS BEST**	**[round, embossed]**
W450	WORLD'S CHEW	[oval, embossed]
W451	**WORLD'S CHEW**	**[octagonal, black on yellow]**
W452	WORLD'S NAVY	[round, blue & black on cream]
W454	WORLD'S LIGHT	[round, embossed]
W456	**WORTH**	**[rectangular, black on red]**
W458	WPP & CO	[round, monogram]
W460	WRIGHT	[rectangular, black on yellow, picture of a man working]
W462	J. WRIGHT & Co.	[round, embossed]
W464	WRIGHT'S EXTRA	[rectangular, embossed]
W468	WRIGHT'S LEADER	[odd rectangular, embossed]
W470	**WYANDOTTE**	**[round, embossed, picture of an Indian, F. & H. Tob. Co.]**

X

Code	Name	Description
X101	**X**	**[die-cut, embossed]**
X102	**X**	**[round, embossed]**
X104	X	[square, large X in diamond]
X106	**X**	**[round, stamped X]**
X108	X	[die-cut embossed Hamilton]
X110	X	[rectangular, yellow X on black diamond & yellow]
X112	**X L TWIST**	**[round, white on red, Ormsby]**
X114	X L	[oval, embossed]
X116	**X T C**	**[die-cut letters, embossed]**
X118	**X X**	**[small die-cut, embossed]**
X120	**X X**	**[large die-cut, embossed]**
X122	X X	[round, embossed]

X124 **X X** **[die-cut embossed, Hamilton]**
X126 X X [rectangular, black on red]
X128 X.X. [round, embossed]
X130 X-X [round, engraved]
X132 **X X X** **[round, embossed]**
X134 X Y C [oval, black on red]
X136 X.Y.Z. (a) [round, black on red, Crews]
(b) [black on yellow]
X138 X BEST [round, red on white]
X139 **X FINE** **[die-cut, embossed Fine]**
X140 X FINE [banner embossed]
X142 X TRA [die-cut shape]
X144 X TRA [round, embossed]
X146 X TWIST [round, brown on yellow]
X148 X X GRADE [rectangular, black on red]
X150 X X NATURAL LEAF [round, brown & red on cream]

Y

Y101 **Y** **[rectangular, black on orange]**
Y102 Y [die-cut letter, red]
Y104 Y Y [rectangular, embossed]
Y106 **YANKEE DOODLE** **[diamond, black on red]**
Y108 YANKEE DOODLE [diamond, black on yellow, Dixson's Tobacco]
Y110 **YANKEE GIRL** **[rectangular, black on yellow, picture of a girl, Scotten, Dillon Co.]**
Y112 **YARBROUGH** **[rectangular, shield, blue on white]**
Y113 YARBROUGH [rectangular, banner embossed]
Y114 YELLOW BELL [round, black on cream, picture of girl]
Y116 **YELLOW HAMMER** **[rectangular, black on red, picture of a bird, L. Ash]**
Y118 YELLOW JACK TWIST [round, black on yellow]
Y120 YELLOW JACKET (a) [round, black on yellow, Crews]
(b) [Crews, Walkertown, N.C.]
Y122 YELLOW JACKET [round, black on orange, picture of a bee]
Y124 YELLOW KID [rectangular, yellow on black, picture of an Oriental boy]
Y126 **YELLOW PINE** **[round, black on yellow, picture of a girl, Leak Bros & Hasten]**
Y127 YELLOW PRINCE [rectangular, embossed]
Y128 YELLOW ROSE [round, yellow on green, picture of a flower]
Y130 YELLOW SAP [octagonal, embossed]
Y132 YELLOW SEAL [round, embossed]
Y134 YELLOW TWIST [rectangular, black on tan]
Y136 YOC-O-MAY [octagonal, red on yellow]
Y138 YOSTS CHOICE [rectangular, odd embossed]
Y140 **YOUNG FRITZ** **[round, black on red]**
Y142 **YOUNG FRITZ** **[oval, black on red]**
Y144 YOUNG JUDGE [rectangular, black on red]
Y146 YUCATAN [oval, black on red]
Y148 YULE [oval, embossed]
Y149 **YUM YUM** **(a) [octagonal red on yellow]**
(b) [black on red]
Y150 **YUM YUM** **(a) [round, scalloped, black on white, picture of a China-man]**
(b) [black on red]
(c) [black on green]
(d) [black on blue]

Z

Z101 **Z** **[round, cut-out Z]**
Z102 Z [round, embossed]
Z104 Z. GRAVELY'S BEST [round, black on yellow, Henry Co., Va.]
Z106 ZACH HAGEDORN [round, black on yellow, picture of a man]
Z108 ZADIE [rectangular, black on yellow, picture of a dog]
Z110 Z BOODLE [die-cut large Z, black on red]
Z112 ZAHN'S OLD NAVY [oval, black on red, picture of a ship]
Z113 **ZANE** **[large die-cut letters]**
Z114 ZAP [oval, embossed]
Z115 ZAP [oval, red on cream]
Z118 ZEBRA [rectangular, black on yellow, picture of a zebra]
Z120 **ZEBRA** **[diamond, black on cream, picture of a zebra running, Hodge]**
Z124 **ZENITH** **[round, embossed]**
Z125 ZEPA [round, red on white]
Z127 ZIEGETS 5 [square, black on red, square deal]
Z128 ZIMMER & CO. [rectangular, black on red, Petersburg, Va.]
Z130 **ZIP** **[diamond, embossed]**
Z132 **ZIG-ZAG** **[rectangular, odd black on gray]**
Z136 ZOO-ZOO [round, picture of a lion's head, Wilson & McCallay]

TIN TAGS ILLUSTRATED

The tags on the following pages are reproduced at approximately actual size.

A-101 A-105 A-109 A-114

A-117 A-121 A-125 A-127

A-130 A-135 A-139 A-142

A-143 A-146 A-147 A-149

A-151 A-163 A-167 A-179

A-180

A-181

A-183

A-186

A-199

A-208

A-209

A-217

A-219

A-232

A-235

A-238

A-242a

A-242b

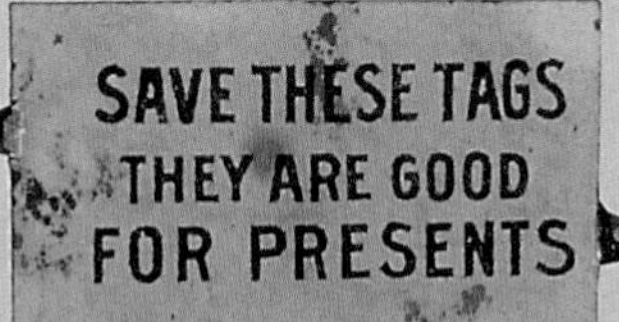

A-242c

A-253

A-255

A-257

A-259

A-261

A-264

A-267

Alto-Cow
200.00

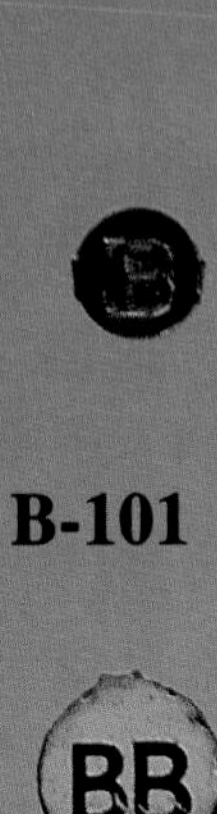

B-101

B-103

B-105

B-108a

B-108b

B-110

B-112

B-111

B-113

B-114

B-119

B-125

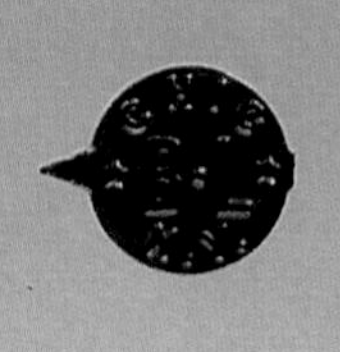

B-127

B-145

B-131

B-132

B-135

B-142

B-137

B-139

B-157

B-158

B-165

B-170

B-181

B-189

B-193

B-194

B-195

B-204

B-209

B-210

B-213

B-214

B-217

B-218

B-225

B-226

B-235

B-237

B-241

B-247

B-253

B-262

B-285

B-287

B-293

B-294

B-301

B-302

B-305

B-303

B-307

B-310

B-311

B-320

B-339

B-341

B-321

B-347

B-349

B-353

B-359

B-361

B-363

B-369

B-355

B-374

B-358

B-382

B-383

B-385

B-386

B-387

B-391

B-397

B-399

B-401

B-403

B-405

B-407

B-411

B-417

B-421

B-427

B-437

B-441

B-443

B-445

B-449

B-451

B-453

B-458

B-462

B-463

B-467

B-471

B-474

B-475

B-477

B-481

B-486

B-495

B-497

B-505

B-507

B-508

B-509

B-511

B-514

B-515

B-522

B-530

B-525

B-544

B-547

B-551

B-555

B-557

B-567

B-572

B-576

B-577

B-581

B-582

B-585

B-593

B-601

B-607

B-608

B-609

B-613

B-619

B-620

B-621

B-627

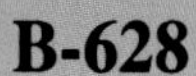

B-628

B-633

B-651

B-653

B-655

B-658

B-659

B-665

B-671

B-673

B-677

B-678

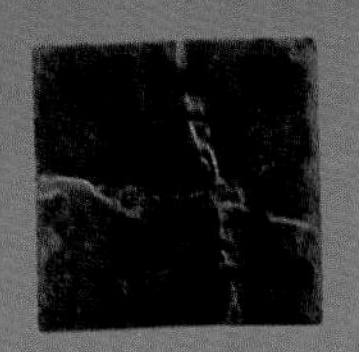

B-679

B-683

B-684

B-685

B-687

B-697

B-700

B-702

B-707

B-711

B-713

B-717

B-721

B-725

B-729

B-730

B-739

B-742

B-747

B-748

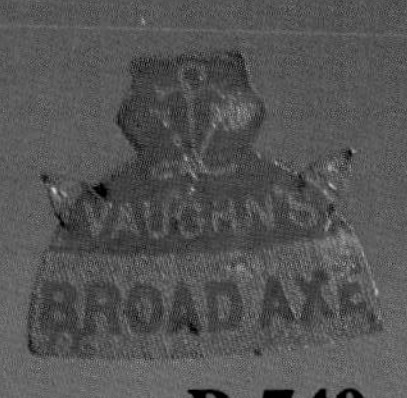

B-749

B-754

B-761

B-755

B-757

B-762

B-766

B-769

B-783

B-784

B-785

B-792

B-795

B-798

B-799

B-803

B-805

B-806

B-809

B-811

B-814

B-815

B-817

B-818

B-819

B-823

B-825

B-833b

B-839

B-842

B-843

B-845

B-853

B-857

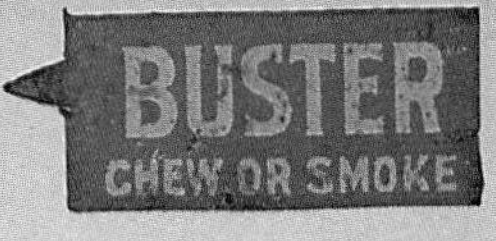

B-861

B-865

B-867

B-877

B-883

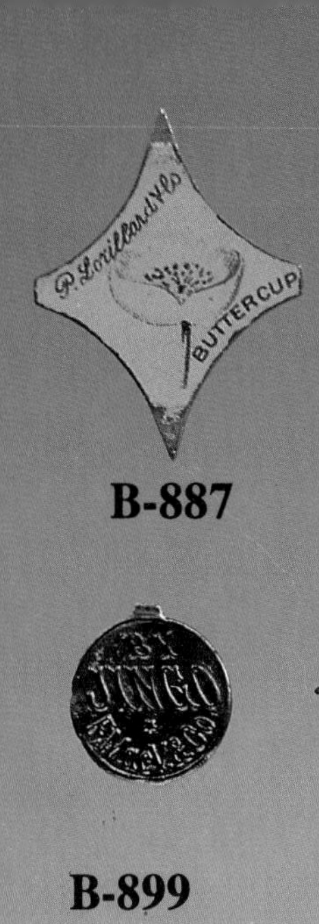

B-889

B-895

BUZZ SAW

B-897

B-887

B-899

B-903

B-905

B-907

C-101

C-103

C-109

C-111

C-115

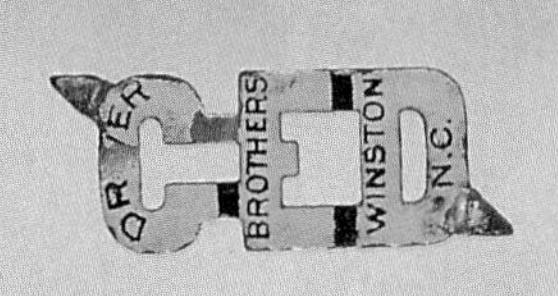

C-119

C-123

C-126

C-128

C-129

C-486

C-137

C-139

C-141

C-145

C-147

C-149

C-153

C-155

C-159a

C-159b

C-161

C-169

C-180

C-181

C-183

C-185

C-187

C-191

C-195

C-203

C-205

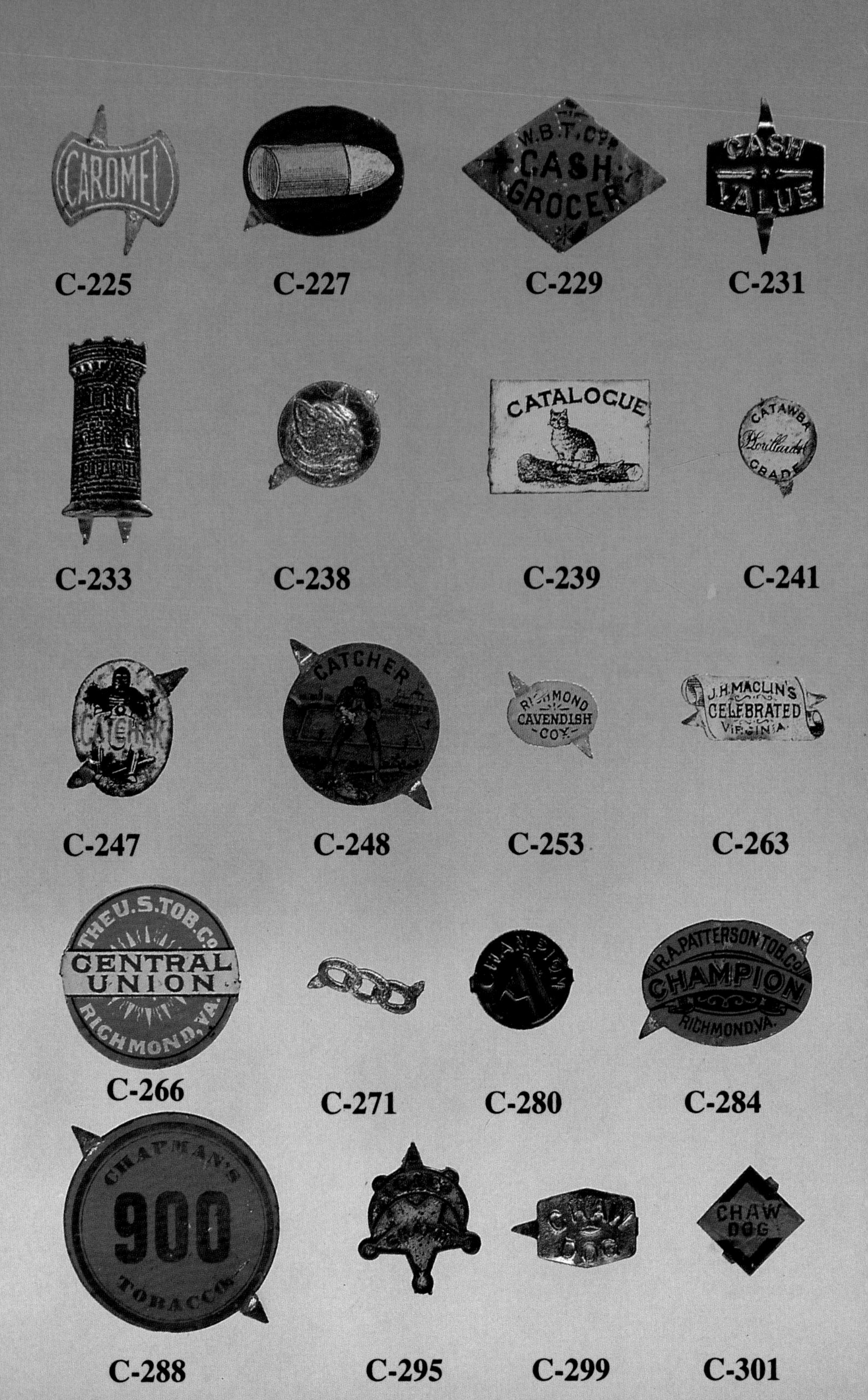

C-225 C-227 C-229 C-231

C-233 C-238 C-239 C-241

C-247 C-248 C-253 C-263

C-266 C-271 C-280 C-284

C-288 C-295 C-299 C-301

C-303

C-305

C-307

C-309

C-321

C-325

C-329

C-330

C-332

C-333

C-323

C-326

C-346

C-349

C-351

C-355

C-357

C-359

C-367

C-371

C-372 C-373 C-377 C-379

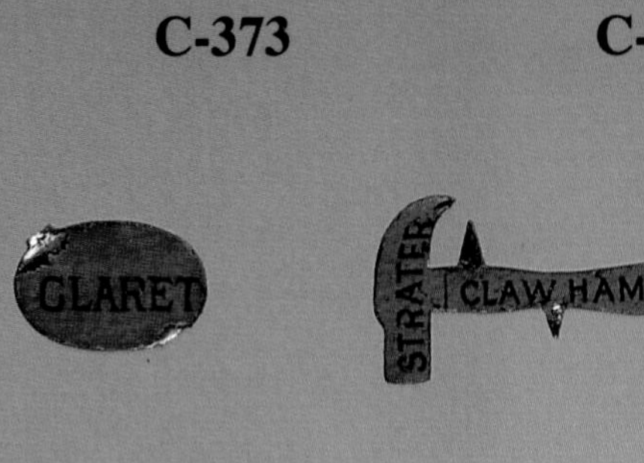

C-389 C-398 C-399 C-407

C-409 C-411 C-415 C-433

C-437 C-443 C-447 C-451

C-453 C-455 C-457 C-465

C-469

C-471

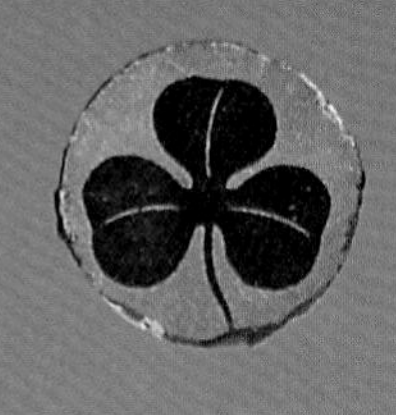

C-472

C-473

C-475

C-477

C-479

C-481

C-488

C-495

C-497

C-501

C-504

C-505

C-511

C-512

C-513

C-515

C-520

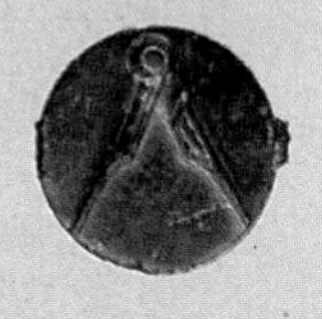

C-526

C-531

C-535

C-538

C-541

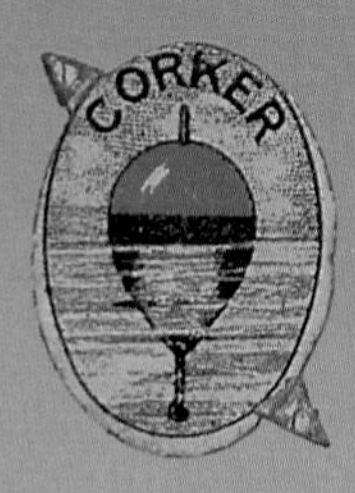

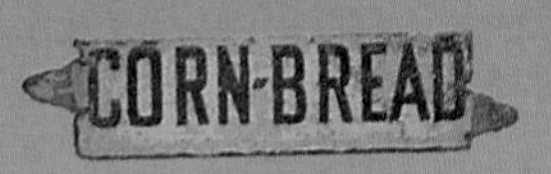

C-557

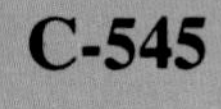

C-545

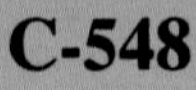

C-548

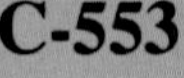

C-553

C-561

C-563

C-571

C-573

C-576

C-579

C-587

C-592

C-596

C-597

C-598

C-607

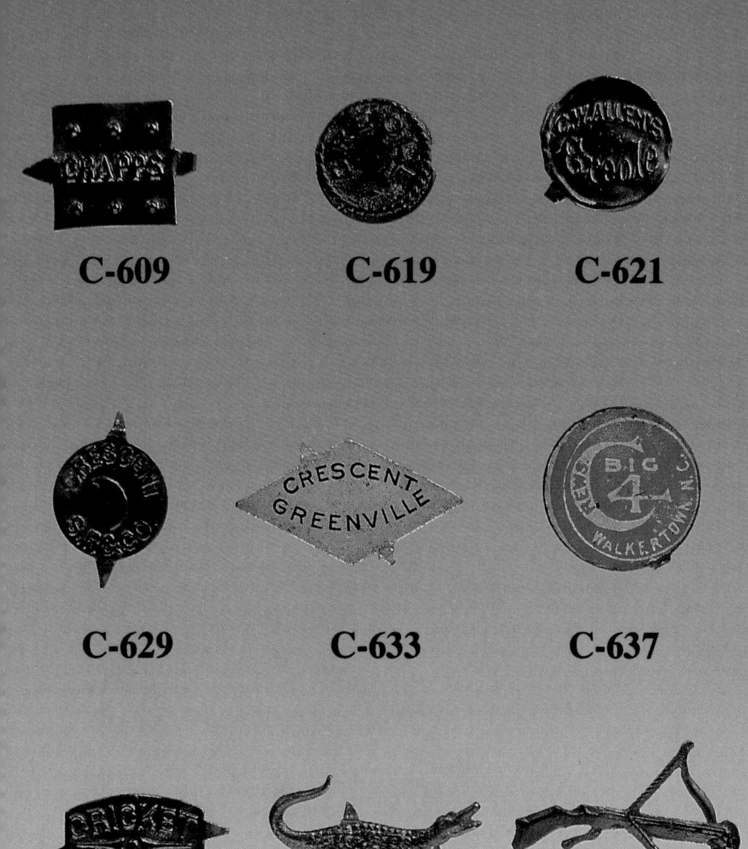

C-609 C-619 C-621 C-627

C-629 C-633 C-637 C-639

C-646 C-647 C-651 C-653

C-655 C-663 C-665 C-669

C-670 C-671 C--677 C-679

C-681

C-683

C-685

C-696

C-699

C-703

C-705

C-707

D-101

D-103

D-105

D-108

D-109

D-121

D-123

D-125

D-127

D-129

D-135

D-137

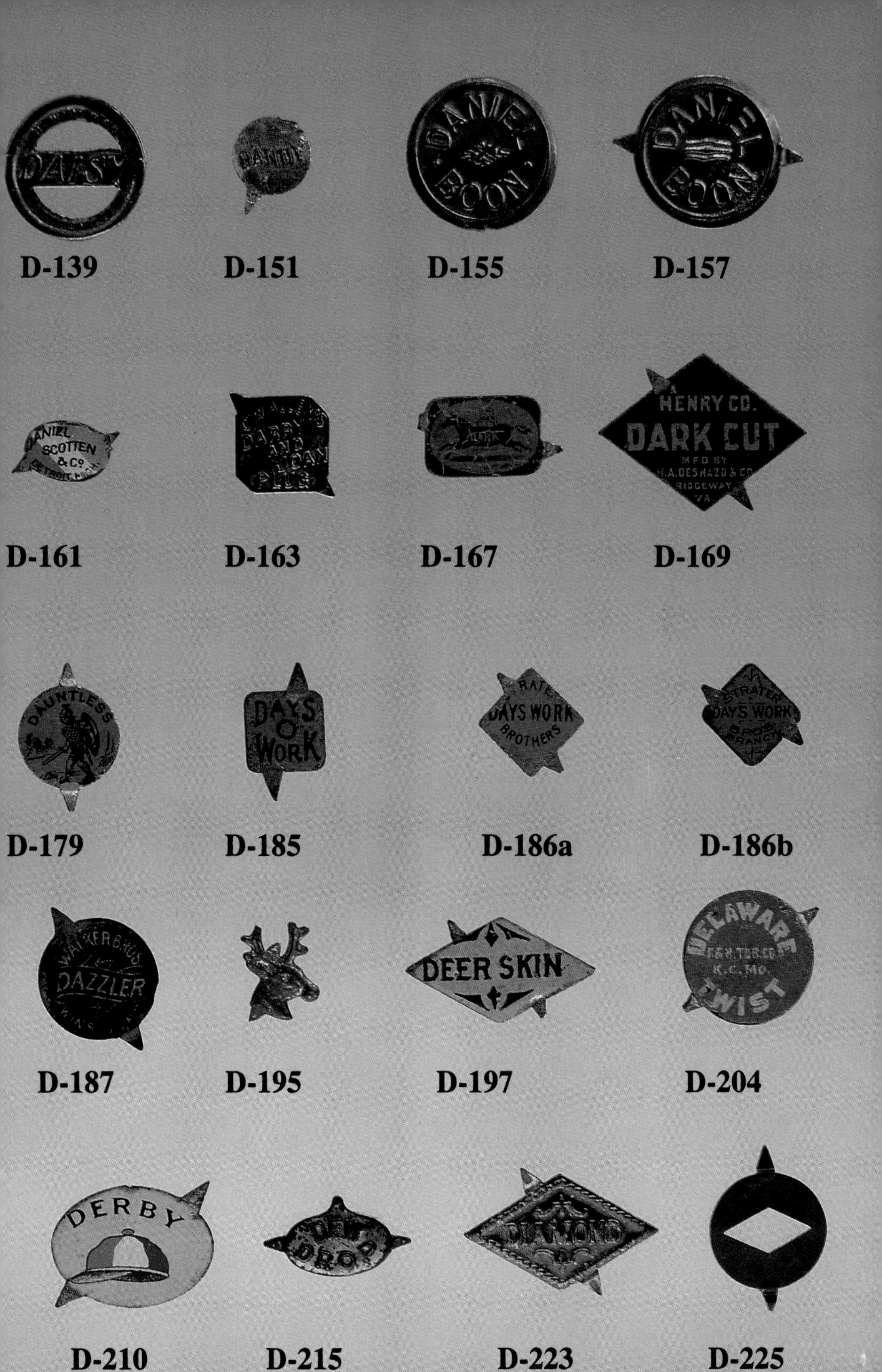

D-139 D-151 D-155 D-157

D-161 D-163 D-167 D-169

D-179 D-185 D-186a D-186b

D-187 D-195 D-197 D-204

D-210 D-215 D-223 D-225

D-229

D-231

D-232

D-237

D-239

D-243

D-245

D-247

D-253

D-259

D-261

D-263

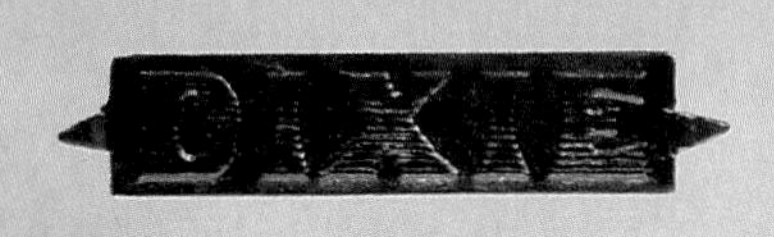

D-269

D-277

D-281

D-289

D-301

D-305

D-306

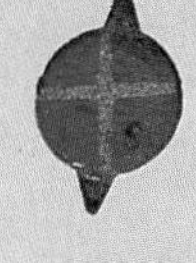

D-307

D-308

D-317

D-319

D-321

D-325

D-331

D-347

D-350

D-351

D-353

D-357

D-359

D-373

E-101

E-103

E-109

E-113

E-117

E-119

E-121 E-123

E-136 E-137

E-143 E-146

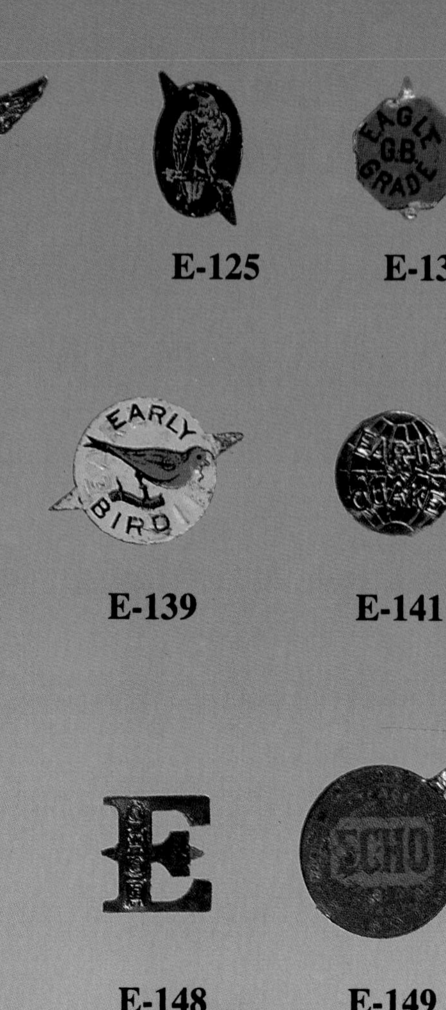

E-125 E-133

E-139 E-141

E-148 E-149

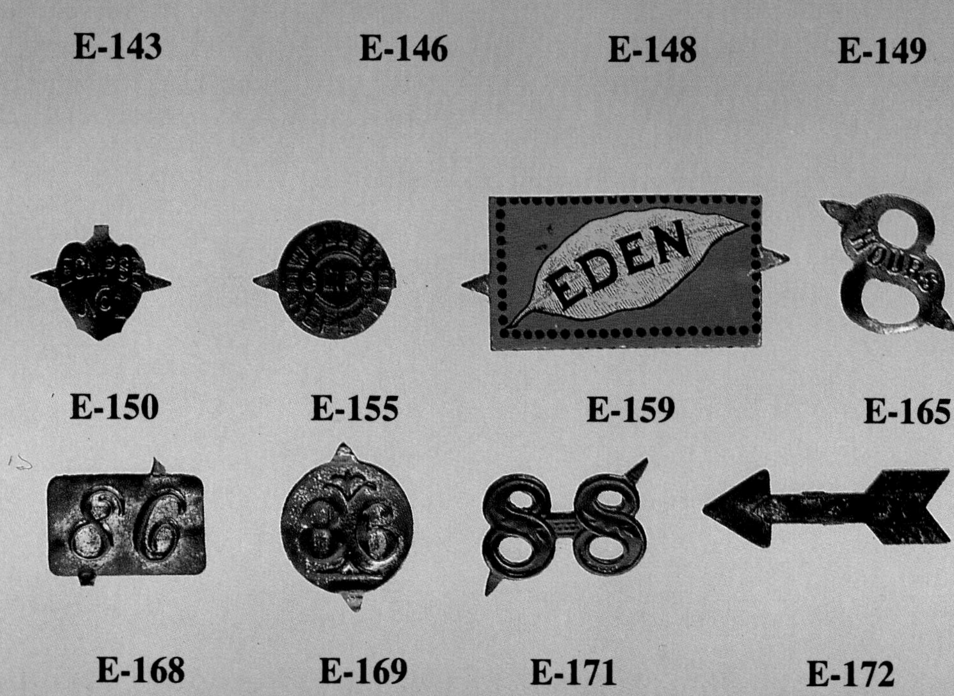

E-150 E-155 E-159 E-165

E-168 E-169 E-171 E-172

E-173

E-177

E-179

E-181

E-182

E-184

E-195

E-211

E-213

E-215

E-219

E-221

E-223

E-225

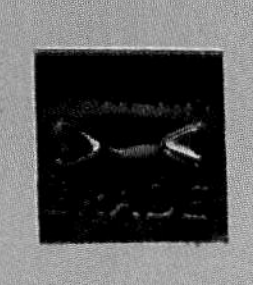

E-229

E-233

E-237

E-239

E-243

E-245

E-250

E-254

E-258

E-259

E-262

E-271

F-101

F-103

F-104

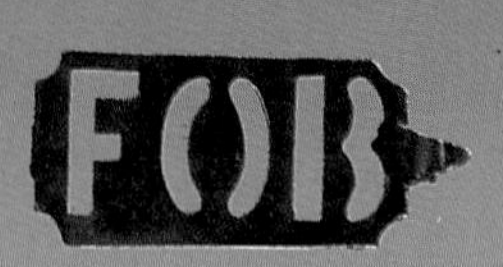

F-114

F-118

F-119

F-123

F-127

F-128

F-130

F-131

F-133

F-137

F-138

F-139

F-151

F-159

F-161

F-167

F-175

F-177

F-182

F-187

F-189

F-195

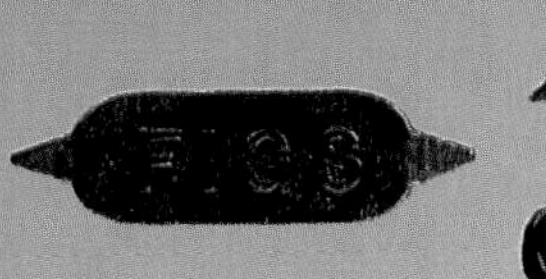

F-203

F-209

F-219

F-223

F-226

F-227

F-234

F-241

F-251

F-255

F-257

F-259

F-261

F-265

F-269

F-271

F-273

F-279

F-283

F-285

F-287

F-291

F-296

F-301

F-302

F-309

F-311

F-313

F-315

F-317

F-318

F-324

F-325

F-328 F-331 F-336 F-339

F-341 F-359 F-363 F-365

F-367 F-373 F-375 F-377

F-382 F-383 F-385 F-386

F-389 F-393 F-397 F-399

G-103

G-105

G-107

G-111

G-118

G-125

G-127

G-135

G-143

G-144

G-147

G-157

G-159

G-169

G-170

G-175

G-183

G-201

G-203

G-205

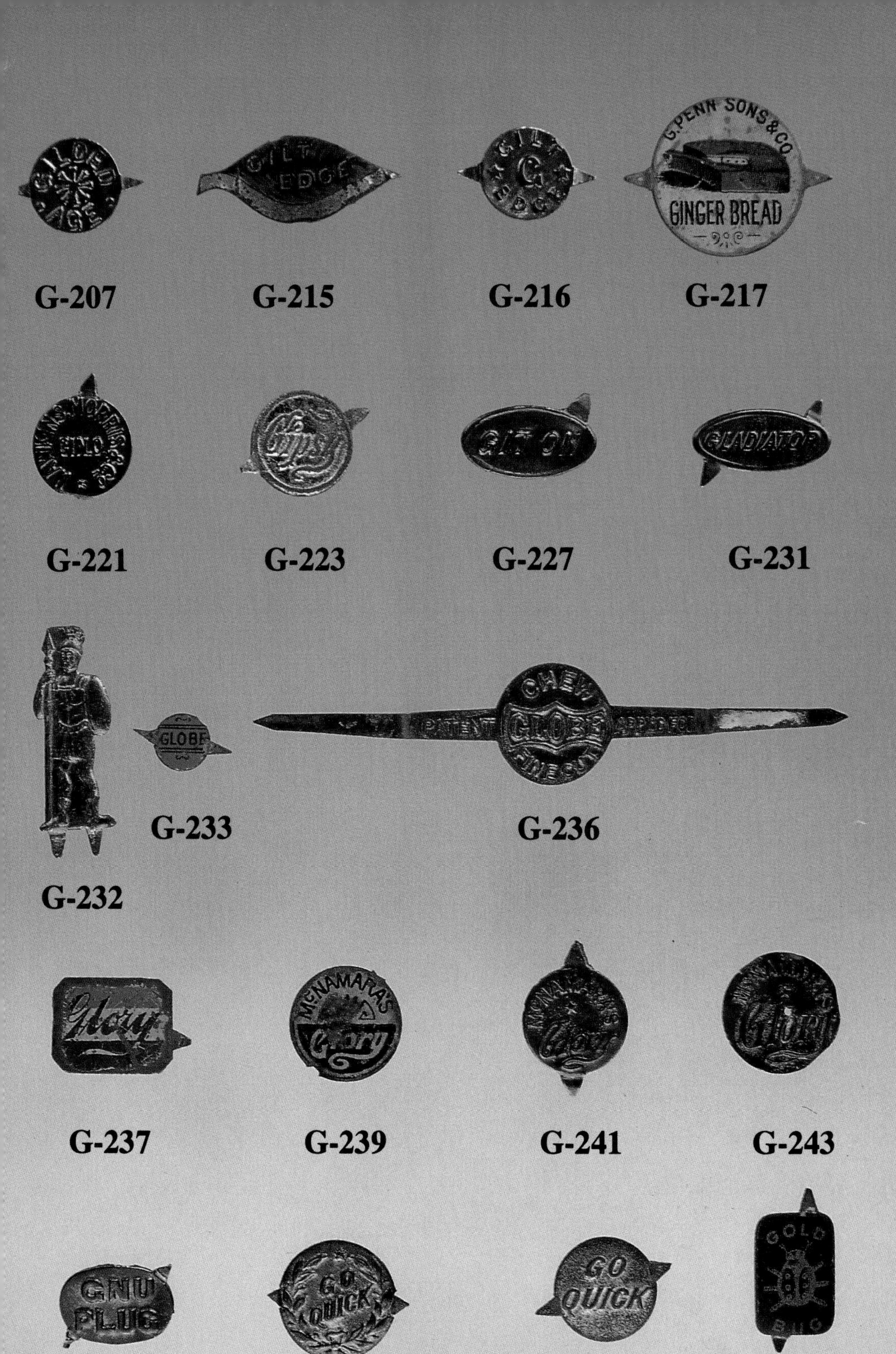

G-207 G-215 G-216 G-217

G-221 G-223 G-227 G-231

G-232 G-233 G-236

G-237 G-239 G-241 G-243

G-247 G-249 G-251 G-261

G-263

G-265

G-267

G-269

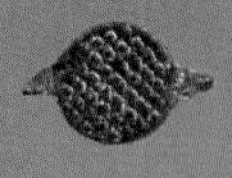

G-275

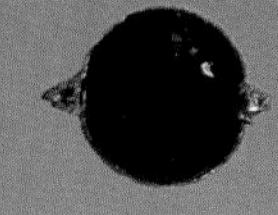

G-277

G-279

G-289

G-292

G-293

G-296

G-299

G-303

G-305

G-309a

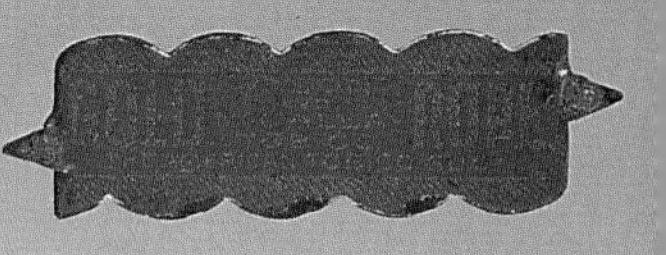

G-309b

G-313

G-317

G-322

G-329

G-331

G-333

G-337

G-347

G-353

G-355

G-357

G-364

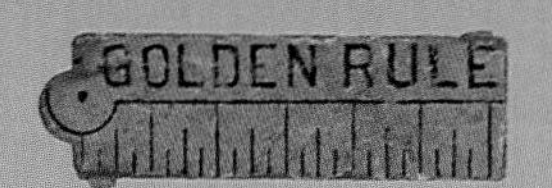

G-367a

G-368

G-371

G-375

G-377a

G-377b

G-379

G-385

G-394

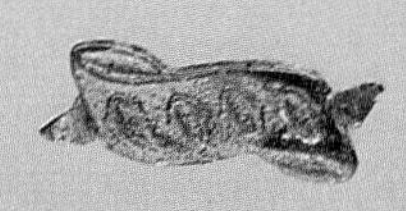

G-395

G-397

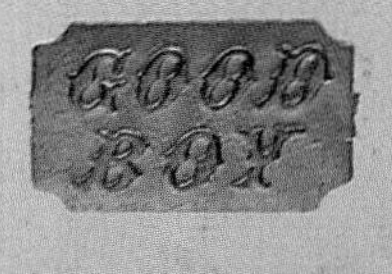

G-401

G-405

G-409

G-410

G-413

G-415

G-417

G-419

G-421

G-429a

G-429b

G-434

G-450

G-455

G-463

G-471a

G-471b

G-477

G-481

G-483

G-491

G-493

G-494

G-495

G-497a

G-497b

G-503

G-507

G-511

G-513

G-515

G-533

G-538

G-539

G-540

G-541

G-543

G-547

G-551

G-561

H-104 H-105 H-111 H-112

H-113 H-114 H-117 H-125

H-137 H-139 H-143 H-145

H-149 H-155 H-161 H-160

H-163 H-166 H-171 H-173 H-193

H-195 H-197 H-199 H-205

H-209 H-211 H-215a H-215b

H-219 H-225 H-227 H-237

H-243 H-245 H-246 H-249

H-251 H-254 H-255 H-.256 H-257

H-266

H-271

H-279

H-287

H-289

H-290

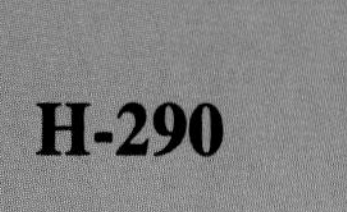

H-299a

H-299b

H-299c

H-299d

H-301a

H-301b

H-306

H-311

H-317

H-321

H-331

HI-YELLOW

H-297

H-333

H-335

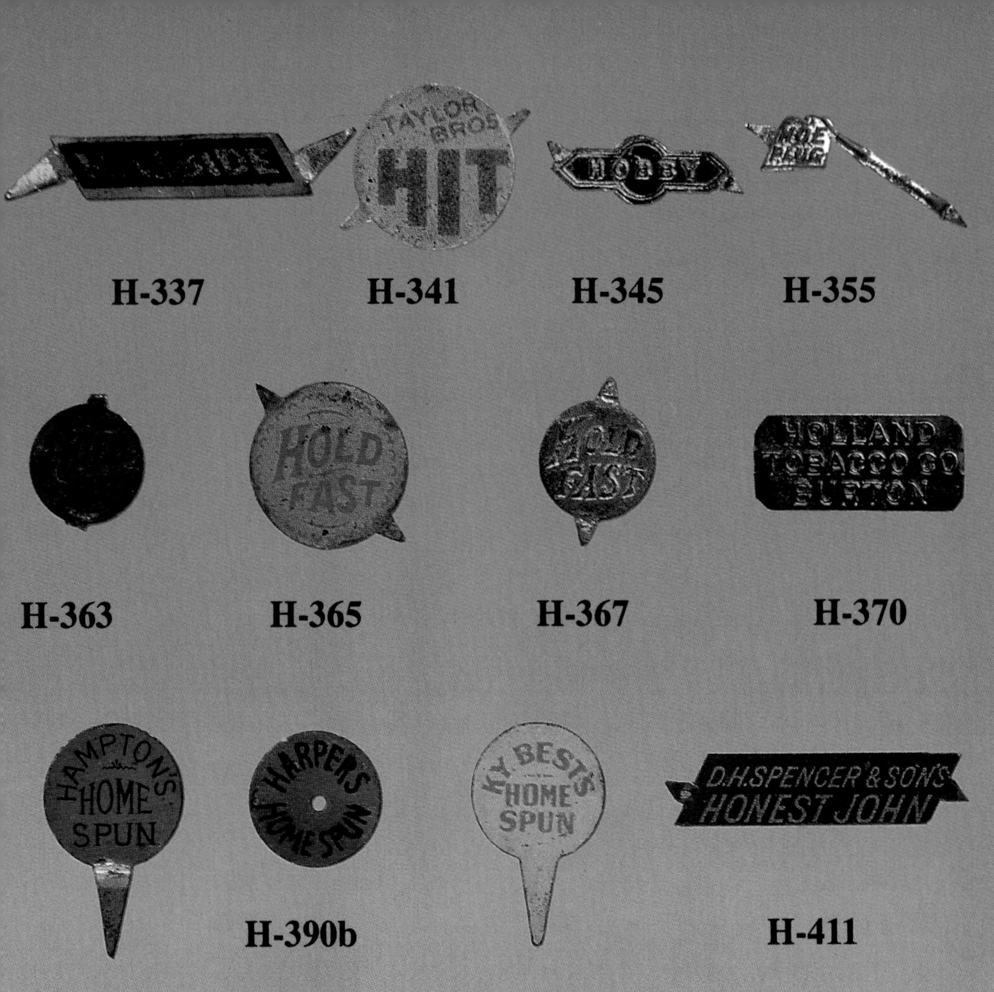

H-337 H-341 H-345 H-355

H-363 H-365 H-367 H-370

H-390b H-411

H-390a H-393

H-413 H-422 H-431a

H-417

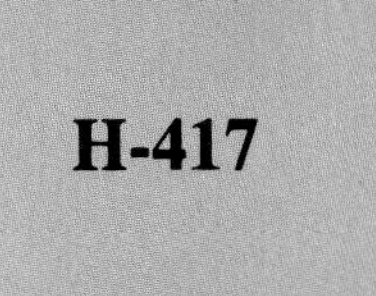

H-439 H-443 H-446 H-447

H-453a

H-453b

H-459

H-469

H-476

H-478

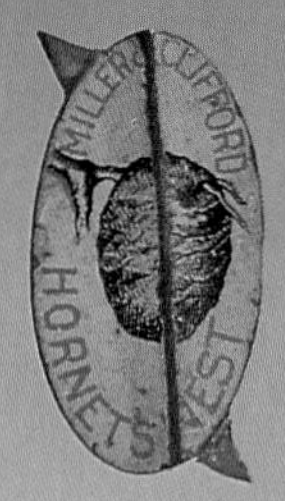

H-487

H-488

H-489

H-491

H-492

H-493

H-495

H-496

H-497

H -499a

H-499c

H-501

H-513

H-515a H521a H-521b H-523a

H-523b H-525 H-527 H-531

H-533 I-102 I-109 I-115 I-123

I-131 I-133 I-135 I-143

I -147a I-147d I-149 I-155

I-156

I-165

I-168

J-103

J-106

J-107

J-108

J-110

J-113

J-117

J-119

J-125

J-131

J-137

J-147

J-157

J-159

J-153

J-161

J-163

J-169

J-173

J-177

J-183

J-186

J-189

J-197

J-203

J-213

J-221

J-226

J-227

J-233

J-237

J-244

J-259

J-261

J-275

J-279a

J-279b

J-288

J-290

J-291

J-292

J-294

J-293

J-298

J-302

J-308

J-309

J-311

J-314

J-319

J-323

J-329

J-331

J-333

K-101

K-103

K-106

K-107

K-115

K-119

K-125a

K-125b

K-133

K-142

K-154

K-156

K-173

K-175

K-177

K-179

K-181

K-189a

K-189b

K-187

K-201

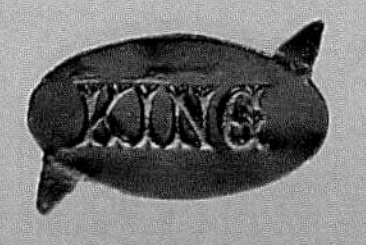

K-207

K-243

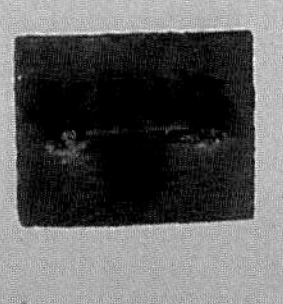

K-212

K-219a

K-219c

K-223

K-235

K-241

K-247

K-249

K-251a

K-251c

K-257

K-259

L-105

L-106

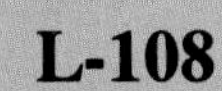

L-108

L-109

L-111

L-119

L-123

L-125

L-134

L-141

L-143

L-153

L-152

L-155

L-163

L-165

L-171

L-173

L-175b

L-185

L-189

L-197

L-207

L-209

L-211

L-213b

L-214b

L-218

L-225

L-229

L-237

L-241

L-249

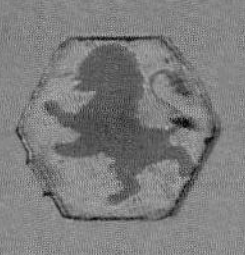

L-251

L-253

L-256

L-257

L-265

L-266

L-269

L-272

L-274

L-287a

L-291

L-304

L-317

L-337

L-341

L-343

L-351

L-353

L-357

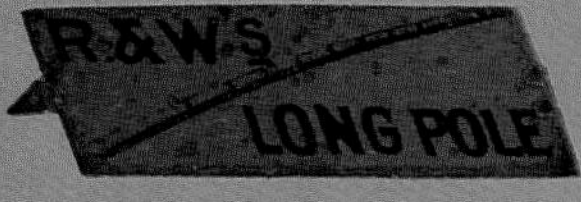

L-359a

L-359b

L-359c

L-367

L-371

L-373a

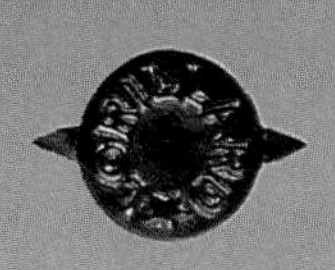

L-373b

L-379a

L-379b

L-383

L-387

L-391

L-395

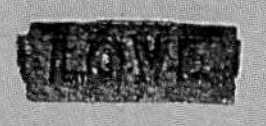

L-401

L-407b

L-417

L-423b

L-429

L-431

L-435

L-437

L-439

L-441

L-445

L-446a

L-449

L-453

L-455

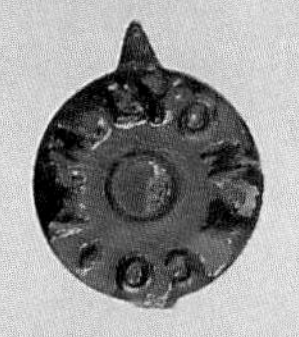

L-462

M-101

M-103

M-117

M-124

M-126

M-129

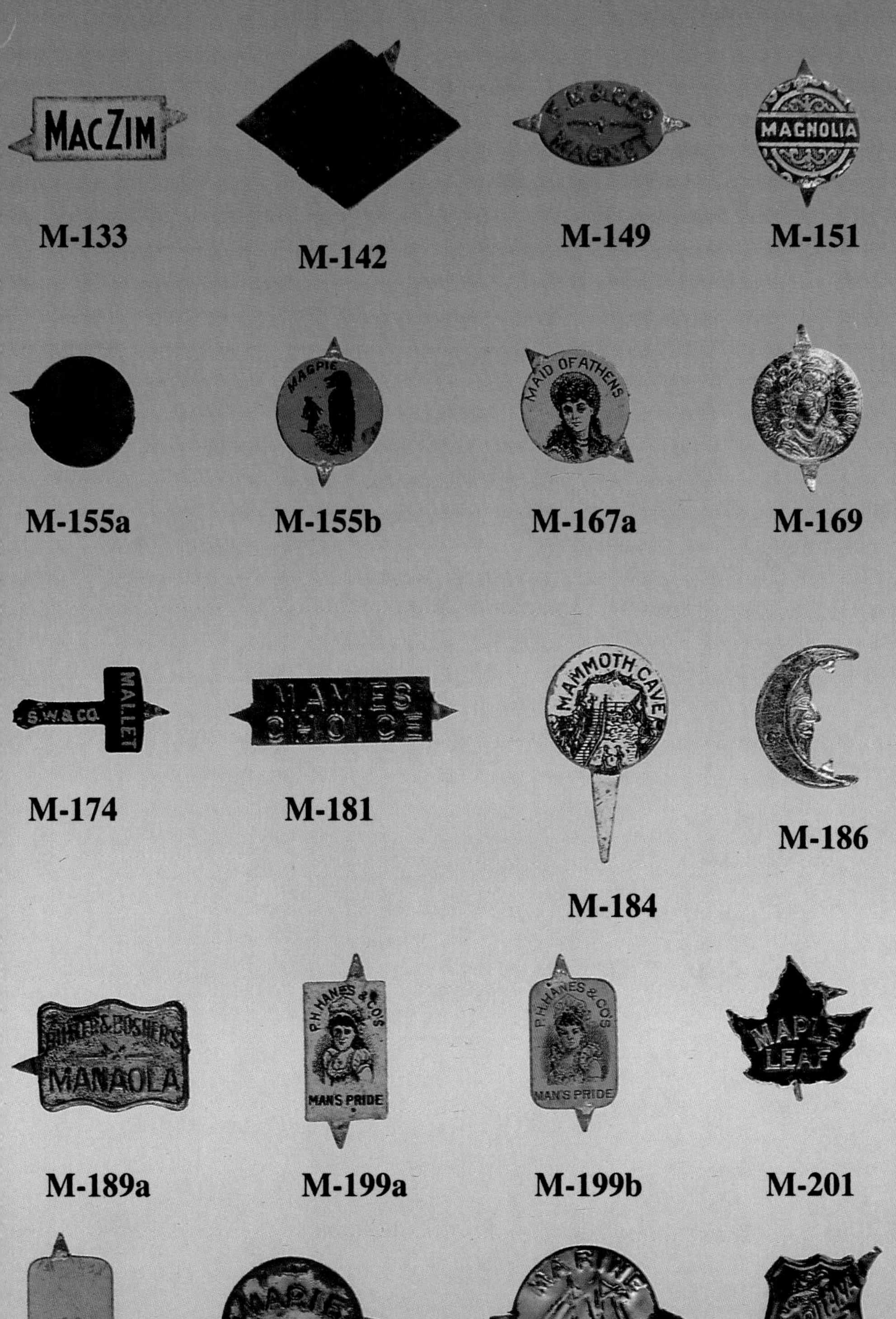

M-133 M-142 M-149 M-151

M-155a M-155b M-167a M-169

M-174 M-181 M-184 M-186

M-189a M-199a M-199b M-201

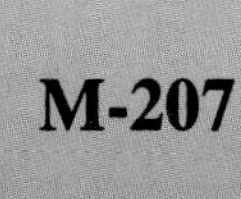

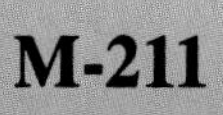

M-207 M-208 M-211 M-213

M-214

M-219

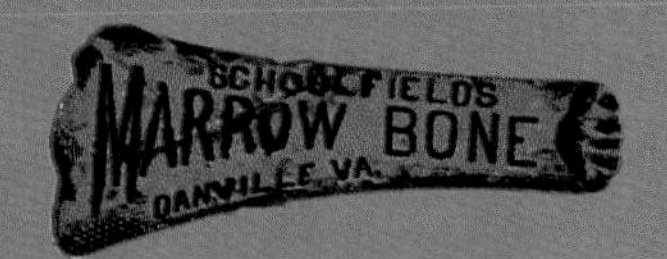

M-221

M-223

M-229

M-233

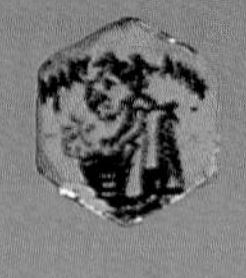

M-237

M-241a

M-242

M-243

M-250

M-251

M-253

M-257

M-262a

M-262b

M-274

M-277

M-282

M-283

M-284

M-286

M-289a

M-289b

M-295

M-301

M-303a

M-303b

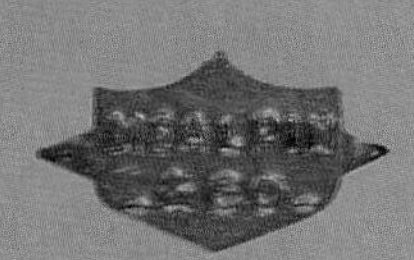

M-303c

M-309

M-316

M-319

M-319

M-341

M-343

M-347

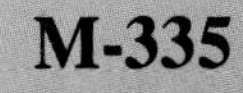

M-335

M-359

M-370

M-373

M-380

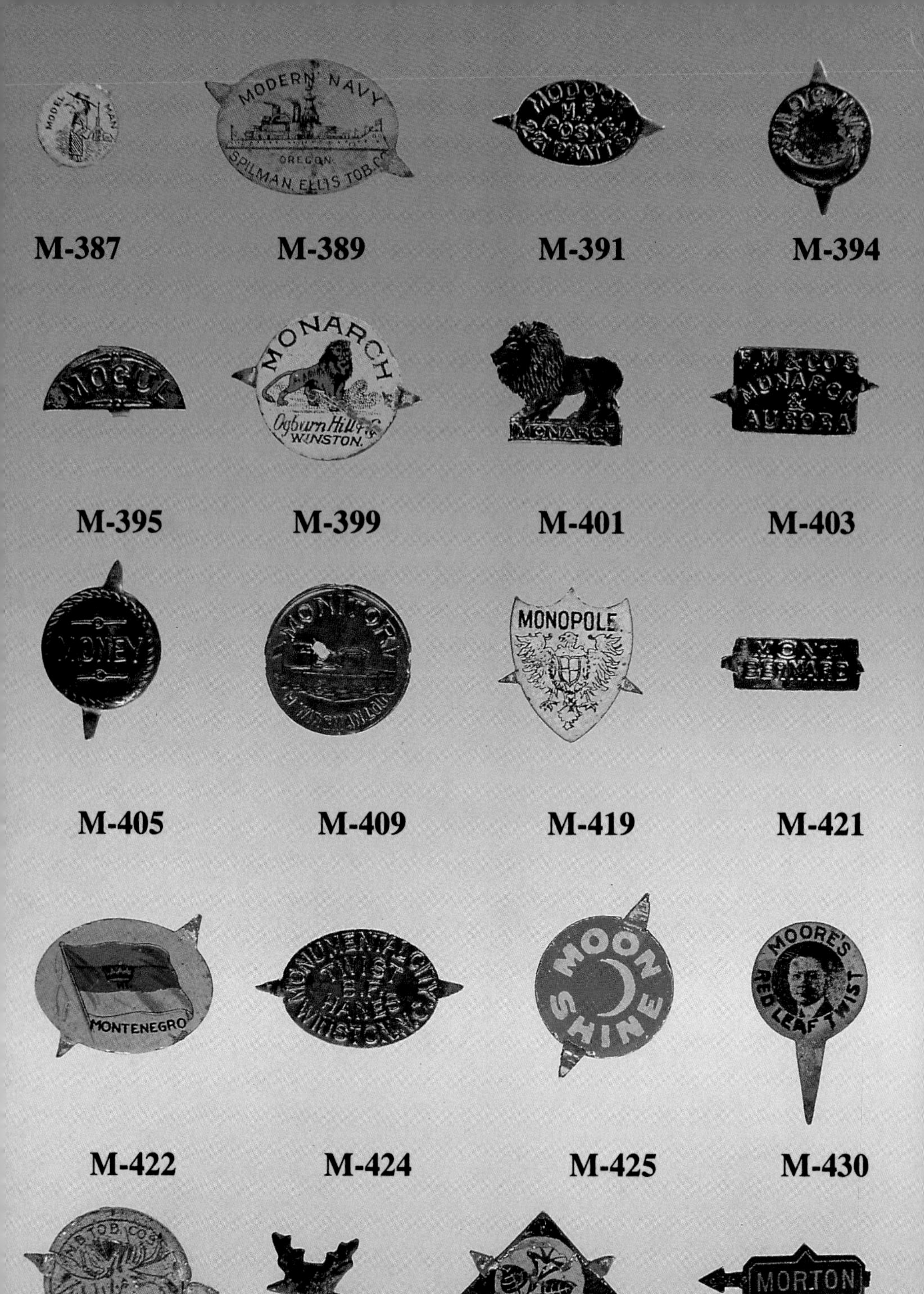

M-432

M-433

M-439

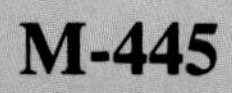
M-445

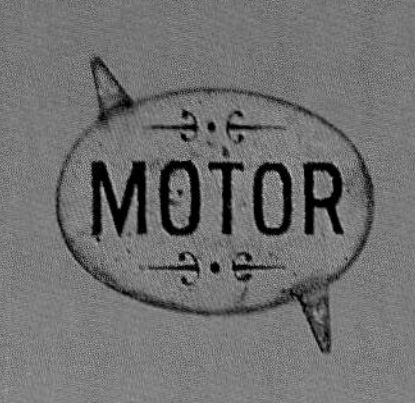

M-451

M-459

M-466

M-470

M-473

M-476

M-481

M-483

M-485a

M-486

M-487

N-101

N-102

N-112

N-115

N-116

N-118

N-119

N-120

N-124

N-128

N-129

N-136

N-140

N-147

N-152

N-160a

N-160b

N-162

N-172

N-173a

N-173b

N-174

N-175b

N-176

N-181

N-182

N-189

N-194a

N-170

N-200

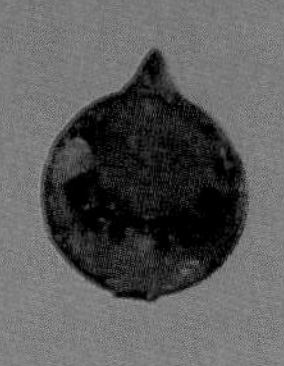

N-216

N-217

N-220

N-222

N-227

N-234

N-236

N-237

N-246

N-252

N-255

N-258

N-259

N-263

N-276

N-280a

N-280c N-289 N-304 N-306

N-318 N-320  N-322 N-333

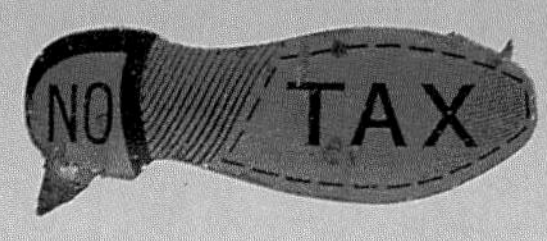

N-340 N-341 N-346 N-347

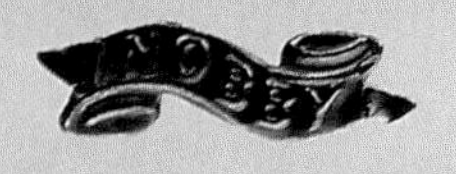

N-349 N-350 N-351 N-353

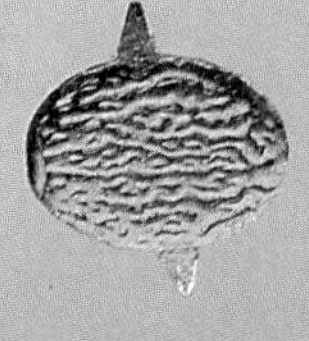

N-362 N-386 N-395 N-396

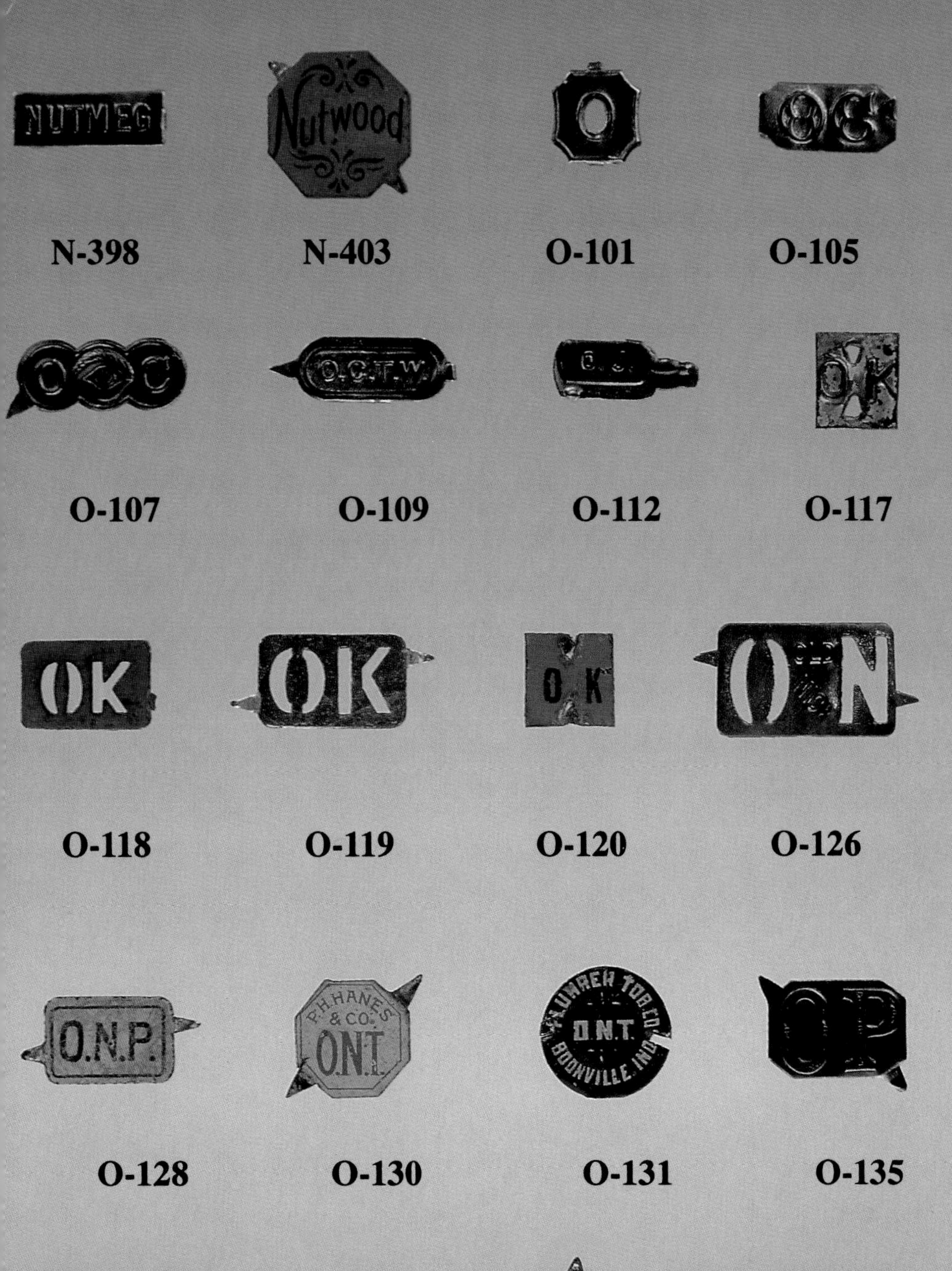

N-398 N-403 O-101 O-105

O-107 O-109 O-112 O-117

O-118 O-119 O-120 O-126

O-128 O-130 O-131 O-135

O-143 O-150a O-162 O-170

O-171

O-179

O-181

O-187

O-191

O-192

O-195

O-209

O-211

O-216a

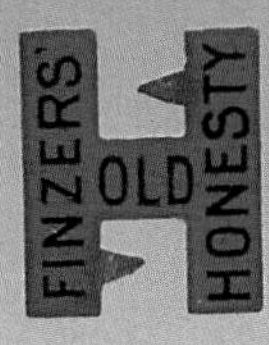

O-216b

O-217

O-220

O-223

O-224

O-229

O-238

O-242

O-254

O-258

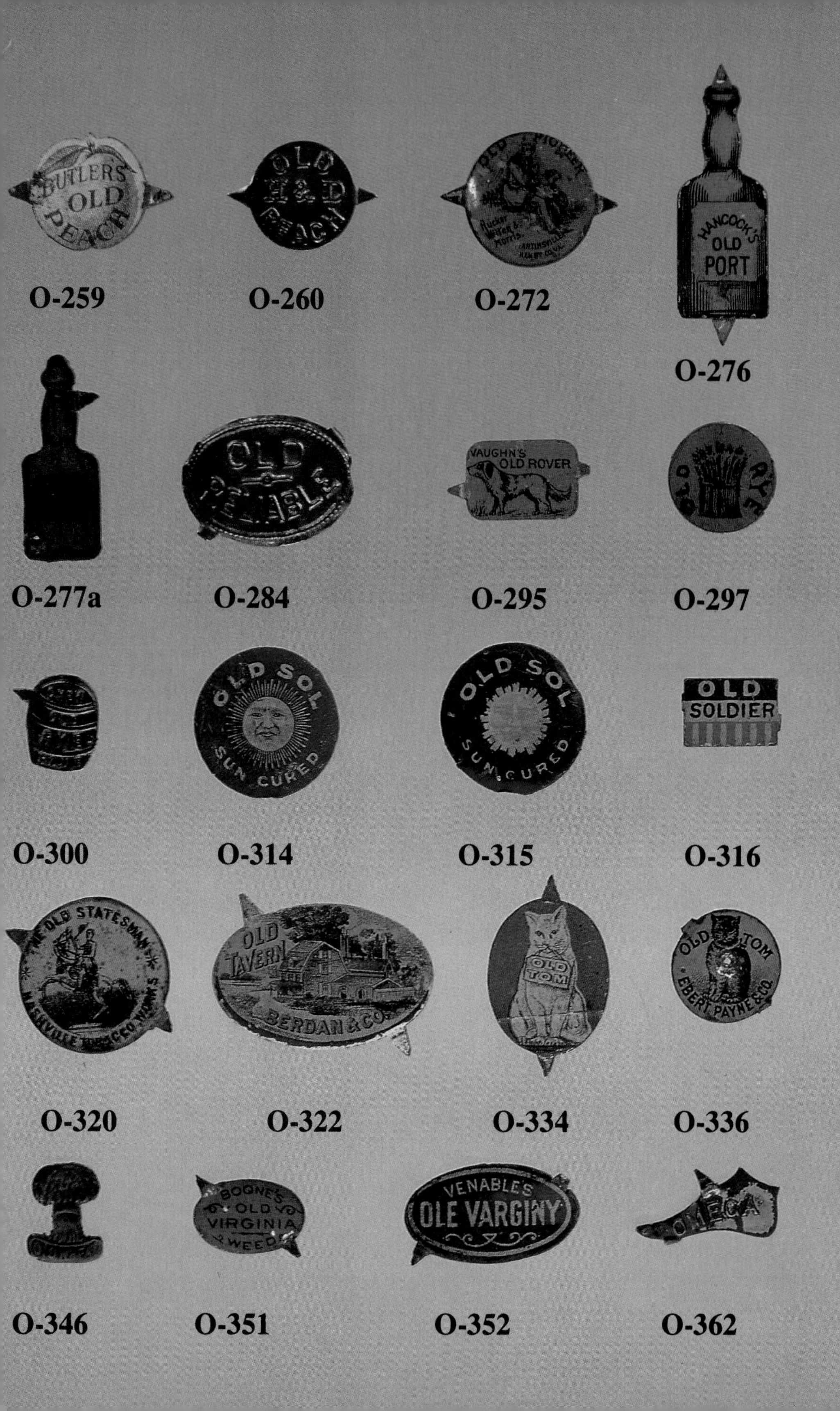
BUTLERS OLD PEACH
O-259
OLD H & D PEACH
O-260
OLD PIONEER
O-272
HANCOCK'S OLD PORT
O-276
O-277a
OLD RELIABLE
O-284
VAUGHN'S OLD ROVER
O-295
OLD RYE
O-297
O-300
OLD SOL SUN CURED
O-314
OLD SOL SUN CURED
O-315
OLD SOLDIER
O-316
THE OLD STATESMAN
O-320
OLD TAVERN
BERDAN & CO.
O-322
OLD TOM
O-334
OLD TOM
EBERT PAYNE & CO.
O-336
O-346
BOGNE'S OLD VIRGINIA WEED
O-351
VENABLE'S OLE VARGINY
O-352
OMEGA
O-362

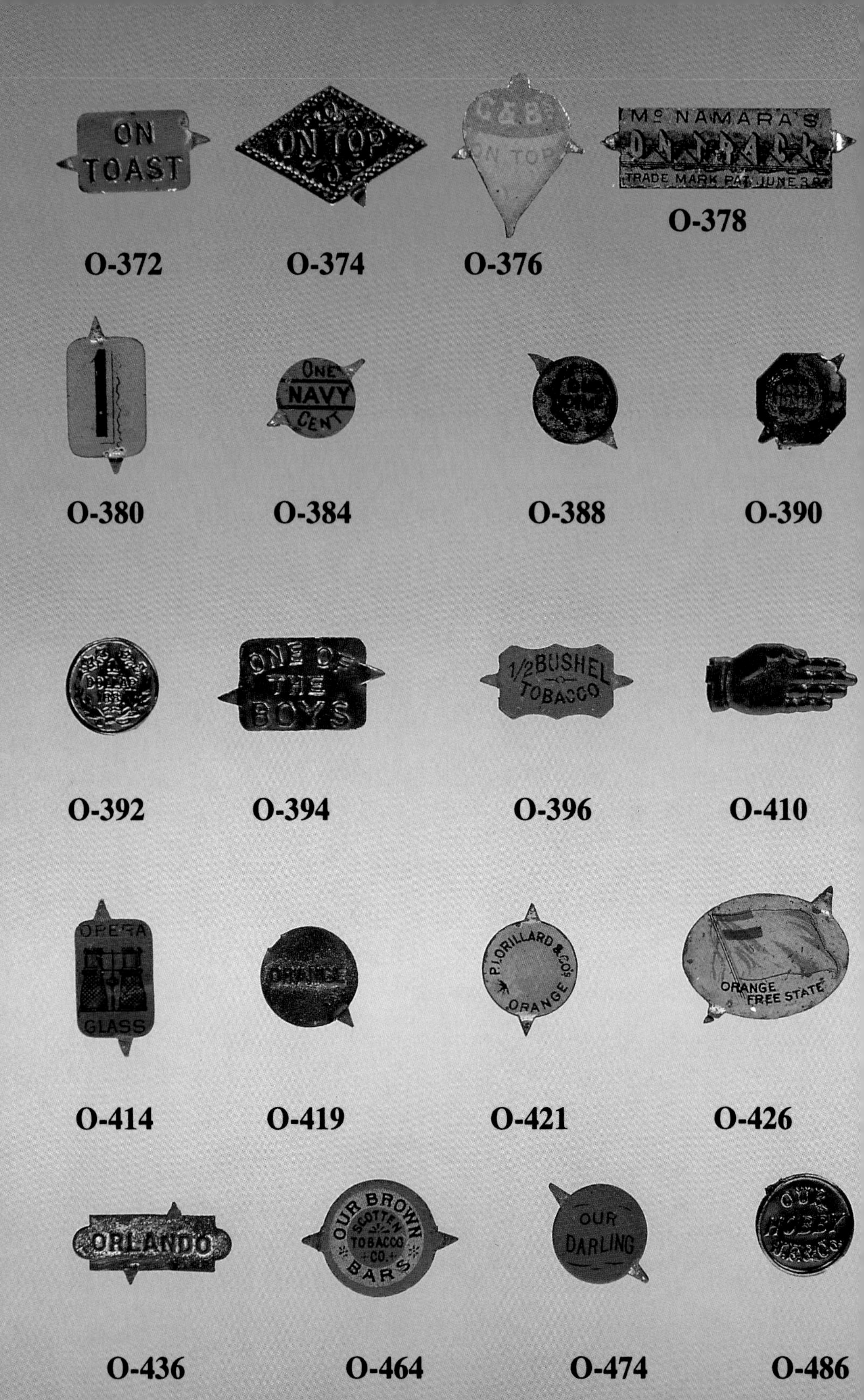

O-372 O-374 O-376 O-378

O-380 O-384 O-388 O-390

O-392 O-394 O-396 O-410

O-414 O-419 O-421 O-426

O-436 O-464 O-474 O-486

O-492

O-500

O-504

O-508

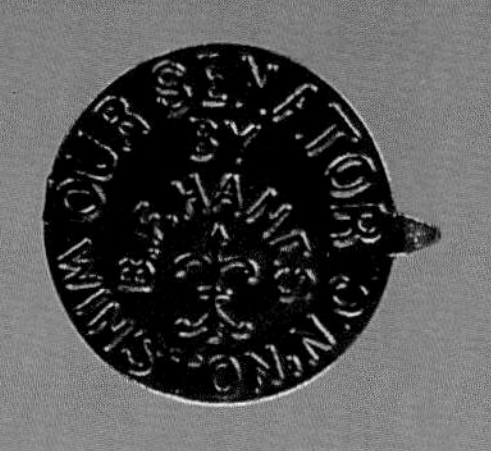

O-514

O-516

O-522

O-524

O-525

O-526

O-527

O-532

O-534

O-539

P-101

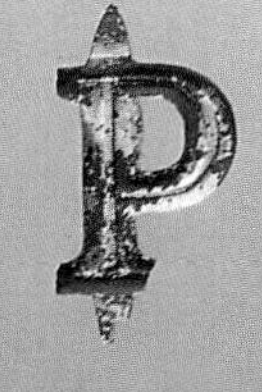

P-103

P-107

P-108

P-112

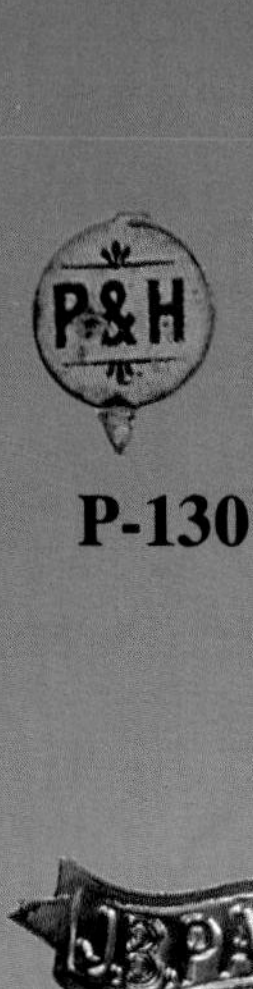

P-130

P-140

P-144

P-160

P-162

P-164

P-168

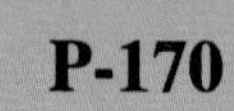

P-170

P-172

P-186a

P-186b

P-188

P-192

P-202

P-204

P-206

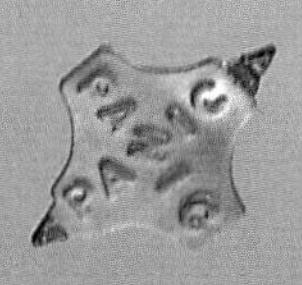

P-208

P-214

P-216

P-224

P-228

P-230

P-240

P-239

P-256

P-258

P-260

P-246

P-268

P-272

P-284

P-286

P-288

P-292

P-300

P-298

P-302

P-308

P-312a

P-312b

P-318

P-336

P-342

P-350

P-352

P-354

P-368

P-376

P-378

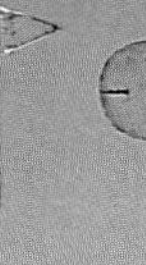
P-380

P-384

P-386

P-388

P-390

P-402

P-404

P-406

P-410

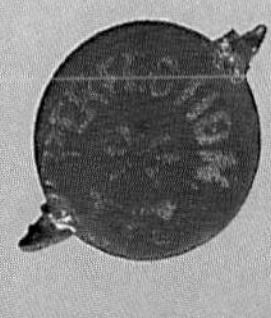
P-412

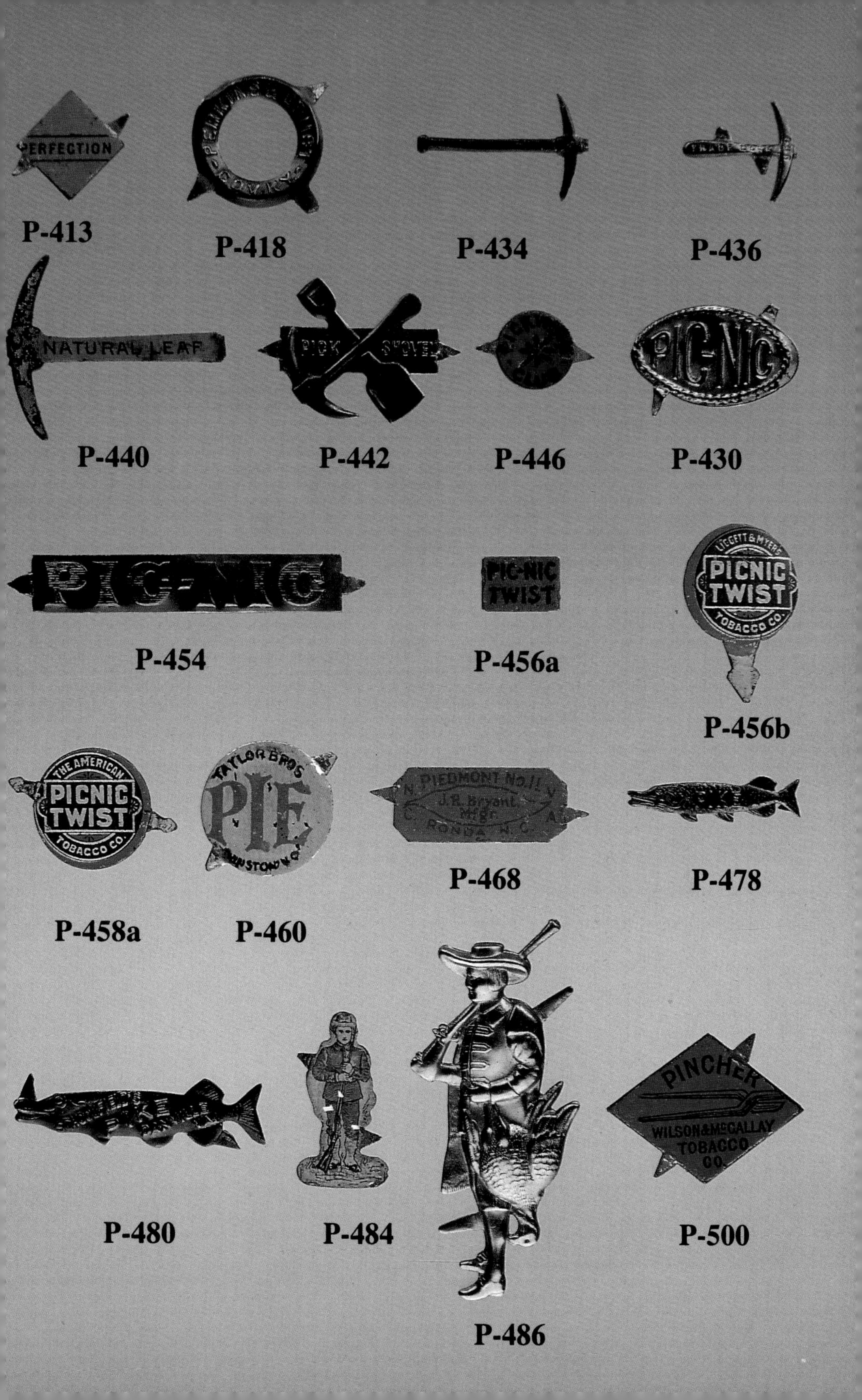

P-413 P-418 P-434 P-436

P-440 P-442 P-446 P-430

P-454 P-456a P-456b

P-458a P-460 P-468 P-478

P-480 P-484 P-486 P-500

P-504

P-526

P-528

P-530

P-538

P-552

P-556

P-557

P-560

P-564

P-568

P-571

P-574

P-580

P-584

P-588

P-590

P-592

P-596

P-600

P-605 P-608 P-610 P-636

P-642 P-644 P-648 P-650a

P-652 P-659 P-664 P-674

P-684 P-686 P-688 P-690

P-692 P-694 P-700 P-702

P-706

P-712

P-718

P-720

Q-103

Q-106

P-628

Q-112

P-632

Q-118

Q-123

Q-126a

Q-126b

Q-130

Q-132

Q-134

Q-142

R-101

P-491

E-263

R-116

R-118b

R-120

R-124

R-130

R-132

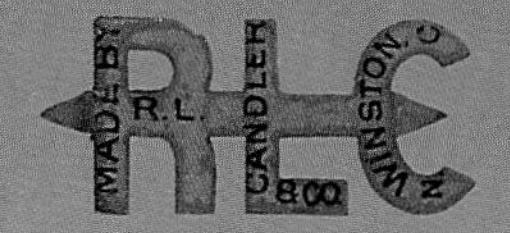

R-136

R-138

R-140

R-148

R-152

R-164

R-176a

R-176b

R-178

R-188

R-190

R-192a

R-192b

R-192c

R-194

R-200

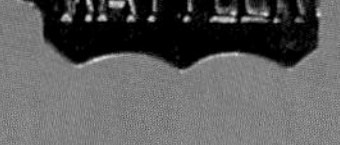

R-202

R-206

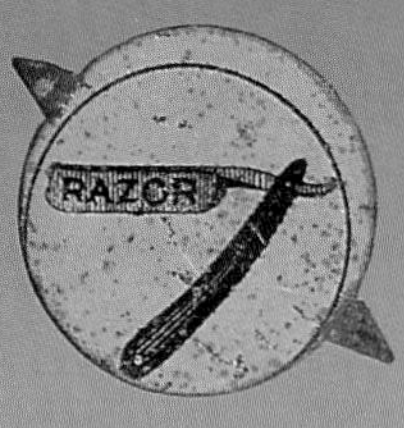

R-212

R-224a

R-224b

R-230

R-236

R-240

R-246

R-250

R-256a

R-258a

R-258b

R-266

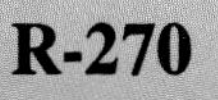

R-270

R-286

R-288

R-290

R-292

R-294

R-302

R-308

R-310

R-312

R-320

R-322

-324

R-330

R-332

R-336

R-338

R-340a

R-340b

R-342

R-344

R-348

R-350

R-352

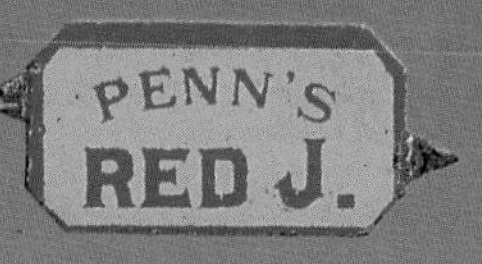

R-356 R-372 R-378 R-386

R-390 R-402a R-402b R-404

R-410 R-416 R-434 R-438

R-454 R-468 R-470

R-444

R-474 R-478 R-480

R-506

R-530

R-536

R-538

R-544a

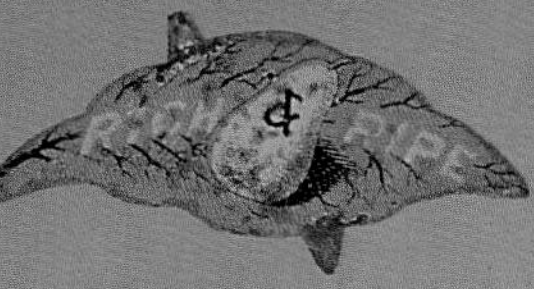

R-544b

R-546

R-574

R-580

R-582

R-590

R-592

R-598

R-600

R-604

R-616

R-624

R-628

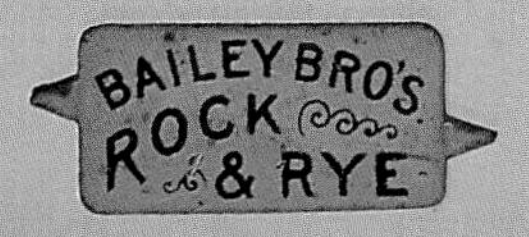

R-629

R-630

R-632

R-636a

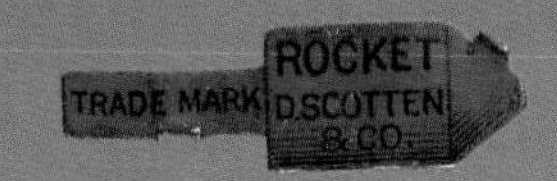

R-636b

R-636c

R-650

R-652

R-658

R-662

R-666

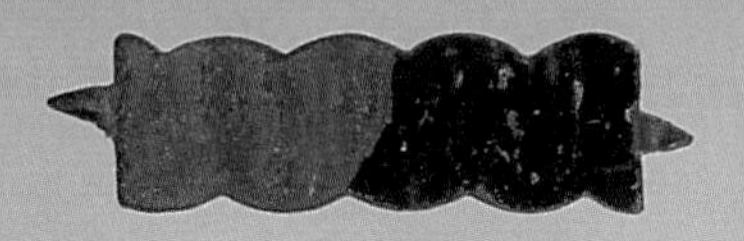

R-668

R-676a

R-676b

R-678

R-682

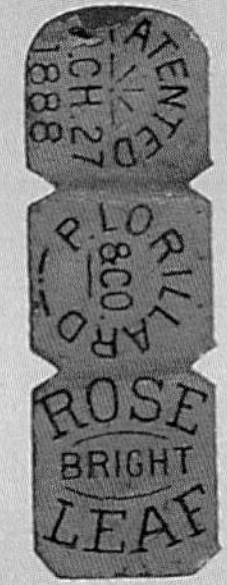

R-688

R-694

R-696

R-712

R-714

R-726

R-732

R-746

R-748

R-750

R-752a

R-752b

R-754

R-760

R-762

R-766

R-768

R-770

R-774a

R-776

R-777

R-778

R-780

R-790

R-794

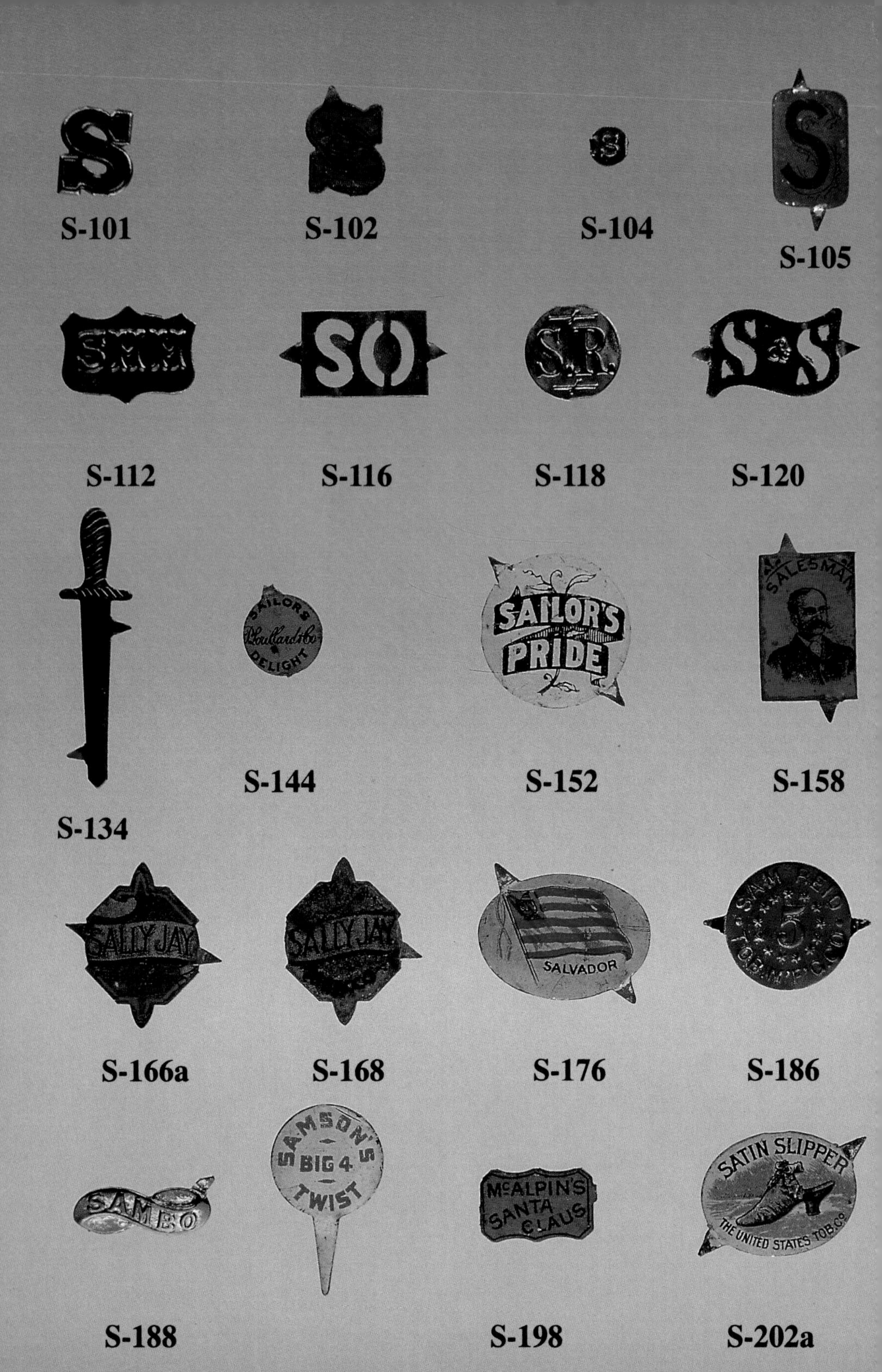

S-101 S-102 S-104 S-105

S-112 S-116 S-118 S-120

S-134 S-144 S-152 S-158

S-166a S-168 S-176 S-186

S-188 S-196 S-198 S-202a

S-204

S-212

S-213

S-216

S-226

S-228

S-246

S-248a

S-250a

S-250b

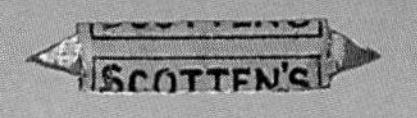

S-250c

S-250d

S-258

S-260

S-264

S-280

S-288

S-290

S-292

S-296

S-302

S-306

S-308

S-314

S-342

S-350

S-354

S-354

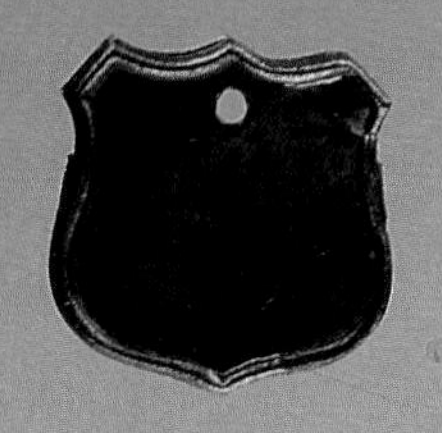

S-358

S-363

S-370

S-382

S-386

S-392

S-394

S-400a

S-408

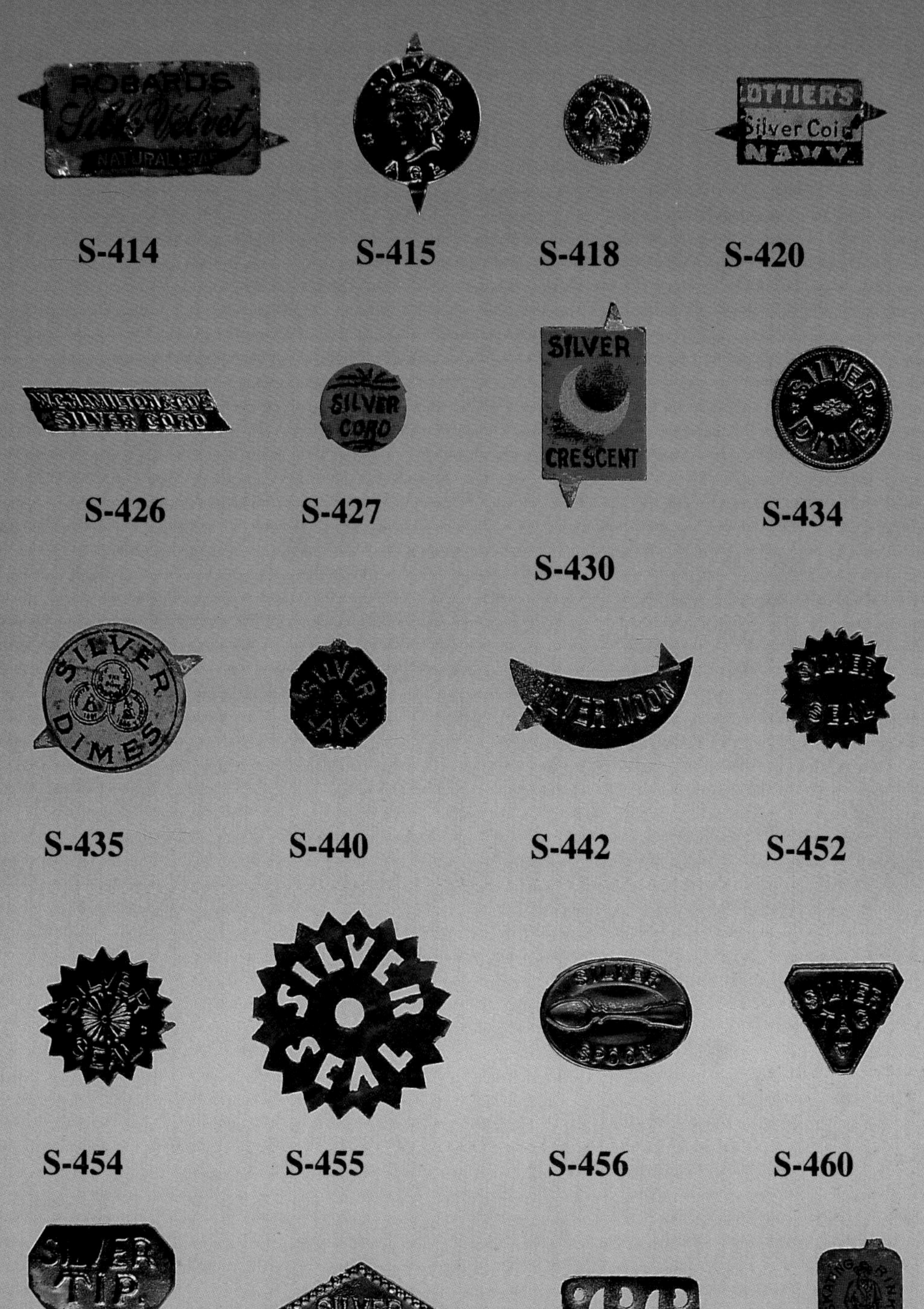

S-414 S-415 S-418 S-420

S-426 S-427 S-434

S-430

S-435 S-440 S-442 S-452

S-454 S-455 S-456 S-460

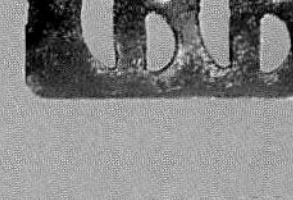

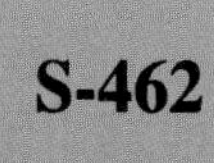

S-462 S-464 S-472 S-474

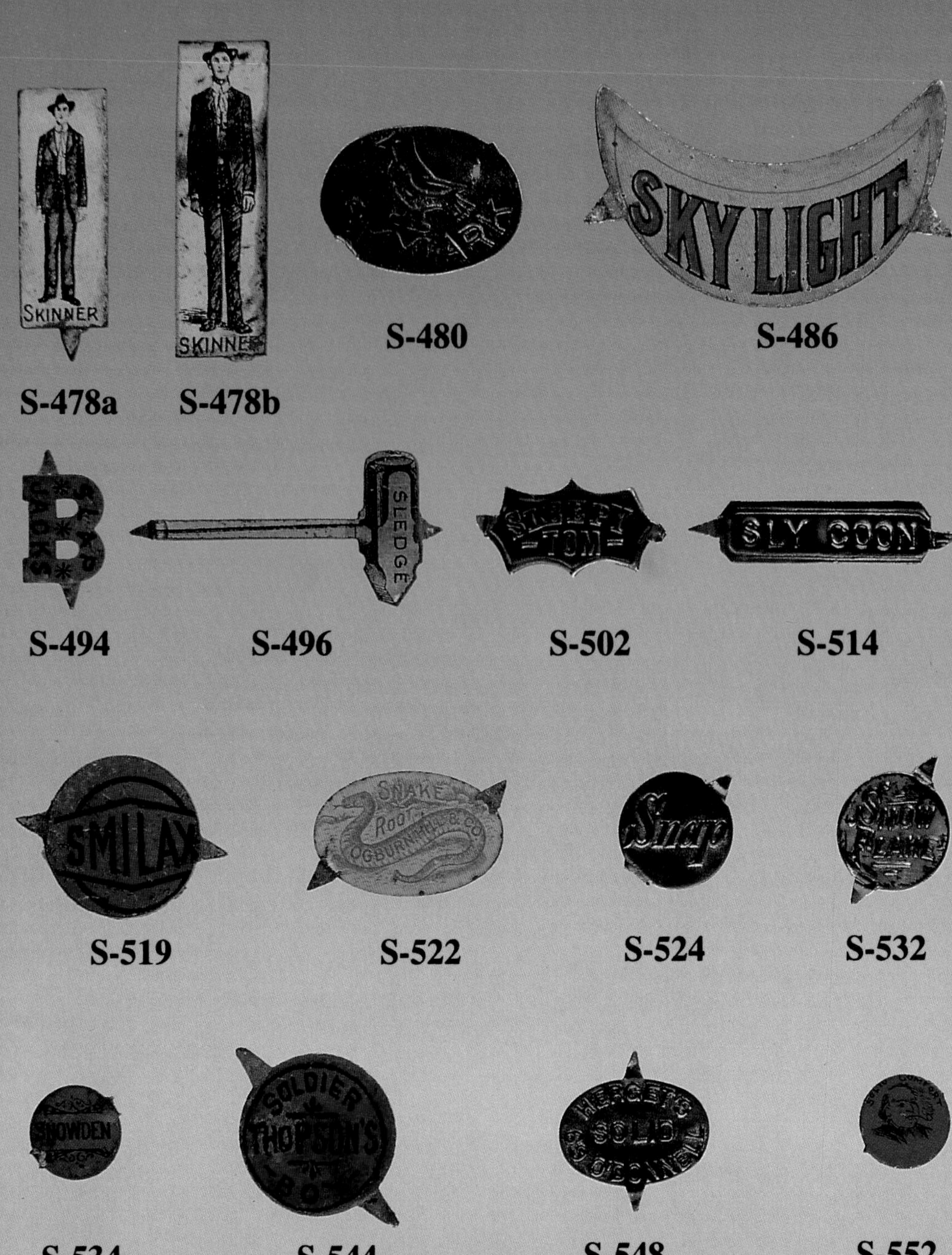

S-478a S-478b S-480 S-486

S-494 S-496 S-502 S-514

S-519 S-522 S-524 S-532

S-534 S-544 S-548 S-552

S-568 S-572 S-576 S-584

S-604a

S-610

S-624

S-626

S-634a

S-634b

S-634c

S-642b

S-646

S-648a

S-648b

S-650

S-652

S-654

S-666

S-668

S-676

S-680

S-688

S-692

S-694 S-696 S-697 S-698

S-699 S-704 S-706 S-710

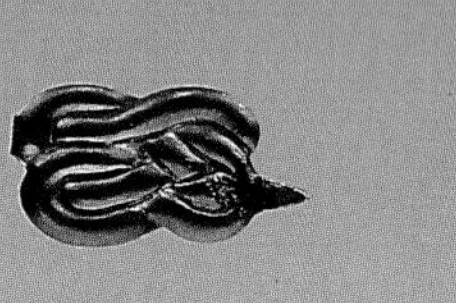

S-710 S-714 S-716 S-718

S-724 S-724 S-725 S-726

S-736 S-738 S-740 S-741

S-742

S-743

S-752

S-754

S-756a

S-756b

S-758

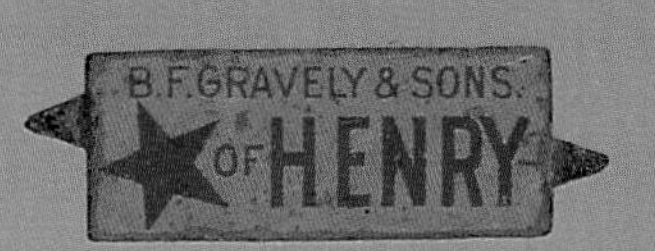

S-766

S-768a

S-768b

S-772

S-790

S-798

S-800

S-802

S-810

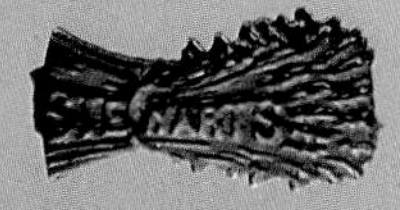

S-814

S-820

S-826

S-776

S-778

S-780

S-830

S-831

S-832

S-784

S-834

S-835

S-850

S-852a

S-853b

S-864

S-866

S-874

S-884

S-886

S-888

S-890

S-891

S-896

S-898

S-899

S-900

S-901a

S-901b

S-904

S-906

S-912

S-914

S-916

S-917

S-920

S-928

S-930

S-932

S-936

S-956

S-959

S-960

S-974

S-980a

S-982

S-984

S-988

S-990

S-993

S-996

S-1012

S-1014

S-1022

S-1024

S-1026

S-1028

S-1034

S-1036

S-1038

S-1064

S-1060

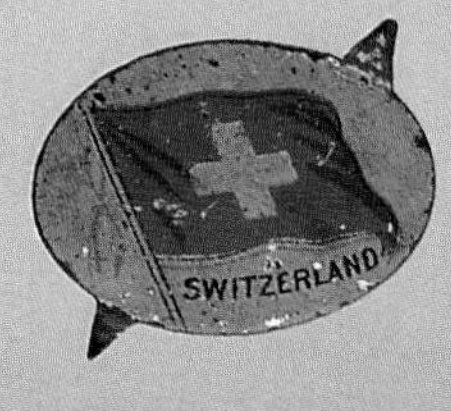

S-1086

T
T-101
T
T-102
TC
T-104
TCJ
T-106
TL
T-120
T-122
T.P.
T-124
T
P A
T-126
TR
T-128
T&W
T-136
T-148
E.J. & A.G. STAFFORD
T-146
T-140
PLUG
T-152
T-156
T-160
TAPE
LINE
T-162
TARGET
T-170
T-172
T-174

T-190

T-192

T-194

T-186

T-184

T-202

T-204

T-208

T-212

T-220

T-228

T-230

T-234

TENNESSEE
— LEAF —

T-238

T-242

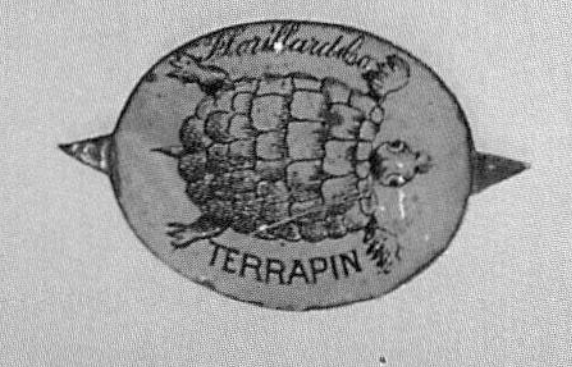

T-246

T-258

T-268

T-286

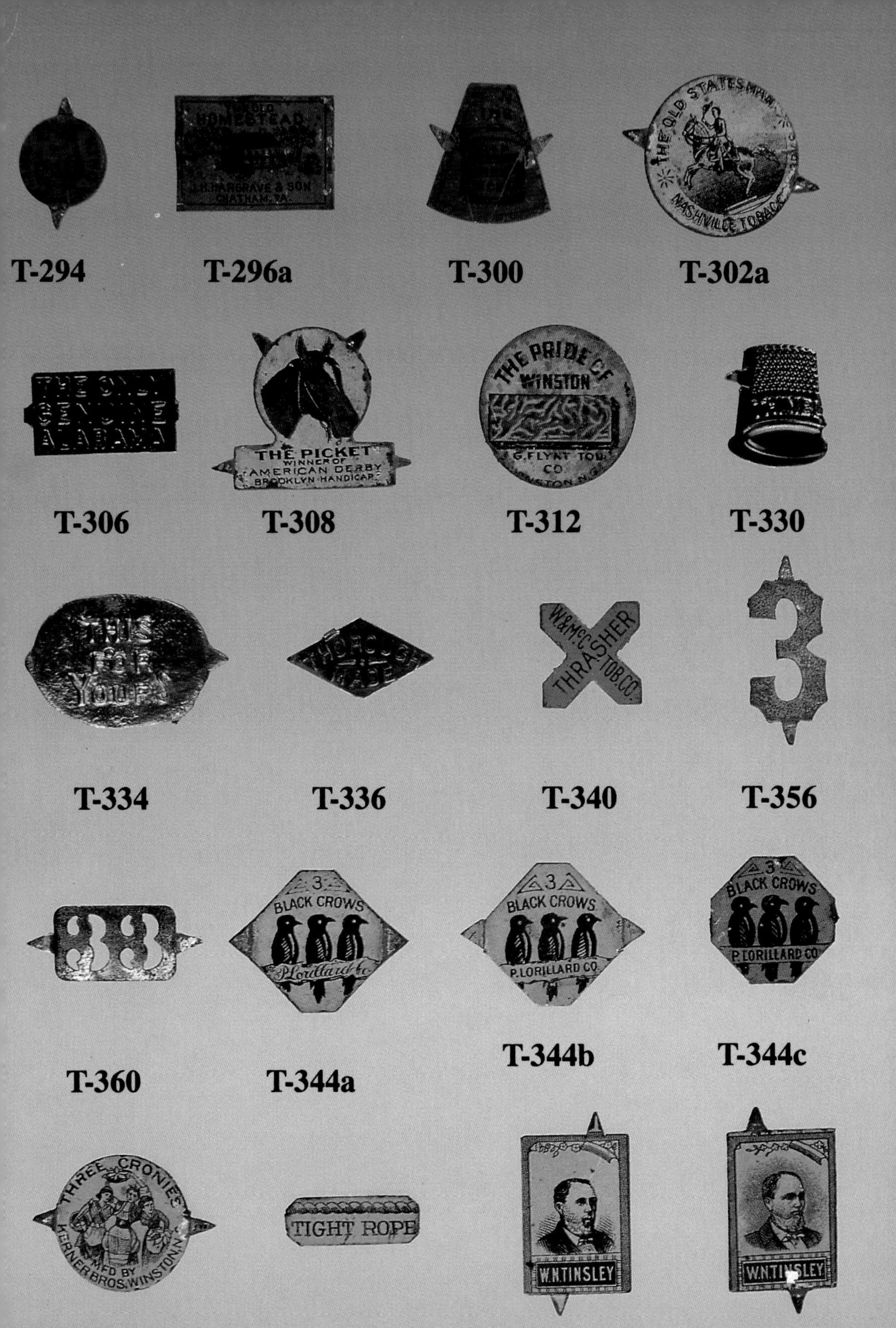

T-362 T-386 T-392 T-392

T-394

T-396

T-400

T-404a

T-410

T-411

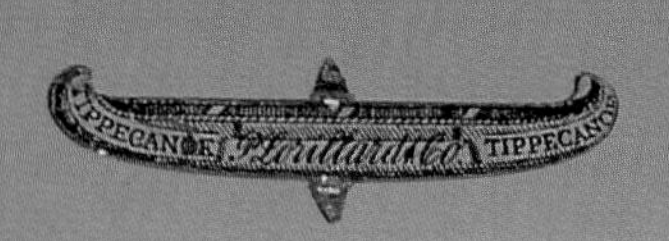

T-414

T-422

T-432

T-438

T-444

T-450

T-460

T-466

T-472

T-474

T-478

T-482

T-484

T-486

T-488

T-490a

T-498

T-502

T-508

T-510

T-518

T-522

T-528

T-530

T-536

T-544

T-554

T-558

T-560

T-582

T-584

T-590

T-592

T-594

T-596

T-607

T-612

T-614

T-630

U-102

U-106

U-108

U-116

U-120

U-124

U-126

U-140

U-146

U-150a

U-150b

U-151

U-158a

U-162

U-166

U-178 U-180 U-186

U-170

U-198

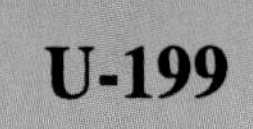
U-199

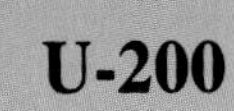
U-200

U-208

U-210 U-214 U-216 U-220

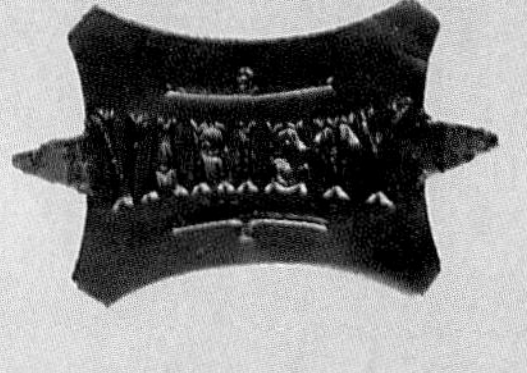

U-226 U-238 U-240 V-112

V-124

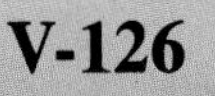
V-126

V-130 V-140

V-142

V-154a

V-156

V-160

V-164

V-166

V-182

V-186

V-194

V-192

W-101

W-112

W-114

W-118

W-124

W-126

W-130

W-134a

W-134b

W-148

W-150

W-152

W-156

W-162

W-170

W-191

W-196

W-198

W-202

W-204

W-208

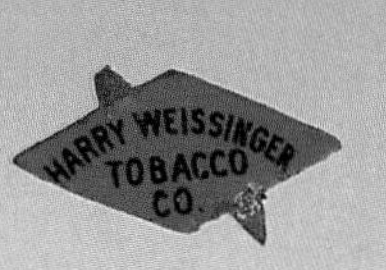

W-212

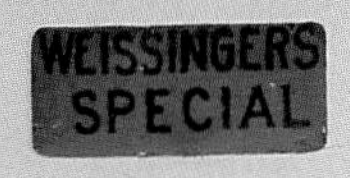

W-214a

W-216

W-224

W-232a

W-232b

W-234

W-238

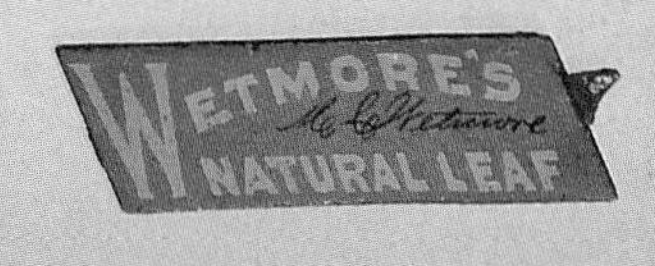

W-240

W-244

W-248

W-250

W-252

W-254

W-255

W-258

W-264

W-268

W-274

W-275

W-284

W-288

W-290

W-294

W-296

W-308

W-310

W-312

W-322

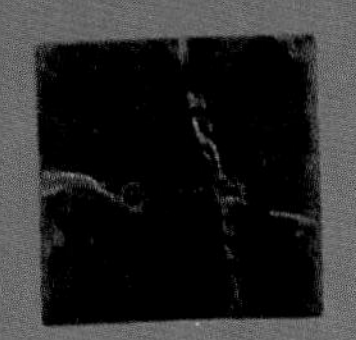

W-323

W-324

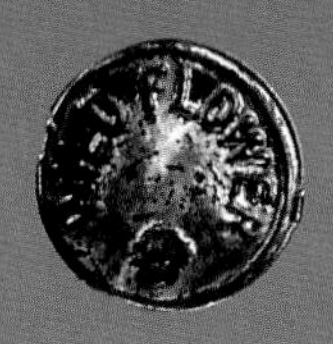

W-332

W-336a

W-356

W-358

W-360

W-364

W-366

W-372

W-374

W-378

W-383

W-387

W-390

W-392

W-396

W-414

W-416

W-418

W-430

W-434

W-444

W-448

W-451

W-456

W-470

X-101

X-102

X-106

X-112

X-116

X-118

X-120

X-124

X-132

X-139

Y-101

Y-106

Y-110

Y-112

Y-140

Y-126

Y-140

Y-142

Y-149b

Y-150

Z-101

Z-113

Z-120

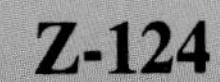

Z-124

Z-130

Z-132

PRICE GUIDE

The values in this book should be used only as a guide. They are for informational purposes only, and are most useful for establishing relative values. They are not intended to set prices. Prices vary from one section of the country to another, and dealer and auction prices vary greatly and are affected by demand as well as condition.

Tobacco tin tags related to other collecting categories and collector interest, such as political, African-American, baseball, and American Indian, are likely to realize higher prices.

Neither the author nor the publisher assumes responsibility for any losses that might be incurred as a result of consulting this guide.

Value Code

Very common tags: made and used in large quantities over a long period of time.

A= $.50-1
B= $1-2
C= $2-4

Many of these tags were made for small tobacco manufacturers and tobacconists, and are quite rare with 1000 or less made, though most tags were made in the tens of thousands.

D= $4-6
E= $6-8
F= $8-10
G= $10-12

Speculative, not based on rarity alone, but also tags theme, popularity, supply and demand.

H= $12-30

A		A145	E	A203	C	A264c	C	A328	H	B109	F	B161	H	B212	E	B268	D
A102	B	A146a	B	A205	C	A265	D	A329	D	B110	C	B163	D	B213	A	B269	F
A103	C	A146b	B	A206	C	A266a	D	A331	C	B111	C	B164	H	B214	C	B270	G
A104	C	A146c	B	A207	C	A266b	D	A333	D	B112	C	B165	D	B215	C	B271	D
A105	B	A147a	C	A208	C	A267	C	A335	D	B113	C	B166	E	B216	F	B272	C
A106	C	A147b	C	A209	D	A268	D	A336	E	B114	B	B167	C	B217	F	B273	D
A107	C	A147c	C	A210	D	A269	C	A337	D	B115	D	B168	E	B218	C	B274	D
A108	B	A149	C	A211	E	A271	D	A338	F	B116	C	B169	F	B219	D	B275	C
A108a	B	A151	C	A213	C	A272	G	A339a	D	B117	D	B170	F	B221a	E	B277	E
A108b	B	A153	C	A215	D	A273	F	A339b	D	B118	C	B171	B	B221b	E	B278	C
A109	C	A155	D	A217	C	A275	D	A340	A	B119	B	B173	C	B223	C	B279	C
A110	D	A157	D	A219	D	A276	C	A341	A	B121	C	B174	C	B224	C	B281	D
A111	D	A158	C	A221	C	A277	C	A342	D	B123	C	B175	D	B225	D	B283	C
A112a	C	A159a	C	A223	E	A279	C	A343	D	B125	C	B176	E	B226	E	B285	C
A112b	B	A159b	B	A225	D	A280	D	A345	C	B126	C	B177	G	B227	C	B286	C
A113	C	A161	H	A227	D	A281	D	A347	C	B127	D	B179	H	B229	D	B287	C
A114	D	A162	H	A229	D	A282	C	A349	D	B129a	C	B181	C	B230	H	B288	C
A115	C	A163	B	A231	D	A283	C	A351a	E	B129b	C	B183	D	B231	E	B289	H
A116	C	A165	C	A232	B	A285	D	A351b	E	B129c	C	B185	C	B232	E	B290	H
A117	A	A167	B	A233	D	A287	D	A351c	E	B131	C	B187	C	B233	D	B291	C
A118	B	A169	B	A234	H	A289	C	A351d	E	B132	C	B188	C	B235	C	B292	C
A119	C	A170	B	A235	D	A291	C	A354	F	B133	D	B189	C	B237	B	B293	B
A120	C	A171	C	A236	D	A293	C	A357	H	B134	C	B190	D	B238	C	B294	C
A121a	B	A173	C	A237	D	A294	B	A359a	C	B135	B	B191	C	B241	D	B295	H
A121b	B	A174	D	A238	E	A295	C	A359b	C	B136	D	B192	D	B242	E	B297	D
A121c	B	A175	C	A239a	C	A296	E	A359c	C	B137	B	B193a	B	B243	D	B298	E
A123	B	A176	E	A239b	C	A297	C	A361	F	B138	D	B193b	B	B244	E	B299	F
A124	D	A177	E	A242a	C	A298	C	A363	C	B139	C	B193c	B	B245	A	B300	D
A125	C	A178	D	A239b	C	A299	C	A365	G	B140	C	B194	C	B247	C	B301a	B
A126	B	A179	D	A239c	C	A301	D	A368	D	B141	C	B195a	C	B249	D	B301b	B
A127	B	A180	C	A239d	C	A302	C	A369	C	B142	B	B195b	C	B251	C	B301c	B
A129	B	A181	F	A247	D	A303a	C	A371	C	B143	E	B197	D	B253	E	B302	C
A130	C	A182	C	A249	C	A303b	C	A373	C	B145a	C	B198	D	B255	E	B303	C
A131	B	A183	H	A251	C	A307	D	A375	C	B145b	A	B199	C	B256	E	B304	B
A133	B	A184	H	A252	D	A308	C			B145c	C	B200	B	B257	G	B305	C
A134	E	A185	C	A253	D	A311	C	**B**		B145d	A	B201	C	B258	D	B306	E
A135	C	A186	C	A255	B	A313	G	B101	B	B151	C	B202	D	B259	D	B307	C
A136	C	A187	D	A256	C	A315	C	B102	D	B153	F	B203	C	B260	C	B308	E
A137	C	A189	D	A257	C	A317	C	B103	C	B155	G	B204	C	B261	F	B309	D
A138	D	A191	D	A259	D	A319	D	B104	D	B157a	D	B207	D	B262	C	B310	C
A139	D	A193	E	A261	D	A321	C	B105	D	B157b	D	B208	H	B263	C	B311	D
A142	D	A199	C	A263	C	A323	G	B106	C	B157c	D	B209	C	B264	D	B312	C
A143	C	A200	G	A264a	C	A325	B	B107	C	B158	F	B210	C	B265	C	B313	C
A144	B	A201	C	A264b	C	A327	D	B108	C	B159	H	B211	E	B267	C	B314	C

B315	C
B316	C
B317	F
B318	C
B319	C
B320a	C
B320b	C
B321	B
B322	C
B323	B
B325	C
B327	B
B329	D
B331	C
B333	C
B335	D
B337	C
B339	C
B340	C
B341a	C
B341b	C
B342	D
B343	C
B344	C
B345	D
B347	E
B348	C
B349	B
B350	E
B351	C
B353a	E
B353b	F
B354	E
B355	B
B357	B
B359	B
B361a	C
B361b	C
B363	B
B364	C
B365	E
B366	G
B367	G
B369	C
B371	F
B373	C
B374	D
B375	D
B376	C
B377	F
B378	C
B379	C
B380	B
B381	C
B382	C
B383	C
B385	D
B386a	D
B386b	D
B386c	D
B387	C
B388	C
B389	D
B390	E
B391a	C
B391b	C
B393	D
B395	C
B397	C
B399	C
B401	D
B402	D
B403	C
B405	E
B407	C
B409	C
B411	D
B412	F
B413	C
B414	D
B415	C
B417	A
B419	C
B420	D
B421	E
B422	C
B423	B
B425	C
B426	D
B427	B
B428	C
B429	D
B431	C
B433	D
B435	C
B437	D
B439	C
B440	C
B441	D
B442	C
B443	B
B444	E
B445	D
B446	C
B447	D
B448	C
B449a	D
B449b	E
B449c	D
B451	E
B453	H
B454	D
B455	C
B456	C
B457	C
B458	B
B459	D
B460	C
B461	C
B462	B
B463	B
B464	D
B465	C
B467	C
B468	G
B469	C
B470	C
B471	C
B472	C
B473	D
B474a	D
B474b	D
B475	C
B477	C
B479	D
B481	B
B482	C
B483	D
B484	F
B485	C
B486	G
B487	G
B488	B
B489	C
B490	C
B491	H
B492	D
B493	D
B494	C
B495	C
B497	C
B499	F
B501	C
B503	E
B505	G
B506	G
B507	C
B508	C
B509	E
B511	D
B512	E
B513	D
B514	C
B515a	D
B515b	D
B516a	C
B516b	C
B517a	C
B517b	C
B518a	D
B518b	D
B519	C
B520	D
B521	E
B522	E
B523	E
B524	D
B525	C
B526	D
B527	D
B528	E
B529	C
B530	B
B531	C
B533	H
B535	E
B537	C
B539	C
B541	H
B543	D
B544	D
B545	D
B547a	B
B547b	C
B551a	C
B551b	C
B553	D
B554	C
B555	E
B557a	E
B557b	E
B557c	E
B559	D
B561	D
B562	D
B563	H
B564	D
B565	C
B566	C
B567	E
B568	D
B569	D
B570	C
B571	E
B572	C
B573	C
B574	C
B575	H
B576	H
B577	H
B579	H
B581	H
B582	D
B583	D
B584	G
B585	B
B586	B
B587a	C
B587b	C
B589	C
B591	C
B593	B
B595	D
B596	D
B597	C
B598a	D
B598b	C
B599	E
B600	C
B601	C
B602	C
B603	H
B604	C
B605	D
B606	D
B607	C
B608	C
B609	D
B610	C
B611	D
B612	E
B613	G
B614	C
B615	C
B616	D
B617	D
B618	C
B619	C
B620	E
B621	C
B622	B
B623	C
B624	H
B625	D
B626	C
B627	D
B628	D
B629	D
B630	C
B631	C
B633	C
B634	B
B635	G
B637	B
B639	G
B641	E
B643	C
B645	E
B646	C
B649	F
B651	B
B653	B
B654	C
B655	B
B656	C
B657	C
B658	B
B659	C
B661	C
B663	C
B664	B
B665a	B
B665b	B
B667	D
B669	C
B671	D
B673	D
B675	C
B677	H
B678	C
B679	C
B689	B
B680	B
B681	B
B682	A
B683	B
B684	C
B685	B
B687a	C
B687b	C
B689	C
B691	C
B692	B
B693	D
B694	D
B695	D
B696	D
B697	C
B698	B
B699	E
B700	D
B701	C
B702	C
B703	G
B704	F
B705	C
B707	B
B709	C
B711a	C
B711b	C
B712	C
B713	G
B714	C
B715	C
B717	B
B718	D
B719	B
B721a	E
B721b	E
B723	F
B725	H
B726	F
B727	G
B728	B
B729a	C
B729b	C
B730	D
B731	C
B735	C
B737	C
B739	C
B741	D
B742a	C
B742b	C
B743	C
B744	H
B745	D
B746	C
B747	C
B748	C
B749	E
B750	C
B751	E
B752	F
B753	C
B754	B
B755	C
B756	B
B757	C
B758	C
B759	D
B760	D
B761	C
B762	C
B763	C
B765	C
B766	D
B767	C
B769a	A
B769b	B
B771	C
B775	C
B776	G
B777	F
B779	C
B781	D
B783	C
B784	D
B785	C
B786	E
B787	C
B788	C
B789a	E
B789b	F
B790	C
B791	B
B792	C
B793	D
B794	D
B795	D
B796	B
B797	F
B798	E
B799	H
B801	E
B803	C
B805	D
B806	F
B807	H
B808	G
B809	E
B811	C
B812	G
B813	E
B814	G
B815	D
B816	G
B817	G
B818	F
B819	C
B821	D
B823	D
B825	C
B827	E
B829	D
B830	H
B831	C
B833a	D
B833b	D
B835	E
B837a	B
B837b	C
B839a	B
B839b	C
B841	D
B842	E
B843	B
B845	C
B847	E
B849	C
B851	C
B852	B
B853	A
B854	A
B855	B
B857	B
B859	C
B861	D
B862	C
B863	D
B864	H
B865	A
B867	A
B869	C
B871	C
B873	C
B875	D
B876	D
B877	C
B879	D
B881a	D
B881b	D
B883	C
B884	C
B885	C
B887	E
B889	D
B891	H
B893	D
B895	E
B896	D
B897	C
B899a	B
B899b	B
B903	D
B904	D
B905	C
B907	C

C

C101a	B
C101b	A
C101c	A
C101d	B
C103	B
C105	B
C107	C
C109	C
C110	C
C111	B
C112	D
C113	D
C114	C
C115	B
C116	C
C117	D
C119a	D
C119b	D
C120	C
C121	B
C123	C
C124	A
C125	D
C126	C
C127	C
C128	C
C129	C
C131	B
C133	D
C135	C
C137	C
C139	B
C141	B
C142	C
C143	C
C144	D
C145	C
C146a	D
C146b	D
C147	C
C148	C
C149	D
C150	E
C151	G
C153	C
C154	C
C155	B
C157	C
C158	C
C159a	C
C159b	D
C161	D
C163	C
C165	D
C167	C
C169	F
C171	C
C172	D
C173	G
C174	C
C175	C
C176	C
C177	D
C179	D
C180	E
C181	D
C182	C
C183	C
C184	D
C185	D
C187	C
C189	E
C190	C
C191	D
C193	C
C194	C
C195	F
C196	C
C197	C
C199	C
C201	C
C202	D
C203	D
C204	C
C205	C
C206	A
C207	B
C208	E
C209	F
C210	G
C211	D
C213	C
C214	B
C215	F
C216	G
C217	C
C218a	C
C218b	B
C219	C
C220	D
C221	C
C220	E
C223	C
C224	C
C225	C
C226	D
C227a	G
C227b	G
C229	E
C231	A
C232	C
C233	F
C234	D
C235	D
C236	D
C237	C
C238	C
C239	H
C241	B
C243	C
C245	G
C246	D
C247	H
C248	H
C249	C
C250	A
C251	E
C252	E
C253	B
C255	A
C257	C
C261	C
C263	D
C264	C
C265	C
C266	D
C269	C
C271	C
C272	C
C273	C
C275	D
C277	C
C279	C
C280	C
C281	H
C282	B
C283	C
C284	C
C285	D
C286	C
C287	C
C288	D
C289	C
C290	E
C291	G
C292	G
C293	C
C294	D
C295	C
C297	C
C299	C
C301	D
C303	B
C305	A
C307	B
C309a	C
C309b	C
C311	C
C312	D
C313	D
C314	D
C315	E
C316	C
C317	C
C319	C
C321	D
C323	C
C324	D
C325	C
C327	G
C326	C
C328	G
C329	C
C330a	C
C330b	C
C331	D
C332	B
C333	D
C335	C
C337	C
C338	C
C339	D
C340	D
C341	F
C342	C
C343	D
C345	F
C346a	C
C346b	C
C347	C
C349	D
C351	D
C353	C
C355	B
C357	C
C358	C
C359	B
C361	C
C362	G
C363	D
C364	A
C365	C
C367	C
C369	C
C370	C
C371	C
C372	F
C373	C
C375	C
C377	A
C378	C
C379	D
C38[illegible]	C
C382	D
C383	D
C384	C
C385	C
C387	C
C389	A
C391	C
C393	C
C395	C
C397	C
C399	D
C401	C
C403	D
C405	C
C407	C
C409	A

Item	Answer
C410	D
C411	C
C412	E
C413	F
C414	E
C415	B
C416	D
C417	D
C418	C
C419	H
C421	H
C425	D
C429	C
C431	C
C433a	A
C433b	A
C435	A
C437	A
C439	C
C441	C
C443	D
C445	C
C446	C
C447a	C
C447b	C
C448	B
C449	A
C451	C
C453	B
C455	B
C457	A
C459	D
C460	D
C461	C
C462	G
C463	G
C465	G
C467	D
C469	B
C471	C
C472	C
C473a	D
C473b	D
C475	D
C477	H
C478	C
C479	B
C480	C
C481	E
C483	D
C485	C
C486	C
C487	C
C488	C
C489	D
C491	D
C492	C
C493	F
C494	G
C495	D
C496	D
C497	C
C499	D
C500	D
C501	D
C503	D
C504	C
C505	D
C507	D
C508	C
C509a	C
C509b	C
C511	C
C512	D
C511	C
C512	D
C513	B
C515	A
C516	B
C517	C
C519	H
C520	C
C521	C
C522	C
C523	C
C524	C
C525	C
C526	D
C527	C
C529	D
C530	C
C531	D
C533	C
C534	C
C535	C
C537	D
C538	C
C539	C
C540	C
C541	C
C543	F
C545	A
C546	C
C547	H
C548	C
C549	E
C551	C
C553	D
C555	B
C557	B
C558	D
C559	A
C560	D
C561	D
C563	C
C565	C
C567	C
C569	C
C571	D
C573a	B
C573b	B
C575	C
C576	C
C577	C
C578	H
C579	D
C581	C
C583	C
C585	D
C587a	B
C587b	C
C587c	B
C589	H
C590	C
C591	C
C592	C
C593	H
C594	E
C595	C
C596	C
C597	D
C598	C
C600	C
C601	C
C603	C
C605	H
C607	G
C609	F
C610	D
C611	E
C613	F
C614	C
C615	C
C617	D
C619	C
C621	B
C623	C
C625	C
C627a	B
C627b	C
C629	C
C631	C
C633	C
C635	E
C637	D
C639	F
C641	E
C642	D
C643	D
C644	C
C645	C
C646	B
C647	C
C649	D
C651	D
C653	B
C654	C
C655	D
C657	D
C658	D
C659	F
C661	D
C663	C
C665	C
C666	C
C667	C
C669	F
C670	C
C671	B
C673	C
C675	C
C677	C
C679	F
C681	E
C683	D
C683	C
C685	C
C687	B
C689	C
C690	G
C691	C
C692	D
C693	D
C694	C
C695	D
C696	A
C697	H
C698	F
C699	B
C701	A
C702	B
C703	D
C705	C
C707	B
D	
D101	B
D103	A
D105	A
D107	C
D108	C
D109	B
D111	C
D113	C
D114	C
D115	D
D116	A
D117	C
D119	B
D121	A
D123	A
D125	C
D127	D
D128a	C
D128b	C
D129	F
D131	C
D133	B
D135	C
D137	F
D139	C
D141	C
D143	D
D145	D
D146	C
D147	D
D149	C
D151	B
D152	C
D153	D
D154	C
D155	C
D157	C
D161	A
D162	C
D163a	C
D163b	C
D163c	C
D167	D
D169	C
D170	C
D171	D
D175	C
D177	F
D179	E
D181	B
D183	A
D184	A
D185	C
D186	A
D187	D
D188	C
D189	C
D190	E
D191	C
D192	F
D193	C
D194	C
D195	C
D197	C
D199	C
D201	B
D203	D
D204	C
D205	D
D206	C
D207	B
D208	C
D209	C
D210	E
D211	E
D213	C
D214	C
D215a	A
D215b	A
D215c	A
D217	C
D218	D
D219	C
D221	D
D222	C
D223	D
D224	C
D225	C
D226	C
D227	D
D229	C
D231	B
D232	C
D233	C
D234	C
D235	G
D236	E
D237	C
D239	C
D241	C
D243	B
D245	F
D246	D
D247	E
D248	C
D249	E
D250	D
D251	C
D253	F
D255	C
D257	D
D258	F
D259	B
D260	D
D261	C
D263	D
D265	D
D267	F
D269	C
D271	D
D273	G
D275	C
D277	C
D279	F
D280	C
D281	C
D283	C
D285	C
D286	D
D287	D
D289	C
D291	C
D293	E
D295	D
D297	C
D299	C
D300	C
D301	C
D302	D
D303	B
D304	C
D305a	A
D305b	A
D305c	A
D306a	B
D306b	B
D306c	B
D307	C
D308	C
D309	C
D310	D
D311	C
D313	D
D314	F
D315	D
D317	B
D319	E
D321	B
D323	C
D325	C
D327	D
D329	C
D330	E
D331a	C
D331b	B
D333	F
D334	D
D335	C
D336	D
D337	C
D338	C
D341	C
D343	G
D344	G
D345	C
D347	G
D349	A
D350	B
D351	C
D353	E
D354	D
D355	C
D357	D
D359	C
D361	F
D362	E
D363	D
D364	C
D365	C
D367	C
D369	C
D371	C
D373	C
D375	D
D377	C
D378	D
D379	G
E	
E101	A
E103	B
E105	C
E107	C
E109	B
E111	B
E113	B
E117	C
E118	C
E119	B
E120	C
E121	C
E123	D
E125	D
E126	C
E127	D
E129	E
E131	C
E133	C
E134	B
E135	D
E136	C
E137	C
E139	C
E140	C
E141a	B
E141b	B
E141c	B
E143	C
E145	C
E146	C
E147	C
E148	C
E149	C
E150	D
E151	C
E153	B
E155	B
E156	C
E157	C
E159	E
E162	D
E163	C
E164	C
E165	B
E167	C
E168	C
E169	C
E170	C
E171	C
E172	C
E173	D
E175	D
E177	C
E179	E
E181	E
E182	B
E183	B
E184	B
E185	A
E186	C
E187	D
E188	C
E189	C
E190	C
E191	D
E192	C
E193	D
E194	C
E195	D
E197	C
E199	H
E201a	C
E201b	C
E202	D
E203	C
E205	D
E207	C
E209	C
E211	D
E213	C
E215	B
E217	C
E219	C
E221	B
E223	F
E224	D
E225	C
E227	C
E229	C
E231	C
E233	C
E237	C
E239	D
E241	D
E243	C
E245	C
E247	D
E249	B
E250	C
E251	C
E252	D
E253	C
E254	D
E255	B
E257	C
E258	C
E259a	E
E259b	F
E260	C
E261	D
E262	D
E263	C
E264	B
E265	C
E266	D
E267	E
E268	D
E269	D
E271	D
F	
F101	B
F103	B
F104	C
F105	C
F107	B
F109	C
F110	E
F112	C
F113	B
F114	B
F115	C
F116	B
F117	C
F118	D
F119	D
F120	C
F121	C
F122	C
F123	D
F124	D
F125	D
F126	E
F127	C
F128	C
F129	C
F130	C
F131	D
F132	D
F133	D
F134	D
F137	C
F138	E
F139	C
F141	H
F142	G
F143	E
F145	C
F147	D
F148	D
F149	D
F150	C
F151	C
F153	G
F155	C
F157	C
F159	D
F160	C
F161	D
F163	C
F165a	C
F165b	C
F166	H
F167	C
F169	C
F171	C
F173	D
F175	C
F177	C
F179	E
F180	C
F181	C
F182	E
F183	D
F185	E
F186	F
F187	A
F189	C
F190	C
F191	C
F192	D
F193	E
F194	C
F195a	F
F195b	F
F196	C
F197	C
F198	B
F199	C
F200	C
F201	C
F202	D
F203	A
F207	C
F209	C
F210	C
F211	C
F212	C
F213	C
F214	D
F215	C
F216	C
F217	C
F219	D
F221	D
F223	D
F225	D
F226	A
F227	A
F231	C
F232	C
F233	E
F234a	E
F234b	E
F235	C
F237	G
F239	F
F241	C
F243	C
F245	D
F247	D
F249	C
F251	C
F253	D
F255	C
F256	D
F257	C
F259	B
F261	B
F262	C
F265	C
F266	C
F267	C
F268	C
F269	C
F271	C
F272	D
F273	B
F275	C
F277	C
F279	D
F280	E
F281	F
F282	E
F283	B
F285	B
F287	C
F289	C
F291	D
F292	D
F293	F
F294	F
F295	C
F296	D
F297	D
F298	C
F301	C
F302	C
F303	E
F305	D
F307	B
F309	D
F311	C
F313	C
F314	C
F315	C
F317	D
F318	D
F319	C
F320	C
F321	C
F322	E
F323	D
F324	B
F325	B
F327	C
F328	C
F329	F
F301	C
F331	C
F333	C
F336	C
F337	C
F338a	D
F338b	D
F338c	D
F339	C
F341	D
F343	D
F345	C
F346	C
F347	C
F349	C
F351	D

F352 C
F353 B
F354 G
F355 C
F357 C
F359 C
F361 B
F362 B
F363 D
F364 F
F365 C
F366 C
F367 D
F371 E
F373 C
F375 C
F377 C
F379 B
F380 C
F381 F
F382 C
F383 C
F384 C
F385a E
F385b E
F386 C
F387 D
F388 D
F389a C
F389b C
F389c C
F390 C
F391 C
F392 D
F393 B
F395 C
F397 C
F399a C
F399b C
F401 D
F402 D
F403 D

G

G101 B
G103 C
G104 B
G105 C
G107 C
G109 C
G111 C
G113 B
G115 C
G116 E
G117 C
G118 D
G119 B
G121 C
G122 D
G123 B
G125 C
G127 B
G129 C
G131 C
G132 E
G133 C
G135 C
G136 D
G137 C
G139 C
G141 H
G143 C
G144 C
G145 D
G147 C
G149 E
G151 D
G153 C
G154 C
G156 D
G157 E
G158 E
G159 C
G161 D
G162 A
G163 C
G164 E
G165a C
G165b C
G166 C
G167 C
G168 C
G169 G
G170 G
G171 F
G172 C
G173 C
G174 C
G175 C
G176 C
G177 H
G181 H
G182 C
G183 C
G185 D
G187 C
G189 D
G190 D
G191 F
G193 C
G195 C
G197 C
G199 H
G201 C
G203 C
G205 C
G207 A
G209 C
G211 A
G213 C
G215 D
G216 B
G217 D
G219 C
G221 D
G223 B
G224 C
G225 D
G227 A
G229 B
G231 C
G232 D
G233 A
G234a C
G234b C
G235 C
G236 D
G237 D
G239 C
G241 B
G243 C
G245 C
G246 C
G247 B
G248 C
G249 B
G251 B
G253 D
G255 C
G256 C
G257 C
G258 D
G259 C
G260 F
G261 D
G263 C
G265 C
G266 D
G267 C
G269 A
G271 C
G273 E
G275 A
G277 A
G278 C
G279 C
G281 D
G283 C
G284 C
G285 D
G287 E
G289 C
G291 C
G292 D
G293 C
G295 D
G296 C
G297 B
G299 C
G301 C
G302 C
G303 C
G305 B
G307 D
G309a C
G309b C
G311 C
G312 C
G313 C
G315 C
G317 B
G318 D
G319 E
G321 C
G322 C
G323 D
G325 E
G327 D
G329 C
G330 E
G331 B
G333 D
G334 D
G335 E
G336 C
G337 C
G339 E
G341 C
G343 C
G344 D
G345 C
G346 D
G347 D
G349 C
G351 D
G353 C
G355 B
G357 C
G359 B
G361 D
G363 D
G364 B
G365 C
G366 C
G367a D
G367b C
G368 C
G369 C
G370 D
G371a C
G371b B
G375 C
G376 C
G377a E
G377b F
G379 C
G381 C
G383 D
G385 F
G387 C
G388 C
G389 D
G391 D
G392 C
G393 C
G394 C
G395 B
G396 F
G397 E
G399 C
G401 B
G403 C
G405 F
G407 B
G409 C
G410 C
G411 B
G413 C
G415 B
G417 E
G419a D
G419b E
G421 C
G423 C
G424 C
G425 D
G427 C
G429a C
G429b D
G431 C
G433 D
G434 C
G435 D
G437 B
G439 C
G441 C
G442 D
G443 C
G444 F
G445 F
G447 C
G448 C
G449 D
G451 D
G452 D
G453 E
G454 E
G455 E
G457 G
G459 D
G461 D
G463 E
G465 C
G466 C
G467 C
G468 C
G469 C
G471a B
G471b B
G474 C
G477a B
G477b B
G481 C
G483 C
G485 C
G487 C
G489 C
G491 D
G492 D
G493 E
G494 C
G495 C
G497a B
G497b C
G498 C
G499 C
G500 F
G501a E
G501b F
G502 C
G503 C
G504 D
G505 E
G506 D
G507 C
G508 E
G509 C
G510 D
G511 C
G513 F
G515 G
G516 C
G517 C
G518 C
G519 B
G521 B
G523 C
G531 D
G532 C
G533 B
G535 C
G537 C
G538 D
G539 C
G540 H
G541 H
G543 H
G544 H
G545 C
G546 B
G547 C
G548 B
G549 C
G551 C
G553 C
G555 C
G557 F
G558 C
G559 D
G561 A
G563 B
G565 C

H

H100 A
H101 C
H102 C
H103 B
H104 C
H105 C
H106a C
H106b D
H107 D
H108 C
H109 C
H110 B
H111 C
H112 B
H113 B
H114 B
H117 B
H119 C
H121 C
H123 C
H125 B
H127 D
H129 G
H131 D
H132 C
H133 C
H135 H
H137 B
H139 B
H141 C
H143 C
H145 C
H146 C
H147 B
H149 C
H151 D
H153 C
H155 C
H157 D
H159 D
H160 C
H161 G
H163 A
H165 B
H166 C
H167 D
H169 F
H171 A
H173a D
H173b D
H175 C
H177 C
H179 C
H183 H
H184 H
H185 H
H186 C
H187 C
H189 C
H190 C
H191 C
H192 D
H193 C
H195 D
H197 D
H199 C
H205 H
H207 C
H209 A
H211 E
H212 C
H213 D
H215a C
H215b D
H217 D
H219 C
H221 D
H223 D
H225 C
H226 C
H227 C
H228 F
H229 C
H231 C
H230 C
H232 D
H233 C
H235 D
H236 D
H237 C
H239 B
H241 E
H242 B
H243 B
H245 D
H246 E
H247 D
H249 C
H251 C
H253 C
H254 C
H255 D
H256 C
H257 B
H258 C
H259 B
H260 D
H261 C
H262 C
H263 E
H264 B
H265 D
H266 D
H267 G
H268 D
H269 F
H270 G
H271 B
H273 D
H275 C
H277 B
H278 D
H279a D
H279b E
H279c E
H281 C
H283 C
H285 C
H286 D
H287 C
H288 A
H289 B
H290 C
H291 D
H293 C
H295 C
H297 C
H299a C
H299b C
H299c D
H299d D
H299e D
H300a D
H300b D
H301a C
H301b C
H302 F
H303 C
H304 C
H305 C
H306 C
H307 D
H308 C
H309 E
H311 E
H312 D
H313 C
H314 D
H315 C
H317 H
H319 G
H321 E
H323 F
H325 G
H327 C
H329 C
H331 D
H332 D
H333 C
H335 F
H337 C
H339 G
H341 A
H343 C
H345 B
H347 C
H348 E
H349 C
H350 C
H351 C
H353 F
H355 D
H357 E
H359 E
H361 D
H363 B
H365 C
H367 A
H369 B
H370 C
H371 D
H372 B
H373 C
H374 C
H375 D
H376 C
H377 D
H378 C
H379 C
H380 C
H381 B
H382 G
H383 C
H384 C
H385 D
H386 F
H387 C
H388 C
H389 D
H390a B
H390b C
H391 C
H392 D
H393 C
H394 C
H395a C
H395b D
H396 C
H397 C
H398 C
H399 C
H400 D
H401 D
H402 C
H403 C
H404 C
H405 D
H406 D
H407 G
H409 C
H411 B
H413 D
H415 D
H417 F
H419 C
H421 D
H422 C
H423 D
H425 G
H427 C
H429 D
H431a C
H431b C
H435 B
H437 D
H439 C
H441 D
H442 C
H443 C
H444 C
H445 B
H446 C
H447 B
H449 B
H451 C
H453a C
H453b C
H454 E
H455 C
H457 C
H459 B
H461 C
H463 C
H464 C
H465 C
H467 C
H469 C
H471 C
H473 D
H475 C
H476 C
H477 C
H478 E
H479 C
H481 E
H483 D
H485 C
H487 G
H488 B
H489 C
H491 E
H492 E
H493 F
H494 F
H495a C
H495b C
H495c D
H496 C
H497 C
H499a A
H499b A
H499c B
H501 C
H502a C
H502b C
H503 D
H505 C
H507 C
H509 C
H511 C
H512 B
H513 B
H515a E
H515b E
H516 C
H517 C
H518 D
H519 D
H520 B
H521a D
H521b D
H521c C
H521d C
H521e D
H522 B
H523a B
H523b B
H525 F
H527 C
H529 B
H531 H
H533 D
H535 C
H537 D
H539 D
H541 C
H543 C
H545 C

I

I100 C
I101 E
I102 D
I103 C
I104 C
I105 C
I106 C
I107 D
I109 B
I110 D
I111 H
I112 F
I113 C
I115 E
I117 C
I121 C
I123 C
I125 D
I127 C
I128 C
I129 G
I130 C
I131 C
I133 D
I135 C
I136 A

Item	Answer
I137	B
I139	D
I141	H
I142	G
I143	D
I145a	D
I145b	D
I145c	D
I147a	D
I147b	D
I149	E
I151	C
I152	D
I153	C
I155	C
I156	D
I157	F
I158	D
I159	D
I160	E
I161	C
I162	C
I163	B
I164	C
I165a	D
I165b	D
I167	G
I168	D
I169	C
I170	D
I171	E
I172	C
I173	C

J

Item	Answer
J101	B
J103	B
J104	D
J105	C
J106	C
J107	B
J108	B
J109	C
J110	C
J111	D
J112	D
J113	B
J115	C
J117	C
J119	D
J121	C
J122	D
J123	C
J124	B
J125	B
J127	B
J129	C
J130	C
J131	C
J133	B
J135	C
J137	B
J138	C
J139	C
J141	D
J143	C
J144	D
J145	C
J147	B
J149	B
J151	B
J153	C
J155	C
J157	B
J159	C
J160	D
J161	C
J163	C
J165	C
J167	A
J169a	C
J169b	C
J169c	C
J171	B
J173	D
J174	D
J175	F
J176	D
J177	C
J178	D
J179	C
J181	C
J183	C
J185	C
J186	C
J187	D
J189	H
J191	C
J193	D
J195	E
J197	C
J199	C
J201	D
J203	H
J204	D
J205	D
J206	C
J207	E
J209	C
J211	C
J212	C
J213	B
J214	C
J215	H
J216	H
J217	H
J218	C
J219	D
J220	C
J221	E
J222	D
J223	C
J224	C
J225	F
J226	E
J227	C
J228	C
J229	D
J230	E
J231	E
J232	C
J233	D
J235	C
J236	C
J237	B
J239	C
J241	E
J242	C
J243	C
J244	D
J245	C
J246	D
J247	D
J249	C
J251	C
J253	C
J255	C
J257	C
J258	D
J259	C
J261	H
J263	E
J265	D
J267	D
J269	C
J270	C
J271	D
J272	D
J273	C
J274	C
J275	F
J276	C
J278	D
J279a	D
J279b	D
J283	E
J285	E
J287	B
J288	C
J289	C
J290	C
J291	B
J292	B
J293	A
J294	C
J295	C
J296	C
J297	D
J298	C
J299	D
J300	B
J301	C
J302	C
J303	B
J305	H
J306	D
J307	D
J308	D
J309	D
J311	D
J313	D
J314	B
J315	C
J317	F
J319	D
J320	B
J321	E
J323	C
J325	D
J326	C
J327	C
J329	D
J331	C
J333	D
J335	C
J337	C
J339	C
J341	D

K

Item	Answer
K101	B
K103	A
K105	C
K106	C
K107	C
K109	C
K111	C
K113	D
K115	B
K117	B
K118	C
K119	E
K120	C
K121	D
K125a	D
K125b	C
K126	C
K127	D
K129	F
K130	C
K131	D
K132	C
K133	C
K134	C
K135	C
K136	D
K137	F
K139	D
K141	G
K142	G
K143	D
K144	C
K145	B
K146	C
K147	C
K149	D
K150	D
K151	D
K152	C
K153	C
K154	D
K155	C
K156	C
K157	C
K158	B
K159	C
K160	E
K161	C
K162	C
K163	D
K165	C
K164	C
K167	D
K169	D
K170	D
K171	C
K172	F
K173	B
K174	C
K175	C
K176	C
K177	C
K179	C
K180	D
K181	D
K182	C
K183	C
K184	C
K185	D
K187	C
K189a	F
K189b	F
K197	C
K201	H
K203	C
K205	D
K207	A
K209	B
K211	C
K212	C
K213	E
K215	D
K219a	B
K219b	C
K219c	C
K220	C
K221	C
K222	C
K223	E
K225	F
K227	C
K229	C
K231	C
K233	D
K235	B
K237	D
K239	G
K241	C
K243	B
K245	C
K247	C
K249	C
K251a	C
K251b	C
K251c	C
K252	D
K253	C
K255	G
K256	E
K257	D
K259	E
K261	C
K263	E
K264	D
K266	D
K267	C
K268	C

L

Item	Answer
L101	B
L103	C
L105	B
L106	C
L107	C
L108	A
L109	A
L111	D
L113	C
L115	D
L117	D
L119	C
L121	C
L123	B
L125	C
L126	C
L127	D
L128	E
L129	C
L130	D
L131	C
L132	D
L133	G
L134	E
L135	H
L136	D
L137	D
L139	D
L140	G
L141	C
L143	C
L145	C
L146	F
L147	D
L148	C
L149	E
L150	C
L151	C
L152	B
L153	A
L154	C
L155	C
L157	C
L159	C
L160	C
L161	C
L162	C
L163	C
L165	C
L166	D
L167	C
L169	D
L171	C
L173	C
L175a	E
L175b	E
L177	C
L179	C
L181	C
L183	E
L185	B
L187	C
L189	B
L191	C
L192	C
L193	C
L195	C
L197	B
L199	C
L200	D
L201	B
L202	C
L203	C
L205	C
L206	F
L207	E
L208	D
L209	B
L211	A
L212	D
L213a	C
L213b	C
L214a	C
L214b	C
L215	D
L218	D
L219	C
L221	D
L223	E
L225	A
L227	A
L229	D
L231	D
L235	C
L237	G
L239	E
L240	C
L241	G
L243	D
L244	G
L245	C
L247	D
L249	C
L250	E
L251	C
L253	C
L255	C
L256	D
L257	C
L258	C
L259	H
L260	B
L261	C
L262	B
L263	C
L265	B
L266	C
L267	E
L268	F
L269	G
L271	D
L272	C
L273	E
L274	B
L275	C
L279a	D
L279b	C
L281	C
L282	B
L295	C
L286	D
L287a	C
L287b	C
L291	D
L293	E
L294	D
L295	D
L296	E
L297	C
L299	D
L300	C
L301	D
L302	B
L303	C
L304	E
L305	C
L306	D
L307	C
L308	C
L309	C
L310	F
L311	E
L313	E
L314	C
L315	C
L317	C
L318	C
L319	D
L321a	H
L321b	H
L323	C
L325	C
L327	C
L329	C
L330	D
L331	C
L333	E
L334	C
L335	E
L337	E
L339	E
L341	D
L343	C
L345	E
L347	G
L348	D
L349	C
L351	D
L353	H
L355	G
L356	C
L357	D
L358	C
L359a	C
L359b	C
L259c	C
L365	C
L367	E
L368	C
L369	C
L370	C
L371	D
L372	A
L373a	A
L373b	A
L374	A
L376	A
L379a	A
L379b	A
L383	D
L385	C
L387	E
L388	C
L390	C
L391	C
L393	C
L394	C
L395	C
L397	D
L399	C
L401	A
L403	B
L404	C
L405	C
L407a	C
L407b	C
L411	C
L413	D
L415	E
L417	D
L419	D
L421	E
L422	G
L423a	B
L423b	B
L425	C
L427	E
L428	D
L429	C
L431	C
L433	C
L435	C
L437	B
L438	C
L439	D
L441	D
L444	C
L445	B
L446a	C
L446b	C
L447	D
L448	E
L449	E
L450	E
L452	C
L453	D
L454	C
L455	C
L456	C
L458	C
L460	E
L461	C
L462	C
L463	C
L465	D
L467	D

M

Item	Answer
M101	B
M103	B
M105	B
M107	B
M109	C
M111	C
M112	B
M113	C
M114	C
M115	C
M117	B
M118	D
M119	C
M120	C
M121	C
M122	C
M123	B
M124	C
M125	E
M126	C
M127	D
M128	E
M129	C
M131	D
M133	B
M135	F
M137	D
M139	B
M140	D
M141	C
M142	E
M143	C
M144	C
M145	D
M147	D
M149	C
M150	C
M151	B
M152	C
M153	B
M154	B
M155a	C
M155b	C
M157a	E
M157b	D
M159	E
M161	C
M163	C
M164	D
M165	C
M167a	E
M167b	E
M169	F
M170	G
M171	E
M172	E
M173	H
M174	D
M175	D
M177	C
M179	C
M181	C
M183	C
M184	C
M185	D
M186	D
M187	C
M189a	C
M189b	B
M191	C
M193	D
M195	C
M197	C
M199a	E
M199b	E
M199c	F
M200	C
M201	C
M203a	D
M203b	D
M204	D
M205	D
M206	C
M207	E
M208	C
M209	C
M211	C
M213	C
M214	C
M215	F
M216	C
M217	C
M219	C
M220	G
M221	F
M223	C
M225	D
M227	D
M229	G
M231	G
M332	E
M233	C
M234	D
M235	B
M236	C
M237	F
M238	C
M239	C
M240	G
M241a	H
M241b	H
M242	H
M243	C
M244	E
M245	C
M246	C
M247	E
M248a	C
M248b	C
M249	D
M250	B
M251	C
M253	C
M255	D
M256	C
M257	C
M258	C
M259	C
M260	E
M261	D
M262a	C
M262b	C
M263	B

M265	D
M267	D
M269	C
M271	C
M273	C
M275	C
M276	D
M277	C
M278	D
M279	C
M280	G
M281	C
M282	D
M283	C
M285	C
M286	C
M287	D
M289a	B
M289b	B
M295	B
M299	C
M301	H
M303a	B
M303b	B
M303c	B
M303d	B
M303e	B
M305	D
M307	D
M309	G
M311	D
M313	E
M314	C
M315	C
M316	C
M317	C
M319	A
M320	C
M321	C
M322	D
M323	D
M325	C
M327	C
M329	C
M331	C
M333	E
M335	F
M337	H
M339	D
M340	C
M341	C
M342	C
M343	B
M347a	C
M347b	C
M349	B
M350	C
M351	D
M352	D
M353	H
M354	C
M355	D
M356	D
M357	C
M358	D
M359	C
M361	B
M363	D
M364	D
M365	C
M366	D
M367	D
M369	C
M370	A
M372	C
M373	F
M374	D
M375	C
M376	C
M377	C
M378	B
M379	D
M380	C
M381	C
M382	C
M383	D
M384	C
M385	A
M387	E
M389	D
M391	C
M393	D
M394	C
M395	C
M396	C
M397	C
M398	E
M399a	D
M399b	E
M401	C
M403	C
M405	C
M406	D
M407	F
M408	E
M409	E
M411	D
M413	E
M415	C
M417	E
M418	D
M419	E
M420	D
M421	B
M422	D
M423	C
M424	C
M425	D
M426	D
M427	D
M428	C
M429	E
M430	D
M431	F
M432	G
M433	E
M434	C
M435	D
M437	C
M438	G
M439	F
M441	C
M443	A
M444	C
M445	B
M449	C
M451	B
M452	C
M453	C
M455	C
M457	C
M459	D
M460	D
M461	C
M463	D
M465	C
M466	C
M467	D
M468	E
M469	C
M470	B
M471	F
M472	D
M473	E
M474	B
M475	G
M476	H
M477	C
M479	F
M481a	B
M481b	B
M482	C
M483	D
M485a	C
M485b	D
M486	D
M487	C
M488	C
M489	F
M491	C
M493	C

N

N101	B
N102	B
N104	G
N105	B
N106	A
N108	B
N110	C
N112	C
N113	C
N115	A
N116	A
N118	B
N119	B
N120	A
N121	A
N122	A
N123	C
N124	B
N126	B
N128	A
N129	D
N130	C
N131a	C
N131b	C
N136	C
N137	D
N139	C
N140	C
N141	E
N142	D
N145	C
N147	D
N149	C
N150	D
N152	C
N154	G
N155	D
N157	B
N158	B
N160a	C
N160b	C
N161	H
N162	C
N163	E
N164	D
N165	C
N166	D
N167	D
N168	E
N169	D
N170	C
N171	C
N172	D
N173a	C
N173b	C
N173c	B
N174	C
N175a	C
N175b	D
N176	B
N177	D
N178	C
N179	C
N180	D
N181	F
N182	D
N183	C
N184	D
N185	D
N186	D
N187a	E
N187b	E
N188	C
N190	D
N191	D
N192	E
N193	C
N194a	C
N194b	D
N195	C
N196	E
N197	C
N198	C
N200	C
N201	D
N204	C
N206	E
N207	C
N208	C
N210	E
N212	E
N214	D
N216	C
N217	B
N218	D
N220	C
N222	D
N223	F
N225	C
N227	B
N227	C
N230	C
N232	A
N234	B
N236	D
N237	C
N239	E
N240	H
N242	E
N244	C
N246	C
N248	F
N250	C
N252	E
N253a	D
N253b	D
N255	D
N256	C
N258	C
N259	D
N261	C
N263	D
N265	C
N267	D
N269	C
N270a	C
N270b	C
N272	F
N274	D
N275	C
N276	C
N278	D
N280a	C
N280b	C
N280c	C
N282	E
N284	C
N286	C
N287	D
N289	F
N290	C
N292	C
N294	E
N296	D
N298	C
N300	F
N301	C
N304	C
N306	B
N308	C
N310	H
N311	H
N314	F
N316	E
N318	A
N320	E
N322	C
N324	C
N325	C
N327	D
N329	E
N332	G
N331	G
N332	D
N333	B
N334	C
N335	B
N336	C
N337	C
N338	C
N340	D
N341	D
N342	C
N343	C
N344	D
N345	C
N346	D
N347	D
N348	C
N350	C
N351	B
N353	D
N355	D
N357	F
N359	C
N360	D
N362	F
N364	G
N365	H
N368	C
N370	E
N372	D
N374	C
N376	B
N378	C
N379	C
N381	D
N382	C
N384	C
N386	C
N388	F
N389	F
N390	D
N392	C
N395	D
N396	C
N397	C
N398	C
N401	F
N403	C

O

O101	B
O102	B
O103	E
O105	C
O107	D
O109	C
O110	C
O112	D
O114	C
O116	C
O117	C
O118	B
O119	B
O120	C
O121	D
O124	C
O125	C
O126	D
O128	C
O130	C
O131	C
O133	D
O135	B
O137	E
O139	C
O141	D
O143	C
O145	G
O147	E
O148	C
O150a	C
O150b	C
O152	D
O154	C
O156	E
O158	D
O160	C
O162	D
O163	E
O165	D
O166	C
O167	C
O169	C
O170	C
O171	C
O173	F
O175	D
O176	D
O177	C
O179	B
O181	C
O183	D
O184	E
O185	E
O187	B
O189	E
O190	B
O191	C
O192	C
O195	C
O197	C
O199	C
O201	C
O202	C
O205	E
O207	G
O209	D
O211	C
O213	C
O214	C
O215	C
O216a	B
O216b	C
O217	A
O218	C
O219	G
O220	F
O221	C
O222	G
O223	B
O224	C
O225	D
O226	C
O227	D
O229	F
O230	D
O232	E
O234	C
O236	D
O238	B
O239	C
O242	B
O244	C
O246	B
O248	D
O250	A
O252	F
O254	C
O256	D
O257	D
O258	C
O259	C
O260	B
O264	C
O268	D
O270	F
O272	G
O274	E
O276	D
O277a	C
O277b	C
O280	C
O282	D
O284	C
O285	B
O286	D
O288	D
O289	C
O290	C
O291	C
O293	E
O295	D
O297	D
O298	C
O300	E
O302	F
O303	E
O304	E
O308	C
O310	D
O312	D
O314	D
O315	C
O316	E
O318	C
O320	E
O322	D
O324a	C
O324b	C
O326	C
O328a	H
O328b	H
O330	B
O332	C
O334	C
O336	G
O340	C
O342	D
O344	C
O346	C
O348	C
O349	C
O350	C
O351	D
O352	C
O354	C
O355	D
O356	C
O358	D
O360	D
O362	C
O364	C
O366	B
O368	C
O370	E
O372	C
O374	C
O376	D
O378	E
O380	C
O382	C
O384a	B
O284b	B
O386	B
O388	D
O390	C
O392	D
O393	C
O394	C
O395	B
O396	D
O398	C
O400	D
O402	D
O404	C
O406	C
O408	E
O410	D
O412	F
O414	E
O415	B
O416	C
O417	C
O418	D
O419	B
O420	C
O421	C
O423	C
O425	D
O426	D
O428	D
O430	C
O432	C
O433	C
O434	C
O436	B
O438	B
O440	C
O442	C
O444	E
O446	D
O450	C
O452	C
O454	C
O456	C
O458	E
O460	C
O462	B
O464	C
O468	G
O470	C
O472	C
O474	C
O476	H
O478	C
O480	F
O482	D
O484	A
O486	B
O487	C
O488	E
O490	B
O492	C
O494	C
O496	D
O498	B
O500	C
O502	D
O504	C
O505	C
O506	C
O507	C
O508	A
O509	C
O510	C
O512	C
O514	C
O515	D
O516	B
O518	D
O520	E
O522	C
O524	C
O525	C
O526	D
O527	C
O528	F
O530	D
O532	C
O534a	E
O534b	F
O534c	E
O536	F
O537	D
O539	C
O542	C
O544	C

P

P101	C
P103	B
P105	A
P107	B
P108	B
P110	C
P111	B
P112	C
P114	B
P116	C
P120	B
P122	E
P124	C
P126	C
P128	C
P130	C
P132	C
P134	C
P136	C
P140	C
P142	C
P144	B
P146	D
P147	D
P149	C
P150	D
P152	C
P153	C
P154	C
P156	B
P158	D
P160	C
P162	B
P164	C
P166	B
P168	C
P170	F
P172	C
P174	C
P176	D
P178	E
P180	D
P182	B
P184	C
P185	C
P186a	D
P186b	C
P188	D
P190	D
P192	C
P194	B
P195	C
P196	C
P198	C
P200	D
P202	C
P203	C
P204	C
P206	A
P208	B
P210	B
P212	C
P214	D
P216	B
P218	D
P220	C

Code	Answer
P222	E
P224	D
P226	D
P228	B
P230	B
P231	D
P232	E
P234	F
P236	C
P238	C
P239	D
P240	D
P242	D
P244	E
P246	D
P247	C
P248	C
P250	F
P252	D
P254	C
P255	D
P256	B
P258	C
P260	C
P262	D
P264	F
P266	E
P268	D
P270	E
P272	C
P274	C
P276	C
P278	D
P280	C
P282	B
P284	C
P286	B
P288	C
P290	C
P292	A
P294	C
P296	C
P297	C
P298	E
P300	C
P301	C
P302	B
P304	E
P306	C
P308	C
P309	D
P310	E
P312a	D
P312b	E
P314	F
P316	D
P318	D
P320	C
P324	D
P326	C
P328	F
P330	D
P332	D
P334	D
P336	C
P338	C
P240	C
P342	C
P244	D
P346	B
P348	C
P350	D
P352	C
P353	D
P354	B
P356	C
P358	C
P360	E
P362	B
P364	C
P366	C
P368	C
P370	C
P372	D
P373	D
P374	C
P376	C
P378	B
P380	B
P382	D
P384	B
P386	C
P387	D
P388	C
P390	C
P392	F
P394	C
P400	E
P402	B
P404	C
P406	A
P407	B
P408	B
P410	B
P412	C
P413	C
P414	B
P415	A
P416	C
P418	B
P419	C
P420	D
P421	C
P422	H
P424	D
P426	A
P428	C
P430	C
P432	D
P434	C
P436	C
P438	C
P440	C
P442	D
P444	D
P446	C
P447	B
P449	C
P452	A
P454	B
P456a	B
P456b	B
P458a	B
P458b	B
P460	B
P462	C
P463	C
P464	C
P468	C
P470	C
P472	D
P474	D
P476	E
P478	D
P480	F
P482	C
P484	F
P486	D
P488	C
P490	C
P492	C
P494	C
P496	C
P500	E
P502	D
P504	C
P506	C
P508	D
P510	C
P512	D
P514	D
P516	C
P517	E
P519	D
P521	D
P522	C
P524	C
P526	C
P528	A
P530	E
P532	D
P534	C
P536	C
P538	C
P540	D
P542	G
P543	G
P546	C
P548	B
P550	C
P552	G
P556	C
P557	C
P559	C
P560	C
P564	C
P566	C
P568	C
P570	E
P571	D
P574	C
P575	D
P576	D
P580	C
P582	D
P584	C
P586	C
P588	C
P590	D
P592	C
P594	C
P596	C
P598	D
P600	D
P602	F
P604	C
P605	C
P606	E
P608	D
P610	D
P612	C
P614	C
P616	C
P618	C
P620	C
P622	D
P624	B
P626	D
P628	B
P630	B
P632	C
P633	C
P634	F
P636	C
P638	B
P640	C
P642	F
P644	G
P646	E
P648	F
P650a	C
P650b	C
P652	C
P653	E
P654	D
P656	D
P658	D
P659	B
P661	C
P664	D
P666	C
P668	D
P670	D
P672	C
P673	D
P674	B
P675	C
P676	C
P680	C
P681	B
P682	C
P684	A
P686	B
P688	C
P690	C
P692	B
P693	B
P694	H
P696	E
P698	D
P700	C
P702	B
P703	D
P704	C
P706	C
P708	D
P710	C
P712	D
P713	A
P714	B
P715	C
P716	C
P718	D
P720	C
P722	C
P724	E
Q	
Q101	C
Q103	B
Q105	D
Q106	D
Q108	E
Q110	C
Q112	C
Q114	D
Q116	C
Q118	C
Q120	E
Q121	E
Q122	G
Q123	E
Q126a	G
Q126b	F
Q128	D
Q130	C
Q132	B
Q134	A
Q136	C
Q138	B
Q140	C
Q142	B
R	
R101	A
R103	B
R106	D
R108	C
R110	C
R112	C
R113	C
R114	C
R116	C
R118a	B
R118b	C
R120	B
R122	D
R124	C
R126	D
R128	C
R130	C
R132	C
R134	B
R136	C
R138	B
R140	B
R142	C
R143	B
R144	C
R146	B
R148	C
R150	C
R152	C
R153	B
R154	C
R156	C
R158	F
R160	E
R162	D
R164	C
R166	E
R168	D
R170	C
R172	D
R174	F
R176a	C
R176b	D
R178	E
R180	F
R182	C
R184	C
R186	D
R188	C
R190	D
R192a	C
R192b	C
R192c	C
R194	B
R196	C
R198	C
R200	C
R202	C
R204	E
R205	C
R206	D
R208	C
R210	C
R212	F
R214	C
R216	B
R218	E
R220	D
R221	C
R222	D
R224a	C
R224b	C
R226	D
R228	C
R230	A
R232	C
R234	D
R235	G
R236	E
R238	C
R240	D
R242	C
R244	D
R246	C
R248	C
R250	E
R252	E
R254	D
R256a	C
R256b	C
R258a	C
R258b	D
R260	E
R262	D
R264	C
R266	C
R268	C
R270	B
R272	C
R274	C
R276	D
R278	D
R280	C
R282	D
R284	E
R286	D
R288	C
R290	B
R292	D
R294	D
R296	C
R298	C
R300	D
R302	B
R304	C
R306	D
R308	C
R310	C
R312	E
R314	D
R318	F
R319	F
R320	C
R322	B
R324	A
R326	A
R328	D
R330	D
R332	D
R334	C
R336	E
R338	F
R340a	D
R340b	D
R342	C
R344	D
R346	D
R347	D
R348	D
R350	F
R352	C
R354	C
R356	D
R358	D
R360	E
R362	G
R364	C
R366	E
R368	F
R369	D
R370	E
R372	B
R374	D
R376	C
R378	C
R380	C
R382	E
R383	F
R384	F
R386	D
R388	D
R390	D
R392	G
R394	C
R395	D
R396	C
R397	C
R398	D
R400	H
R402a	D
R402b	D
R404	E
R406	D
R408	G
R410	C
R412	D
R414	G
R416	D
R417	C
R418	E
R420	C
R422	C
R424	D
R426	D
R428	E
R430	C
R432	F
R434	G
R436	G
R438	C
R440	E
R442	E
R444	D
R446	C
R448	C
R450	D
R452	E
R454	C
R456	D
R458	D
R460	C
R462	E
R464	D
R466	D
R467	C
R468	D
R469	F
R470	B
R472	E
R474	C
R476	D
R477	D
R478	C
R480	C
R482	D
R484	C
R486	E
R488	C
R500	C
R502	C
R505	C
R506	D
R508	F
R510	C
R512	D
R514	C
R516	D
R518	F
R520	C
R522	G
R524	F
R526	D
R527	B
R528	C
R530	C
R532	D
R534	C
R536	B
R538	C
R539	C
R540	B
R541	D
R542	B
R544a	C
R544b	C
R546	B
R548	D
R550	D
R552	D
R554	E
R556	D
R558	C
R560	C
R562	E
R564	D
R566	C
R568	C
R570	B
R572	C
R574	C
R576	D
R578	D
R580	D
R581	D
R582	D
R584	D
R586	C
R588	E
R590	C
R592	B
R594	D
R596	F
R598	C
R600	C
R604	C
R606	D
R608	B
R610a	D
R610b	D
R612	C
R614	B
R616	C
R618	D
R620	C
R622	C
R624	C
R626	C
R628	C
R629	C
R630	C
R632	D
R634	D
R636a	D
R636b	D
R636c	D
R636d	E
R637	D
R638	C
R640	D
R642	D
R644	C
R646	C
R648	E
R650	C
R652	D
R654	C
R656	D
R658	B
R660	D
R662	C
R664	E
R666	B
R668	C
R670	C
R672	D
R673	C
R674	D
R675	C
R676a	C
R676b	D
R677	C
R678	B
R680	C
R681	D
R682	C
R683	C
R684	E
R686	E
R688	E
R690	E
R692	C
R694	C
R696	D
R697	C
R700	E
R702	C
R703	C
R706	D
R707	C
R708	C
R710	H
R712	C
R714	D
R716	H
R718	C
R720	C
R722	C
R724	E
R726	D
R728	D
R730	C
R732	D
R733	G
R734	C
R736a	C
R736b	C
R738	B
R740	C
R742	C
R744	E
R746	G
R748	C
R750	C
R751	B
R752a	C
R752b	C
R754	B
R756	C
R758	E
R760	C
R762	B
R764	C
R766	E
R768	F
R770	C
R771	B
R772	C
R773	B
R774a	D
R774b	D
R776	C
R777	C
R778	D
R780	C
R782	D
R784	C
R786	C
R788	C
R790	B
R792	C
R794	A
S	
S101	B
S102	A
S104	B
S105	C
S106	D
S107	C
S108	C
S110	B
S112	C
S114	D
S116	B
S117	B
S118	B
S120	C
S122	C
S124	C
S125	C
S126	C
S128	D
S130	C
S131	G
S132	G
S134	E

S136	D
S138	C
S140	H
S141	H
S142	C
S144	A
S146	B
S148	D
S150	C
S152	E
S154	C
S156	E
S158	D
S160	C
S162	C
S164	C
S166a	C
S166b	C
S168	D
S170	C
S172	D
S174	C
S176	E
S178	D
S180	E
S182	D
S184	E
S186	D
S188	C
S190	C
S192	C
S194	C
S196	C
S198	D
S200	C
S202a	F
S202b	E
S203	B
S205	B
S206	D
S208	C
S210	C
S212	E
S213	D
S214	D
S216	D
S218	E
S219	C
S220	F
S222	D
S223	C
S224	D
S226	C
S228	C
S230	H
S232	E
S234	D
S236	C
S238	D
S240	C
S242	E
S244	C
S245	D
S246	A
S248a	A
S248b	A
S248c	A
S248d	A
S250a	B
S250b	B
S250c	B
S250d	B
S252	D
S254	B
S256	C
S258	B
S260	C
S262	C
S263	C
S264	B
S266	E
S268	C
S270	C
S272	D
S274	E
S276	G
S278	C
S280	C
S282	D
S284	C
S286	H
S288	H
S290	B
S291	C
S292	C
S294	B
S296	C
S298	D
S300	C
S302	C
S304	C
S306	C
S308	C
S310	C
S311	B
S312	C
S314	D
S316	D
S318	C
S320	B
S321	E
S322	D
S324	E
S326	B
S327	D
S328	F
S330	D
S332	D
S334	D
S336	D
S338	E
S340	D
S342	C
S344a	G
S344b	F
S346	G
S348	C
S350	B
S352	C
S354	D
S356	H
S358	C
S360	C
S362	D
S364	C
S366	E
S368	B
S370	B
S372	D
S374	C
S376	D
S378	C
S380	B
S382	B
S383	G
S384	D
S386	C
S388	D
S390	D
S392	C
S394	C
S396	E
S397	B
S398	C
S400a	C
S400b	B
S402	C
S404	B
S406	B
S408	D
S410	F
S412	D
S414	C
S416	C
S418	C
S419	C
S420	B
S422	D
S424	B
S425	C
S426	A
S427	C
S428	C
S430	E
S432	C
S434	C
S435	D
S436	D
S437	D
S438	D
S440	C
S442	D
S444	C
S448	D
S450	E
S452	A
S454	B
S455	C
S456	C
S458	F
S460	B
S462	B
S464	D
S466	E
S468	E
S469	C
S470	C
S472	B
S474	D
S476	C
S478a	F
S478b	G
S480	C
S482	C
S484	E
S486	E
S488	C
S490	C
S492	C
S494	D
S496	D
S497	C
S499	G
S500	D
S502	C
S504	E
S506	F
S508	C
S510	C
S512	C
S514	C
S516	D
S518	C
S519	C
S520	D
S521	C
S522	G
S523	F
S524	B
S525	C
S526	C
S528	D
S530	C
S532	B
S534	B
S536	C
S538	B
S540	D
S542	E
S544	C
S546	H
S548	D
S550	B
S552	D
S554	C
S556	C
S558	C
S560	B
S562	D
S564	C
S566	E
S568	C
S570	C
S572	A
S574	B
S576	C
S577	C
S578	C
S580	B
S582	E
S584	C
S586	C
S588	C
S590	F
S592	C
S594	D
S596	D
S598	C
S600	C
S601	D
S602	G
S604a	C
S604b	C
S606	C
S608	E
S610	C
S612	C
S614	G
S616	D
S618	D
S620	D
S622	C
S624	B
S626	B
S628	E
S630	E
S631	F
S632	E
S634a	A
S634b	A
S634c	A
S636	A
S638	D
S640	B
S642a	C
S642b	D
S643	A
S644	C
S645	C
S646	B
S648a	E
S648b	E
S650	B
S652	C
S654	C
S656	C
S658	D
S660	D
S662	C
S664	C
S665	C
S666	D
S667	E
S668	E
S670	D
S672	C
S674	C
S676	H
S678	D
S680	D
S682	E
S684	C
S686	H
S688	C
S690	H
S692	A
S694	E
S696	G
S697	C
S698	C
S699	G
S700	C
S702	C
S704	B
S706	C
S707	C
S710	C
S712	C
S713	C
S714	C
S716	C
S717	C
S718	B
S720	D
S722	B
S724	C
S725	E
S726	D
S728	C
S730a	H
S730b	H
S732a	H
S732b	H
S734	G
S736	B
S738	B
S739	D
S740	A
S741	D
S742	C
S743	C
S744	C
S746	C
S748	C
S750	E
S752	D
S754	C
S756a	B
S756b	A
S756c	A
S758	B
S760	E
S762	C
S764	F
S766	D
S768a	E
S768b	E
S770	D
S772	E
S774	A
S776	B
S778	B
S780	A
S782	C
S784	C
S785	C
S786	D
S788	C
S790	C
S792	C
S794	C
S796	C
S798	B
S800	E
S801	C
S802	G
S804	C
S806	C
S808	C
S810	D
S812	C
S814	D
S816	C
S818	D
S820	C
S824	H
S825	H
S826	C
S828	C
S830	A
S831	B
S832	A
S834	C
S835	C
S836	C
S837	C
S838	C
S840	D
S842	E
S844	C
S846	G
S848	C
S850	D
S852a	C
S852b	C
S854	C
S856	B
S858	D
S860	F
S862	C
S864	B
S865	C
S866	B
S868	D
S870	G
S872	F
S874	E
S876	E
S878a	C
S878b	D
S880	E
S882	D
S883	C
S884	C
S886	E
S888	C
S890	C
S892	B
S891	C
S893	D
S894	C
S895	C
S896	B
S897	C
S898	C
S899	C
S900	B
S901a	C
S901b	B
S902	C
S903	E
S904	B
S906	C
S908	C
S910	D
S912	D
S914	D
S915	C
S916	C
S917	C
S918	D
S920	C
S922	C
S924	C
S926	C
S928	B
S930	D
S932	C
S934	C
S936	A
S938	B
S940	D
S942	C
S943	D
S944	C
S946	C
S948	G
S949	G
S950	E
S951	B
S952	A
S954	C
S956	C
S958	D
S959	D
S960	C
S962	D
S964	D
S966	D
S968	D
S970	E
S972	C
S974	C
S976	C
S977	D
S978	C
S980a	D
S980b	C
S982	C
S984	C
S985	C
S986	C
S988	B
S990	C
S992	D
S993	C
S994	D
S995	C
S996	C
S998	G
S1000	E
S1002	C
S1004	C
S1006	F
S1010	E
S1012	E
S1014	D
S1015	C
S1016	C
S1018	E
S1020	G
S1021	C
S1022	C
S1024	D
S1026	F
S1028	E
S1030	B
S1032	G
S1034	C
S1036	B
S1038	C
S1040	F
S1041	F
S1042	G
S1043	F
S1044	E
S1046	C
S1047	C
S1048	D
S1050	C
S1052	B
S1054	E
S1056	E
S1058	C
S1060	C
S1062	D
S1064	F
S1066	C
S1068	C
S1070a	C
S1070b	C
S1071	D
S1072	C
S1073	C
S1074	D
S1075	D
S1076	E
S1078	C
S1080	D
S1084	C
S1086	D
S1090	D

T

T101	B
T102	C
T104	C
T106	B
T108	C
T110	C
T112	C
T114	D
T116	C
T118	C
T120	B
T122	C
T124	C
T126	D
T127	B
T128	C
T130	C
T132	B
T134	C
T136	C
T138	C
T140	B
T142	B
T144	C
T146	G
T148	B
T150	E
T152	E
T154	C
T156	C
T158	D
T160	C
T162	C
T164	D
T166	C
T168	B
T170	C
T172	B
T174	C
T176	C
T178	C
T180	D
T182	F
T184	D
T186	C
T188	C
T190	C
T192	B
T194	A
T196	C
T198	C
T200	D
T202	E
T204	B
T206	E
T208	C
T210	C
T212	C
T213	C
T214	B
T216	D
T218	C
T220	D
T222	C
T224	C
T226	D
T228	G
T229	E
T230	F
T232	C
T234	C
T236	C
T238	C
T240	C
T242	C
T244	C
T245	C
T246	G
T247	C
T248	E
T250	C
T252	C
T254	G
T256	E
T258	B
T260	B
T262	C
T264	B
T268	C
T270	C
T272	H
T274	C
T276	D
T278a	G
T278b	G
T280	D
T282	C
T284	C
T286	C
T288	E
T290	C
T292	C
T294	C
T296a	E
T296b	F
T298	D
T300	E
T302	H
T304	F
T306	C
T308	H
T310	D
T312	D
T314	D
T316	C
T318	C
T319	C
T320	D
T324	E
T326	C
T328	C
T330	C
T334	C
T336	B
T338	C
T339	C
T340	C
T342	F
T344a	C
T344b	B
T344c	B
T346	C
T348	D
T350	C
T352	C
T354	C
T356	B
T358	B
T360	C
T362	F
T364	C
T366	C
T368	G
T370	C
T372a	D
T372b	E

T374 E
T376 C
T378 C
T380 D
T382 D
T484 C
T386 D
T387 D
T388 C
T390 G
T392 D
T394 C
T396 A
T398 C
T400 B
T402 C
T404 C
T406 D
T408 B
T410 C
T411 C
T412 C
T414 D
T416 C
T418 C
T420 C
T422 B
T424 C
T426 E
T427 D
T428 C
T430 C
T432 B
T434 C
T436 D
T438 D
T440 D
T442 E
T444 H
T446 C
T448 D
T450 C
T452 D
T454 B
T456 C
T458 H
T460 C
T462 C
T464 D
T466 E
T468 D
T470 C
T472 C
T474 C
T476 C
T478 C
T480 C
T482 D
T484 C
T486 C
T488 B
T490a C
T490b C
T492a C
T492b C
T494 D
T495 C
T496 D
T498 B
T499 B
T500 C
T502 B
T504 C
T506 D
T507 D
T508 C
T510 C
T512 B
T514 C
T516 C
T518 B
T520 C
T522 B
T524 D
T525 C
T526 C
T528 C
T530 C
T532a E
T532b F
T534 D
T536 D
T538 C
T540 E
T542 C
T543 C
T544 D
T546 D
T548 C
T550 C
T552 C
T554 C
T556 C
T558 D
T560 C
T562 C
T564 F
T566 C
T568 C
T570 B
T572 C
T574 E
T576 G
T578 F
T580 E
T582 B
T584 C
T586 D
T588 C
T590 B
T592 C
T594 G
T595 E
T596 F
T598 D
T600 D
T602 F
T604 C
T606 D
T607 D
T608 D
T610 D
T612 B
T614 C
T616 F
T618 C
T620 B
T622 C
T624 C
T626 D
T628 E
T630 C
T632 D

U

U101 A
U102 B
U104 C
U106 B
U108 B
U110 C
U112 C
U114 C
U116 C
U118 C
U119 B
U120 C
U121 B
U122 C
U124 B
U126 D
U128 A
U130 E
U132 H
U134 H
U136 C
U138 C
U139 C
U140 C
U142 D
U144 H
U146 G
U148 C
U150a C
U150b C
U151 F
U152 E
U154 H
U156 D
U158a D
U158b D
U160 E
U162 A
U164 B
U166 D
U168 C
U170 C
U172 C
U174 C
U176 E
U178 C
U180 D
U182 D
U184 D
U186 D
U188 C
U190 C
U192 D
U194 C
U196 C
U198 B
U199 B
U200 C
U202 E
U204 C
U206 C
U208 D
U210 D
U212 C
U214 B
U216 C
U220 H
U222 C
U224 C
U226 E
U228 C
U230 C
U232 C
U234 F
U236 D
U237 D
U238 D
U240 C
U242 E
U244 C

V

V101 C
V102 D
V104 E
V106 D
V108 C
V110 D
V112 C
V114 C
V116 E
V118 C
V120 B
V122 A
V124 B
V126 C
V128 D
V130 C
V132 C
V134 E
V136 E
V138 C
V140 B
V142 A
V144 C
V146 C
V148 C
V150 H
V152 C
V154a A
V154b B
V156 D
V158 D
V160 C
V162 C
V164 C
V166 C
V168 D
V170 C
V172 E
V174 D
V176 C
V178 B
V180 C
V181 C
V182 C
V184 C
V186 D
V188 D
V190 C
V192 C
V194 C

W

W101 B
W102 C
W104 B
W106 D
W108 D
W110 B
W112 C
W114 C
W116 C
W118 C
W119 C
W120 D
W121 B
W122 C
W124 B
W126 B
W127 C
W128 C
W130 C
W132 D
W134a C
W134b B
W134c C
W136 D
W138 C
W140 B
W142 C
W144 C
W146 B
W148 B
W150 C
W152 C
W153 C
W154 C
W156 D
W158 D
W160 C
W162 C
W164 C
W166 C
W167 C
W170 C
W172 B
W174 E
W176 D
W178 C
W180 C
W182 D
W184 C
W186 D
W188 F
W189 C
W190 D
W191 C
W192 C
W194 B
W196 B
W198 D
W200 C
W202 C
W204 B
W206 C
W208 C
W210 D
W211 C
W212 C
W214a B
W214b C
W216 C
W217 D
W218 C
W220 C
W222 C
W224 C
W226 C
W228 E
W230 D
W231 F
W232a E
W232b E
W234 B
W236 C
W238 C
W240 C
W242 C
W246 C
W248 G
W250 C
W252 D
W254 C
W255 C
W256 F
W258 C
W260 E
W262 E
W263 C
W264 B
W268 C
W269 C
W270 C
W271 C
W272 C
W274 D
W275 C
W276 C
W278 C
W280 D
W282 C
W284 F
W286 D
W288 B
W290 F
W292a E
W292b E
W294 B
W296 C
W297 D
W298 C
W300 D
W301 D
W302 D
W304 C
W306 C
W308 C
W310 D
W312 C
W314 D
W316 C
W318 H
W320 E
W322 C
W323 D
W324 D
W326 C
W328 B
W330 C
W332 B
W336a C
W336b C
W338 C
W340 C
W341 G
W342 D
W343 D
W346 H
W348 B
W350 C
W352 C
W354 F
W356 C
W358 C
W360 C
W362 F
W364 C
W366 E
W368 F
W370 E
W372 D
W374 C
W376 C
W378 C
W380 C
W382 C
W383 C
W384 D
W385 C
W386 D
W387 C
W388 C
W389 D
W390 C
W392 D
W393 C
W394 C
W395 D
W396 C
W398 D
W400 C
W402 D
W404 C
W406 D
W408 F
W410 D
W412 E
W414 C
W416 C
W417 D
W418 C
W420 C
W422 F
W424 D
W426 D
W428 D
W430 C
W432 C
W434 E
W436 D
W438 C
W440 C
W442 C
W444 C
W446 F
W448 A
W450 B
W451 C
W452 D
W454 C
W456 B
W458 C
W460 F
W462 C
W464 B
W468 C
W470 F

X

X101 B
X102 B
X104 C
X106 B
X108 C
X110 D
X112 C
X113 B
X114 B
X116 C
X118 A
X120 B
X122 C
X124 B
X126 D
X128 C
X130 C
X132 C
X134 C
X136 C
X138 C
X139 C
X140 C
X142 D
X144 C
X146 D
X148 C
X150 E

Y

Y101 C
Y102 B
Y104 C
Y106 C
Y108 D
Y110 C
Y112 D
Y113 D
Y114 G
Y116 F
Y118 E
Y120a C
Y120b D
Y122 E
Y124 D
Y126 E
Y127 C
Y128 E
Y130 C
Y132 C
Y134 D
Y136 E
Y138 B
Y140 C
Y142 B
Y144 C
Y146 C
Y148 C
Y149a C
Y149b C
Y150a D
Y150b D
Y150c D
Y150d D

Z

Z101 B
Z102 C
Z104 C
Z106 F
Z108 E
Z110 C
Z112 F
Z113 G
Z114 C
Z115 C
Z118 F
Z120 G
Z124 B
Z125 E
Z127 G
Z128 B
Z130 C
Z132 D
Z136 E